THE DRONE AGE

THE
DRONE AGE

HOW DRONE TECHNOLOGY WILL CHANGE WAR AND PEACE

MICHAEL J. BOYLE

OXFORD

UNIVERSITY PRESS

Oxford University Press is a department of the University of Oxford. It furthers
the University's objective of excellence in research, scholarship, and education
by publishing worldwide. Oxford is a registered trade mark of Oxford University
Press in the UK and certain other countries.

Published in the United States of America by Oxford University Press
198 Madison Avenue, New York, NY 10016, United States of America.

Library of Congress Cataloging-in-Publication Data
Names: Boyle, Michael J., 1976– author.
Title: The drone age : how drone technology will change war and peace /
Michael J. Boyle, La Salle University.
Other titles: How drone technology will change war and peace
Description: New York, NY : Oxford University Press, [2020] |
Includes index.
Identifiers: LCCN 2019047301 (print) | LCCN 2019047302 (ebook) |
ISBN 9780190635862 (hardback) | ISBN 9780190635886 (epub) |
ISBN 9780197501788 (online)
Subjects: LCSH: Drone aircraft. | Drone aircraft–Political aspects.
Classification: LCC UG1242.D7 B69 2020 (print) |
LCC UG1242.D7 (ebook) | DDC 358.4/183–dc23
LC record available at https://lccn.loc.gov/2019047301
LC ebook record available at https://lccn.loc.gov/2019047302

1 3 5 7 9 8 6 4 2

Printed by LSC Communications, United States of America

CONTENTS

———⟫•⟪———

ILLUSTRATIONS

———◆◆◆◆———

ACKNOWLEDGMENTS

This book would not have been possible without the support of many institutions and people.

I would like to first thank the Smith Richardson Foundation for awarding me a Strategy and Policy Fellowship in spring 2015. This fellowship enabled me to conduct interviews for this book and to take a semester away from teaching to work on the research and writing. I also wish to thank Alan Luxenberg and Michael Noonan of the Foreign Policy Research Institute (FPRI) who recommended that I apply for this fellowship. The FPRI has also been kind enough to invite me to present this work in a number of ways and I have benefited from the discussion and feedback that I received at each presentation.

I would also like to thank the Leaves and Grants committee of La Salle University for granting me a one semester research leave to work on this project. For their support of my proposal, I would particularly like to thank former dean Thomas Keagy, Professor Emeritus Michael R. Dillon, and Professor Miguel Glatzer. I would also like to thank all of the staff of the Connelly Library for their help.

I would also like to thank the faculty and staff of Perry World House at the University of Pennsylvania, where I was a visiting fellow in fall 2016. I am particularly indebted to William W. Burke-White and Michael C. Horowitz for arranging my visit and making my stay there such a pleasant one. I was also fortunate to have an opportunity to present my work at their research seminar and I thank all of the participants at the university for their feedback.

I would like to thank all of the interviewees that have spoken to me for this project. I would particularly like to thank the US Air Force Academy in Colorado for facilitating a very useful visit to their campus.

Elsewhere, I have presented this work at the American Political Science Association (APSA) and International Studies (ISA) conferences. I also presented versions of the chapters at the John Jay College at the City University of New York (CUNY), the Jenkintown Lyceum and the Valley Forge Freedom Foundation. I am grateful to all of these audiences for their feedback. I would like to thank in particular friends and colleagues who read or commented on some early drafts of the proposal and the chapters, especially William W. Burke-White, Michael C. Horowitz, Emma Leonard Boyle, Michael Noonan, Nicholas Staffieri, Dominic Tierney, and the anonymous referees. I am also grateful to my research assistant, Selena Bemak, who did a wonderful job fact-checking the manuscript and helping with its last stages. I would also like to thank Julie Shawvan for indexing the book. Of course, I am wholly responsible for any errors or omissions in the book.

I would also like to thank Oxford University Press for permission to reprint some amended text from my original article, "The Costs and Consequences of Drone Warfare," *International Affairs* 89:1 (2013), pp. 1–29 and MIT Press for the right to reprint some parts of my correspondence in Michael J. Boyle, Michael C. Horowitz, Sarah E. Kreps, and Matthew Fuhrmann, "Debating Drone Proliferation," *International Security*, 42:3 (Winter, 2017/18), pp. 178–182.

I am also very grateful to my agent, Andy Ross, who assisted with the proposal and sharpening the argument of this book. I would also like to thank David McBride and Oxford University Press for their patience as this book was finished.

Finally, I would like to thank my family for their love and support for all of the long days and weekends lost writing this book. My son, George, arrived during the writing of this book and his laughter in the room next to me provided just the right impetus to get it done. I am also thankful for my dog and (according to my son) research assistant Millie who patiently sat in my study as I worked. But the book is dedicated with love to my wife, Emma, who has shown patience and a bit of bemusement as I wrestled with yet another book project. I thank her for everything she has done to enrich my life in so many ways.

I

The Drone Age

ON A LATE SEPTEMBER morning in 2011, the Yemeni-American preacher Anwar al-Awlaki was finishing breakfast with a group of men in the Jawf province of southwestern Yemen. This desolate region, once known for its fine Arabian horses and beautiful oases, was now a chaotic no man's land far outside of the control of the central government in Sana'a. For the last decade, the government of Ali Abdullah Saleh had struggled to restore order to the southern regions of the country amid a rising tide of tribal unrest and Islamic militancy. Like many other Islamists, Awlaki had retreated to the ungoverned spaces of the countryside because they provided free movement outside of the reach of Yemen's ruthless security services. He had good reasons to fear another run-in with them. Awlaki spent eighteen months as a prisoner in Yemen's jails after being arrested without charge in 2006 and knew that he would face a return to jail—or worse—if he were caught again.[1]

This breakfast meeting was not out of the ordinary for Awlaki, a leader in al Qaeda in the Arabian Peninsula (AQAP), an increasingly powerful offshoot of the terrorist organization started by Osama bin Laden. Since returning to Yemen in 2004, Awlaki had become the public face of this organization through his charismatic sermons, both in Arabic and in English, which criticized US foreign policy and inspired people to join al Qaeda's jihad. By 2009, he moved beyond inflammatory speeches and became "operational," according to US counterterrorism officials.[2] He had been linked to a mass shooting at Fort Hood in November 2009 by a disaffected US Army psychiatrist, Major Nidal Malik Hasan, as well as a plot by the young Nigerian Umar Farouk Abdulmutallab to destroy a Delta airlines flight over Detroit weeks later on Christmas Day. In the US media, he was being dubbed as the "next bin Laden" even before Osama bin Laden was killed by a US Navy SEAL team in May 2011.[3]

For the Obama administration, Awlaki posed a more vexing problem than bin Laden because he was a US citizen. Born in Las Cruces, New Mexico, he spent several years as a child in the United States when his father, Nasser, was on a Fulbright scholarship. After living his teenage years in Yemen, he moved back to the United States to study civil engineering at Colorado State University in 1991. He was later active in mosques in Denver and San Diego and moved to Falls Church, Virginia, shortly before the September 11 attacks. Although the FBI investigated his links with the September 11 hijackers, they found nothing incriminating, but nevertheless continued to keep him under surveillance. For a brief time, Awlaki took on a public role as a spokesperson on interfaith tolerance. He was even invited to the Pentagon as part of an Islamic outreach program and to an interfaith prayer service at the US Capitol.[4] After he moved to London in 2002, his sermons became more strident in their criticism of US foreign policy. He was jailed more than a year after his return to Yemen for unspecified charges, and only released due to family pressure on the Saleh government.[5] By 2008, he was highly influential among jihadi groups around the world. His natural charisma and his ability to reach a global audience with his English language sermons praising jihad against the United States made him a dangerous enemy. What concerned the CIA was his global reach: Awlaki's sermons were popular and widely quoted online and his influence was evident in al Qaeda's English-language magazine *Inspire*.[6] But he remained a US citizen, with all of the constitutional protections that this status implied, and all US government agencies were forbidden from conducting assassinations abroad under an executive order signed by President Gerald Ford in 1976.[7]

Even if that were not the case, a decision to assassinate Awlaki would hardly be expected from President Barack Obama. Obama had come into the presidency as a critic of the seemingly endless wars in Iraq and Afghanistan and wanted the United States to focus more on what he called "nation building at home."[8] By training and temperament, he was skeptical about foreign policy overreach, particularly under the banner of the war on terror. As an expert in constitutional law who had taught the subject at the University of Chicago, Obama knew well what due process rights a citizen like Awlaki would have in the United States and that killing him could be a violation of federal law.[9] He also knew it was morally problematic. Obama's thinking was shaped by scholars like Reinhold Niebuhr, who warned that evil was always present in the world and that a foreign policy committed to destroying every enemy abroad could be dangerous and even morally corrupting.[10] Yet Obama also recognized that, as president, he had an obligation to defend US citizens from terrorist attacks, even when those attacks were orchestrated by one of their own. The specifics of the Awlaki case—that he was a member of a terrorist organization, not a recognized foreign army, and that he was living abroad in a country where the United States was not legally at war—would need to be swept aside in the face of that obligation.

Supreme Court justice Oliver Wendell Holmes once remarked that lawyers spent a great deal of their time "shoveling smoke" to obscure issues that are well established within the law. To solve his Awlaki dilemma, President Obama turned to his own smoke-shovelers in the Office of Legal Counsel (OLC) at the Department of Justice. They dutifully produced two memoranda that concluded that killing a citizen like Awlaki was permissible because he was participating in a war as part of al Qaeda and his activities could be considered an imminent threat.[11] Notably, their analysis emphasized that killing Awlaki would constitute neither murder nor assassination, thus sidestepping the two actions that a US president could never officially authorize. Instead, the Obama administration's lawyers embraced a legal gray area called "targeted killing," which some influential legal scholars argued was permissible under conditions of armed conflict.[12] According to this school of thought, targeted killings were different from murders because they were legal, and from assassinations because they were conducted against active combatants in a declared war. Meanwhile the Pentagon and CIA ramped up their search for Awlaki, using satellites, intercepts of electronic communication, and local spies. The chief way that they hunted for him in the distant regions of Yemen was with Unmanned Aerial Vehicles (UAVs), more commonly known as drones. The manhunt involved "near continuous coverage" of Yemen by powerful Predator drones flown from US bases in Djibouti and Saudi Arabia.[13]

In May 2011, only days after bin Laden's death, the Pentagon's Joint Special Operations Command (JSOC) located Awlaki driving through the Shabwa province in the south of Yemen.[14] With President Obama's approval, JSOC deployed manned Harrier jets, a Special Operations Dragon Spear aircraft and Predator drones to kill Awlaki and his companions. Another drone, a massive surveillance aircraft called the Global Hawk, flew above the strike team and relayed a live video feed to JSOC commanders. This would be a drone strike which US officials could watch in real time, giving instructions to the pilots as needed. Within hours of their launch, they found him. Using the Predator's laser guidance system, Harrier Jets launched three Hellfire missiles against his convoy, but they missed. His Suzuki truck raced through the dust and smoke to evade the US drones. He radioed for help from two sympathetic Yemeni brothers, Abdallah and Musa'd Mubarak, both of whom were widely known in the AQAP community.[15] Taking advantage of the confusion, Awlaki quickly switched cars with them, speeding away in their Suzuki truck toward safe haven in a nearby cave. The Mubarak brothers took off in the other direction, followed by US drones, and were killed by the final series of missile strikes. Awlaki reported seeing a flash of light behind him and feeling a shockwave as his SUV sped away from the missile strike.[16] Only later did it become clear to JSOC that Awlaki survived.[17]

For the rest of the summer, Awlaki lived as a hunted man. He knew his days were numbered. He warned his colleagues that the next set of missiles might be more accurate, reportedly saying that "this time eleven missiles missed [their] target, but the next time the first rocket may hit."[18] He became resigned to his fate, saying that no human being died until they reached their "appointed time."[19] He saw his upcoming death in religious terms, remarking that "the tree of martyrdom in the Arabian Peninsula has already got ripe fruits on it, and the time for reaping has come."[20] He moved around every few days for much of the summer but eventually settled for a longer period in a mud house in Khashef, allegedly planning to make another video exhorting jihad.[21]

Only a few months later, on a bright September morning, Awlaki and three companions again heard the familiar whirring noise of a US drone overheard and scrambled for their nearby jeeps to escape.[22] But they were confronted not by one drone but by three: two Predators and a larger Reaper model equipped with Hellfire missiles.[23] For their second attempt to kill him, the Obama administration relied on drones to fire the missiles rather than manned aircraft. As the Predator drone affixed a laser target on the jeeps, the Reaper drone blasted a series of missiles at Awlaki and his companions, instantly killing them and burning their bodies beyond recognition. What was left of their bodies was later cut out of the jeep by local officials and given an Islamic funeral.[24] Amid the burning wreckage, villagers found a traditional Yemeni rhinoceros-horned dagger called a jambiya, known to belong to Awlaki.[25] Along with two Yemeni members of AQAP, another US citizen, Samir Khan, a North Carolina–born jihadi responsible for al Qaeda's English-language magazine *Inspire*, was also killed in the strike.[26] In this drone strike, the Obama administration had killed two of its own citizens—one intentionally, one accidentally—in a country in which the US government was not at war.

Shortly after the strike, President Obama offered a carefully worded acknowledgment of the killing, noting that Awlaki "was killed" in Yemen and calling the successful strike "a tribute to our intelligence community and to the efforts of Yemen and its security forces who have worked closely with the United States."[27] He made no mention of Samir Khan, although Obama administration officials later described him as "collateral damage."[28] He also made no mention of drones or of the crucial role that this technology played in enabling the deliberate killing of US citizens on foreign soil. He did not tell the US public that buried in a CIA archive was a strike video made in real time of the killing of Awlaki.[29] The strike was immediately controversial, as civil liberties advocates demanded to know the legal rationale behind the president's decision to kill a US citizen without trial or criminal charges. At the same time, drone technology itself remained widely popular with the US public, despite the fact that its features—speed, precision, adaptability, and remoteness—provided the essential foundation for President Obama's decision.[30] What was sometimes overlooked in the debate that followed was that it was the rise of drone technology itself that had opened the door for a global expansion of targeted killing. While the policy decision was justly controversial, it was the technology that allowed President Obama to make life-or-death decisions about individuals in foreign lands from the quiet remove of the Oval Office. That technology—and now that capability to kill around the world—would be given to President Donald Trump and all those who held that office in the future.

In this case, President Obama did not believe this was a difficult choice, calling the decision to kill Awlaki an "easy one."[31] But drones and their messy human consequences would continue to haunt him. In early September 2011, Anwar al-Awlaki's sixteen-year-old son, Abdulrahman, went looking for his father in southern Yemen. Born in Denver and also a US citizen, Abdulrahman had not seen his father for two years, but he knew that Awlaki was on a "kill list" drawn up by the US government. Abdulrahman stayed with family and friends in Yemen and learned of the death of his father while there. Two weeks later, he was spending the evening with friends in a café in Shabwa when a US drone strike killed him and five of his friends.[32] Abdulrahman was not deliberately targeted and was not involved in militant activity. US officials had received bad intelligence that suggested that an Egyptian al Qaeda operative, Ibrahim al-Banna, was in that café and struck the site to eliminate him. Abdulrahman and several of his cousins were killed instantly, with their shredded bodies left unrecognizable to their family. The US government refused to comment on the case despite an ACLU lawsuit and anguished calls by Nasser al-Awlaki for the US government to explain the deaths of his son and grandson.[33]

What Are Drones?

How did President Obama go from being a skeptic about the use of force, and mindful of his limits under the Constitution, to authorizing the extrajudicial killing of a US citizen on foreign soil? Is there something about drone technology that altered his choices or made his decision easier? The United States has long had the ability to see enemies around the world with satellites and to strike them with manned aircraft and precision-guided missiles, but only since the rise of drone technology has it done so on an almost daily basis. Equipped with what one former CIA director called an "exquisite weapon," Obama wound up authorizing hundreds of targeted killings, including of US citizens, during his time in office.[34] Yet even as this happened, he acknowledged that drones could affect his strategic choices and worried that they could be abused or misinterpreted as a "cure all for terrorism."[35]

To understand how drones might have influenced President Obama's strategic choices, it is important to begin by defining drones and identifying what makes them different. Drones are aircraft of varying size that do not have a pilot on board and are instead controlled by someone on the ground. Unlike missiles and other aerial projectiles, they are typically intended to be reused rather than simply rammed into a target. Most drones are not in and of themselves weapons, although they can be equipped to carry weapons of different sizes.

While most people use the term "drone" broadly, the language around drones is controversial. Within the US Air Force, for example, the terms "drones" and "UAVs" are generally rejected because they obscure the fact that all models are controlled by human beings even if the pilot is not physically located in the vehicle itself. These terms can be seen to imply that drones are flying in an autonomous way with relatively little control by a human on the ground. Although autonomous drones are on the horizon, this is not the reality today. The present generation of military drones is under direct human control, while the more sophisticated models—for example, the MQ-9 Reaper or RQ-4 Global Hawk—are no more automated than a commercial airliner. In fact, most large military drones are useless without a ground control station where pilots are located and a supply chain to ensure that the vehicle remains in the air.[36] Because pilots are so essential, there is a vast bureaucratic infrastructure around the training and supervision of pilots during flight. For this reason, some have argued that drones are better described as remotely piloted aircraft (RPA) or unmanned aerial systems (UAS), as these terms emphasize the human control over the technology and the degree of control and oversight that they feature. Others have argued that retaining the term "drone" is important, because other formulations are not popular and too "bland and bureaucratic" to portray the reality of what drones do.[37] Either way, it is important not to be too fixated on the term "unmanned," as in reality it is the interaction between humans—the pilots, but also those supplying their operations on the ground and setting policy—and the technology which determines what drones are capable of doing.

This book will employ the term "drone," to follow popular use, but it is important to bear in mind that not all drones are alike. Among the key characteristics that most drones share is that: (1) they are flown remotely; (2) they are capable of flight maneuvers; and (3) they are typically intended to be reused, unlike missiles and other disposable projectiles.[38] But on many other criteria—range, endurance, speed, and ability to carry a payload—drones can vary so substantially that they hardly look like the same technology. The most famous drone, the now-retired MQ-1 Predator, is closer to a military jet like an F-16 than to the small drones employed by hobbyists.[39] The Predator model MQ-1B, for example, has a wingspan of 55 feet, carries 2,250 pounds of weight upon takeoff and flies for 770 miles.[40] Its larger cousin, the Reaper (MQ-9), is similar, although it has a slightly longer range (1,150 miles).[41] These models can be flown by pilots located in ground control stations hundreds or even thousands of miles from the drone itself and are linked by satellites to an intelligence infrastructure that enables their use.

Compare this to the ordinary quadcopter drones employed by hobbyists and small companies for basic photography and delivery of goods. The popular Phantom 3 model drone, produced by the Chinese company DJI, has a weight of 2.64 pounds and can fly at 52.5 feet per second. At most, it can be flown 1.242 miles from the operator.[42] The Phantom 3 has a sophisticated camera, but it cannot carry a significant payload and is small enough to be carried in its operator's hand. In general, drones like the Phantom 3 must remain within the line of sight of the operator and have a standard radio link to that person. In many respects, they are closer to a hobbyist's model airplane than to an F-16 or any other type of manned aircraft. The term "drone" encompasses all of these models. To date, the public debate has largely focused on the Predator and Reaper drones, due to their prominence in the targeted killing program, but has paid comparatively less attention to the rapid diffusion of small quadcopter drones in other areas of life.

There have been a number of attempts to classify drones, but a classification scale originally produced by the US Air Force focused on two general characteristics of drones: their altitude and endurance.[43] In broad terms, some models are capable of only low altitude (such as several hundred feet in the air) while others can fly at the same height as commercial aircraft (approximately 30,000 feet). Some military surveillance drones can fly at even higher altitudes, such as 60,000 feet. The endurance for each model also varies substantially, with some small hobbyist drones capable of flying only a few hours and other, military models capable of remaining in air for more than a day. Table 1.1, drawn from that classification scheme, identifies the five tiers of drones and provides some examples of each.

Aside from the differences in the altitude and endurance of these drones, their functions vary across the different tiers. Drones have two distinct functions: (1) the *collection* and recording of images and data from the external environment; and (2) the *delivery* of payloads, ranging from food, medicine, and commercial packages to missiles. Some drones are capable of both, while others can do only one.[44] Most small drones are equipped to collect images and scientific data and can be used for a wide range of development activities, such as monitoring crops and environmental damage. They can also be used for reconnaissance on the battlefield. Today, commercial retailers like Amazon are working on building small drones that are capable of delivering parcels and other commercial packages, but this effort remains a work in progress. Multipurpose medium-altitude long-endurance (MALE) drones, like the Predator and Reaper models, can be used for both intelligence collection and delivery, as evidenced by their role in delivering the missiles that killed Awlaki. The high-altitude long-endurance (HALE) drones, such as the Global Hawk (RQ-4), are generally used for sophisticated collection tasks, such as advanced imagery surveillance,

TABLE 1.1 US Air Force UAV Classification System

Tier	Altitude	Endurance	Examples
Tier N/A	Very Low	Very Low	Hobbyist Drones, Micro-Drones
Tier I	Low	Prolonged	GNAT 750
Tier II (MALE)	Medium	Prolonged	Predator (MQ1), Reaper (MQ-9)
Tier II+ (HALE)	High (30,000 feet)	Prolonged	Global Hawk (RQ-4)
Tier III	Very High (60,000 feet)	Prolonged	Dark Star (RQ3)

Source: http://www.hse-uav.com/military_uav_platforms.htm, accessed July 27, 2015

and they involve technology not widely available on the commercial market. Both collection and delivery models can be used for peaceful purposes (such as surveillance, scientific research, and commerce) and for military purposes (such as reconnaissance and targeted killing). In the debate over their use, it is important to keep in mind the diversity of models and functions of drones, as those most widely discussed types and missions—the Predators and Reapers used for targeted killing—are hardly typical.

It is equally important to recognize what is new—and what is not—about drones. Many of the most widely discussed characteristics of drone technology—for example, the distance between the operator of the weapon and the target and the depersonalized nature of the violence itself—have arisen before with other military innovations, such as artillery, aerial bombardment, and nuclear weapons.[45] Drones are slower than most forms of manned aircraft and can be less precise than a cruise missile; they are no more automatic or dependent on computers than commercial airliners. Drones can see things on the ground just as satellites can, but not necessarily at a higher level of detail. Drones are less adaptable and nimble than manned aircraft and, at present, many military drones, such as the Reaper or Global Hawk, are less likely than a manned aircraft to evade enemy fire.

What makes drones different is that they combine characteristics seen elsewhere into a single technology that can be deployed at low cost. It is the amalgamation of these characteristics in a single technology that makes drones distinctive. They combine the speed and precision of a cruise missile with the durability and responsiveness of a manned aircraft. They convey images that are strikingly real and accurate like a digital camera or a satellite but are neither as small as cameras nor as removed from their subject as satellites. By combining features like speed, precision, adaptability, and remoteness into a single technology, drones become what is sometimes known as a "disruptive technology" and alter the strategic choices of those that have them.[46]

The Drone Age

It is hard to overstate the speed with which drone technology has emerged as a force in global politics over the last twenty years. The concept of a remotely piloted aircraft has long been a dream for militaries and aviation enthusiasts, but the technology was difficult to develop and early prototypes failed more often than they succeeded. As late as 2000, the US military had only a handful of drones, most of which were seen as unreliable.[47] By 2014, the total US military inventory had increased to over 10,000 drones.[48] Inside the military, the demand for drone overflights is insatiable. In August 2015, the Pentagon announced plans to increase drone overflights in conflict zones by 50% by 2019.[49] In December 2015, the US Air Force unveiled a plan to double the number of drone pilots, but even that may not be enough to meet demand.[50] Other countries are following the lead of the United States in embracing drone technology. The United Kingdom, for example, announced that at least one third of the Royal Air Force (RAF) fleet will be drones by 2030.[51] According to the Pentagon, China plans to produce as many as 42,000 land- and sea-based drones by 2023.[52] Between 2004 and 2011, the number of states with active UAV programs doubled from forty to more than eighty.[53] By 2017, according to an estimate by the Center for New American Security, ninety countries had developed some kind of unarmed drone technology.[54] Of these, thirty had programs for armed drones, but many more had the latent capabilities to do so.[55]

Governments are not the only organizations to embrace the drone age: international organizations, non-governmental organizations (NGOs), law enforcement, private companies, and even universities are also seeking drones for a variety of purposes. The United Nations and an array of NGOs are beginning to employ them for peacekeeping and humanitarian operations. Surveillance drones have proven useful in crisis mapping, search and rescue following natural disasters in Haiti and the Philippines, and monitoring refugee camps in the Democratic Republic of Congo and the Central African Republic.[56] They are also being used in economic development and environmental conservation. For example,—small surveillance drones are now being deployed to deter the poaching of elephants in Malawi.[57] Private companies are using drones to encourage economic growth in the developing world by monitoring crops, tracking weather events, and moving goods and essential supplies to hard-to-reach areas.

In the United States and elsewhere, law enforcement and emergency response organizations are discovering new purposes for drones almost daily. Local police across the United States have begun to deploy them to track suspects, to record videos of events, and even to produce heat maps which can help to rescue stranded hikers.[58] One Federal Aviation Authority (FAA) estimate in 2013 suggested that there would be 30,000 drones in US skies by 2020.[59] This turned out to be a wild underestimate. By 2016, the FAA predicted that "small, hobbyist UAS purchases may grow from 1.9 million in 2016 to as many as 4.3 million by 2020. Sales of UAS for commercial purposes are expected to grow from 600,000 in 2016 to 2.7 million by 2020. Combined total hobbyist and commercial UAS sales are expected to rise from $2.5 million in 2016 to $7 million in 2020."[60] Another estimate suggested that 200,000 drones are sold worldwide each month.[61] The FAA has estimated that the global market for drones will be worth approximately $90 billion between 2013 and 2023.[62]

The rapid pace of innovation in UAV technology has also led to dozens of commercial applications. Among the most famous of these is the proposal by Amazon to deliver packages by drones. Amazon's founder, Jeff Bezos, has promised that seeing one of their drones in the future will be "as common as seeing a mail truck."[63] Small surveillance drones have now been used by real estate agents to photograph properties and by electricity companies to check for downed power lines. Farmers are also using drones to monitor their crops and check for disease. In 2015, a small Australian company called Flirtey undertook the first successful delivery of a package by drones, dropping medical supplies at a rural clinic in Virginia. While this flight was relatively short—the drone carried the package only 1 mile—Senator Mark Warner (R-VA) called it a "Kitty Hawk" moment in drone aviation.[64] While the FAA has been slow to approve commercial drone deliveries, various lobbies and major corporations are pressuring the US government to open the skies and to allow drones to do even more.

One of the many reasons why drones are spreading so rapidly to so many domains is cost. For many governments, private companies, and individuals, drones are a way to take to the skies at a remarkably low cost. For example, the top-of-the line Predator model costs approximately $10.5 million, compared to the $150 million price tag of a single F-22 fighter jet.[65] While new models such as the Reaper are more expensive to fly ($12–13 million) and to maintain ($5 million per year), their cost still compares favorably with manned aircraft.[66] In general, the drones purchased for military use are vastly more expensive than those on the commercial market. Smaller commercial models, such as quadcopters capable of modest tasks like the collection of imagery, can be bought for hundreds of dollars through commercial retailers like Amazon or Best Buy. As the price drops every year, more organizations get the ability to fly, record images, and collect data. Due to their low cost and suitability for routine tasks, small drones are spreading with astonishing speed to a growing array of users.

What impact will the rapid diffusion of drones have? It is difficult to say because of how suddenly all of this change has come. Consider the comparison with manned aircraft. The first flight was conducted by the Wright brothers in 1903 in North Carolina with a rudimentary machine that could fly only a small distance. Although a press release was issued at the time, relatively little attention was paid to their experiment until the commercial and military implications of the technology became clear. Early planes were relatively crude, often unreliable, and had a limited ability to carry freight or passengers over a significant distance. In the words of one aviation historian, "the world still regarded airplanes as toys, and those who dared to fly them were looked upon as crackpots or madmen."[67] World War I changed all of that. Recognizing that aircraft could be an effective weapon, the British, French, and German militaries poured resources into making aircraft more reliable and lethal.[68] By 1918, fifteen years after the Wright brothers announced the first heavier-than-air manned flight, world-changing advances in the technology enabled dozens of new uses for aircraft including mail delivery.[69] By 1919–1920, the first steps toward passenger aviation had been taken, although large-scale commercial flying would not become a reality for another decade.[70] The full consequences of the Wright Brothers' innovation were hardly clear twenty years after the technology first made the news. It was another several decades before the extent to which the world had changed due to manned aircraft was known.

We are in a similar situation today with drones, although they do not represent as dramatic a change as the beginning of manned flight. It has been approximately twenty years since the Pentagon first developed models that could regularly be used for counterterrorism. Those initial drones, like the aircraft of the Wright brothers and other aviation pioneers, were prone to crashes and often far less effective than hoped. They were also widely ignored, with few realizing the vast commercial and scientific potential that they represented. As in the case of manned aircraft, it was war which spurred more research into drones and gave rise to new applications for their use. Once the military had adopted drone technology, allowing expertise to be established, commercial applications began to flourish. Today, this has resulted in a Wild West atmosphere in the field, with dozens of actors—governments, NGOs, private companies, even individuals—throwing out new ideas for their use. While many of these ideas will fail, a few will succeed and begin to quietly change our daily life. Over time, drones will be seen as manned aircraft are today: as a technological achievement once seen as extraordinary but now so ubiquitous that most people take them for granted.

What is different about the diffusion of drones is that they have wound up in the hands of a greater number of users than traditional aircraft ever did. We will soon be living in a world in which everyone can get their hands on a drone and take to the skies, for good or bad reasons. The spread of drones has effectively democratized the air: allowing many actors (individuals, terrorist groups, NGOs) into the skies for the first time, while enabling others long in the air (governments, private companies, the United Nations) to do things that they never thought possible. By expanding the number of people who can take to the skies, drones disrupt the traditional "knowledge monopolies" over information gleaned from the air held by the military and commercial aviation authorities.[71]

As a result, drones will empower a range of states and non-state actors, such as civilians, terrorist groups, and even humanitarian organizations, and change how they interact across the world. The net effect of a world full of drones will change over time. In the short term, the availability of drones will entrench the advantages of powerful countries like the United States, which has a vast fleet of drones that allows it to fight in a growing number of conflicts without risking the lives of pilots. At home, the availability of drones will initially work to the advantage of law enforcement and big companies like Amazon, who will use their resources to scale up drone use and take advantage of open airspace. But in a drone age these advantages will not last forever. In the long run, latecomers to the drone age will catch up. In time, we will see weak states use drones to strike against stronger ones, and other equally matched states begin to deploy drones against each other in a war of nerves. We will see non-state actors like rebel groups and terrorists use drones to attack governments and even civilian targets from the air. We will see authoritarian governments deploy drones to watch and even suppress their own population. We will also see drone technology become folded into the arsenals of humanitarian organizations and peacekeepers, amplifying their ability to understand the dangers facing and protect the vulnerable in war zones and other disaster areas. Over time, all of these organizations will behave and interact differently because of the impact of drone technology on their choices.

What Drones Do

The Awlaki case gives us some glimpse of what drones can do. Barack Obama was not the first US president to face an enemy abroad who meant harm to the United States. He was also not the first president to try to eliminate his country's enemies with the most advanced technology available. From far-fetched assassination plots against Fidel Castro to bombing raids against Muammar Qaddafi, US presidents have long sought to kill their enemies with a variety of tactics that sidestep legal prohibitions against murder and assassination. But they had to rely on costly weapons (e.g., cruise missiles) and ones that had a prohibitively long delay between seeing a target and being able to strike it. President Obama was the first to have sustained access to a low-cost technology that enables killing to occur almost immediately, while at the same time facing no risk to the lives of US personnel. Drone technology did not create the practice of targeted killing, but it enabled it to the degree that it is barely noticed by most of us when it is reported in the newspapers each day. Inside the government, the availability of drone technology quietly changed the doctrines and practice of the US military and intelligence agencies and led to the creation of a "kill list" of potential targets. It has also led to a series of shadow wars in places like Pakistan, Yemen, Somalia, and Libya, which together are challenging the conventional models of counterterrorism and blurring the lines between war and peace. All of this is possible because drone technology produced a stepwise change in what states are able, and now willing, to do.

The chief argument of this book is that drone technology alters the strategic choices of its users, governments, and non-state actors alike, in two ways. The first is by transforming their risk calculation. Perhaps the most salient fact about drones is that they enable their users to get into the sky without risking the life of a pilot. This is naturally attractive for all of the obvious reasons: everyone wishes to do things with the least risk possible. But this change in the calculation of risk is especially important today because we live in what Ulrich Beck has described as a "risk society"—that is, one in which finding ways to avoid the hazards of life and misfortune is of paramount importance.[72] As an example, consider the ways in which the growing aversion to risk has affected the way in which militaries fight their wars. In 1994, the strategist Edward Luttwak made the controversial argument that the great powers, especially the United States, Britain, and France, were increasingly eager to avoid battlefield casualties because of their low birth rates.[73] Societies with fewer children, he argued, are less inclined to see death in battle as heroic and more inclined to consider it as a personal or even national tragedy. The result is that they seek ways to use alternatives, like technology, to fight their wars rather than risk the lives of their sons and daughters.[74] Similarly, in 2001, Michael Ignatieff observed that many governments want to fight what he calls "virtual" wars: clean ones, conducted from the air, and with the loss of almost no lives on their side.[75] Today, many governments try to protect their personnel from harm in war zones by insisting on strict rules for force protection and engagement with the enemy.[76] This fact has led some scholars to suggest that we are entering an era of post-heroic warfare in which traditional notions of honor have given way to a more complex calculation of risk and reward.[77] If war is reconceived as an exercise of risk management, as some have argued, it follows that governments will prioritize and exploit technologies which mitigate or even eliminate risk.[78]

From this perspective, it is not hard to imagine why drones are so attractive: they appear to offer a solution to avoiding casualties on the battlefield because they keep pilots out of harm's way, often thousands of miles away from the scene of the fighting. Most US military drone pilots are working from small bunkers on bases across the continental United States and are able to go home to their families at night. This avoids a significant "footprint" of military personnel on the ground, which means that there is little risk for those supporting the operations and, at least in theory, a reduced risk of backlash from the local population.[79] To governments, drones appear to be almost a magic bullet: a way in which one can fight wars without incurring the risk of deaths that warfare normally implies.

On some level, drones appear to minimize risks for civilians as well. US officials have argued that drones provide governments with the ability to wage war with such high levels of precision that they minimize civilian casualties to an extraordinary degree.[80] With targeted killings, drones can follow prospective targets for a long time and determine if their activities constitute involvement in combat, thus making them a fair target under the international law of armed conflict. Even more, drone pilots can consult with an array of experts, such as lawyers and intelligence specialists, to ensure that the target is a legitimate one.[81] They can also wait to strike the target until civilians have cleared out of the area. Given this, some argue, drones might actually be morally preferable to ground combat.[82] At a minimum, they might allow for a more deliberative form of warfare where the combatants are given time to anticipate and reflect on their decisions without fearing for their own lives in the midst of battle. The political advantages are also clear: because drones appear to offer a clean form of warfare which minimizes civilian casualties, they may make longer military campaigns more sustainable for a casualty-sensitive public.

The problem with these arguments is that they derive from the myth that drones eliminate risk, or at least radically reduce it. In reality, drones do not eliminate risk but rather redistribute it. A closer look at how drones operate with targeted killings illustrates this point. First, military drone pilots are not as insulated from the costs of war as they are assumed to be. Some have reported higher levels of post-traumatic stress disorder (PTSD) because they view the horrors of warfare more directly than traditional pilots do.[83] Drone pilots often follow their targets for hours and watch them interact with their wives and children. When their targets are killed, the drone's camera lingers for a long time over the strike zone, allowing the pilots to see the bloodshed that their actions caused and the horrified reactions of family and first responders. This does not happen once, but continuously. As one pilot remarked, this is "war at a very intimate level."[84] An Air Force study found that drone pilots struggled with psychological distress, including emotional numbness and distance from families, as well as trouble sleeping caused by "images that can't be unseen."[85] Drones may reduce ordinary physical risk for pilots but they also may increase the psychological risks, including that of depression and burnout, which all pilots face.

Second, drones may allow the United States to conduct targeted killings in places like Yemen, but they put US troops at risk of similar attacks in other theaters of war. For years, US military officers in foreign wars did not need to fear attacks from the sky because the United States established air superiority almost immediately in any war it fought. But today terrorist organizations are adapting commercial drones to take to the skies and to attack US troops with improvised explosive devices (IEDs). Here the pace of change is stark. In 2014, the Islamic State had just begun to employ rudimentary commercial models for surveillance over battle sites in Syria.[86] Within a year, the Islamic State had mastered the use of drones for reconnaissance and propaganda over the battlefields in Syria and Iraq. By 2017, US forces had come under attack from the air for the first time in years. Near al-Tanf, Syria, US advisors and their Syrian allies found themselves shot at by a drone, likely made by Iran and operated by pro-Assad rebels, in an airborne attack that would have been unthinkable only a few years before.[87] Although the United States destroyed that drone with an F-15 fighter, it nevertheless represented a dramatic illustration of how drones are shifting risks long thought virtually eliminated back onto US troops. As a result, the Pentagon is pouring money into anti-drone technology to prevent its soldiers from being attacked from the air, a risk it has not regularly faced for decades.

Third, drones can shift risks to people on the ground. When the war on terror started in 2001, people living in the tribal areas of Pakistan and Yemen were not being watched by drones and did not face the risk of being accidentally caught up in drone strikes. Today, their reality is very different. US drones offer nearly continuous coverage of conflict zones around the globe as they seek out al Qaeda, Islamic State, and other affiliated forces. Even with the precision offered by drones, civilians are accidentally killed, and even those left untouched by drone strikes must deal with the negative psychological effects of living with the risk of drone strikes. The US journalist David Rohde described the psychological effect of living under drones after he was captured by the Taliban and held hostage for seven months in Afghanistan in 2008.

> The drones were terrifying. From the ground, it is impossible to determine who or what they are tracking as they circle overhead. The buzz of a distant propeller is a constant reminder of imminent death. Drones fire missiles that travel faster than the speed of sound. A drone's victim never hears the missile that kills him.[88]

In the words of a Yemeni child later killed by a CIA drone strike, living under them has "turned our area into hell and continuous horror, day and night, we even dream of them in our sleep."[89] This fear also has political consequences: it leads people to turn on each other for allegedly cooperating with the CIA and to become disenchanted with their own government for allowing these strikes to take place.[90] All of these risks are effectively offloaded onto the local population in the favor of reducing physical risks to US drone pilots.

Finally, and perhaps most importantly, the use of drones shifts the risk calculations of decision-makers and produces a danger of moral hazard: a condition in which someone increases their risk-taking behavior because they feel insulated against the consequences. Drones may allow for precise, careful wars, but they may also make wars easier to fight and thus more frequent.[91] Because drone technology is low risk for their own soldiers, governments may be more willing to use force rather than seek other, non-violent means of addressing the problem. This can be seen in the expansion of the shadow wars beyond Pakistan and Yemen to more locations worldwide. Drones may not lead pilots to become heartless and indifferent to the lives of civilians killed by a push-button war, but they may lead politicians to become more risk-taking and to fight wars on an ever-greater number of fronts. These might be slow-burning, remote wars, but they have real consequences for those living where they are fought. To be watched by drones on a nearly constant basis and perhaps to be killed suddenly in a blinding strike, is a life that that none of us would want. Seen from this perspective, drones are producing a novel disjuncture in the calculation of risk, with those under the gaze of drones directly acquainted with their risks but governments more removed, and thus perhaps more risk-taking with the decision to use them.

The second major way that drones affect strategic choice has to do with goal displacement: the process by which an organization enlarges its ambitions and begins to substitute alternative, sometimes more expansive, goals for the ones that it originally had. This process, known to the military as "mission creep," can be driven by two interrelated factors. The first is cost. When technology like drones are cheap, governments and other actors can afford to use them on an ever-greater number of tasks, including those only indirectly related to their original one. The reasoning behind this—that drones are cheap, so we can throw them at a problem and see what happens—can be clearly seen in expansion of the US targeted killing program. Under the Obama administration, the original purpose of the targeted killing program was to target al Qaeda and "associated forces," although that phrase was left undefined. Over time, the United States began to use drones not just to destroy al Qaeda and its immediate allies, but also to strike an array of other Islamist groups around the world, including the Haqqani network and Tehrik-e-Taliban in Pakistan, al Qaeda in the Arabian peninsula and its allies in Yemen, and even al-Shabaab in Somalia. It also began to involve itself in eliminating tribal enemies of the governments in these countries.[92] Because they were so inexpensive and expendable, US drones drifted into becoming the "counterinsurgency air force" of these governments, implicating the United States in conflicts that it did not fully understand.[93] Over time, drone strikes also became a substitute for a political strategy, with the United States abjuring traditional diplomacy to resolve conflicts in favor of building drone bases around the world to police them. What began as an effort to remove specific individuals from a battlefield drifted into an effort to conduct aerial policing over larger swaths of the globe.

Goal displacement can also come about because of the nature of the technology itself. Drone technology not only produces a rich stream of information, including photographs, video, and other types of data, but it also allows its users to see the world in a more vivid, textured detail than ever before. Drones also offer the promise of being able to monitor a place not just for a limited time, like a satellite, but rather on continuous basis. For governments, this is often a good thing, as a greater degree of information generally has a beneficial effect on decision-making.[94] No one would ever want to know less about a crisis that they face. But technology that yields vast increases in information and data can sometimes lead to false confidence among its recipients about their ability to master their external environment and to tame the uncertainties that they face. This false confidence can, in turn, feed on itself and encourage its users to pursue new goals and to lose sight of their original ones. Just as the sketching of maps laid out new worlds for imperialists to conquer, technology that allows us to know more can also produce a corresponding impulse to do more and, in some cases, to control more.

This dynamic can be seen with the US military's desire to deploy drones for "dominant battlespace knowledge" over present and future battlefields. In less than two decades, the United States has gone from using Predator drones selectively over a few conflict zones to building a vast surveillance system and fielding ninety combat air patrols (CAP) of drones per day.[95] Each CAP consists of typically four drones, as well as nearly two hundred people supporting them.[96] Although some officials recognize that the demand for drones is unsustainable, it has not stopped the appetite for drone imagery from increasing. The United States has now developed a drone surveillance system called Gorgon Stare—named after the figure from Greek mythology whose look could turn someone to stone—designed to continuously watch whole cities and record video of all movements on the ground for thirty days.[97] The ability of the United States to see everything with drones has produced a cultural change within the military, prioritizing the acquisition of information as an essential goal in and of itself. Aside from fueling false confidence that the United States may eventually control the battlefield, banish risk, and even transform the nature of war, it has also led the United States to displace its original goal—to fight al Qaeda more effectively—in favor of a larger one of knowing, and possibly even controlling, greater portions of the earth than it had previously imagined possible.

This argument—that drones alter strategic choice through the recalculation of risk and goal displacement—does not imply that drones are necessarily good or bad. It accepts that drones, like all technology, are subject to human misuse.[98] It also does not suggest that these changes in risks and goals are exclusive to drones; other forms of technology may produce similar effects. But this book seeks to show that the introduction of drones has fundamentally changed how we understand the strategic choices that we face in war and peace. These changes are often subtle: just like manned aircraft once did, drones are altering the field of vision of those that employ them and changing the habits of mind of their users. The French sociologist Jacques Ellul long ago argued that technology is accompanied by "technique," a mode of thinking that emphasizes efficiency and instrumental rationality, but also dehumanizes those subject to the technology and banishes discussions of its moral dimensions.[99] Some elements of this can be seen in the US government's defense of the targeted killings, which emphasizes the efficiency of drones while sidelining the essential moral questions their use raises. As Ellul noted, technique can even go as far as to redefine what certain words mean and recast controversial behavior away from morally loaded terms (assassination) toward more conducive ones (targeted killing).[100] Through altering our risk calculations and goals, drones go beyond being a mere tool, but rather have distinct political qualities that shape how we act.[101] It is these qualities which helped to turn Barack Obama from being a skeptic of the US war on terror into the first president to authorize the targeted killing of a US citizen without a trial.

Consequences for War and Peace

This book explores six ways in which drones affect risk calculations and goals of their users, government and non-state actors alike, and have consequences for war and peace in the twenty-first century. First, drones undermine the long-standing legal and ethical prohibitions on assassination and extrajudicial violence outside of wartime. Chapter 3 will trace the emergence of the practice of targeted killing from its origin to its embrace by the United States after the September 11 attacks. It shows how the United States adopted drones alongside the practice of targeted killing to control risks as it fought a new war against al Qaeda, but found itself gradually drifting into more conflict zones and fighting new enemies. While the United States used drones to protect its pilots from physical risk, it altered the nature of the risks that they faced and created new risks for the population who live under the drones' watch. Drone technology also subtly changed how the United States wages its wars, making it more willing to countenance killing people off traditional battlefields, while undermining the standards of accountability and transparency that it has traditionally employed. Over the last twenty years, multiple administrations have shielded the targeted killing program from Congressional, judicial, and public scrutiny, thus pushing the United States' embrace of drones deeper into the shadows. But the world has paid attention. Today, as more countries are getting drones, they are experimenting with targeting killings of their own and eroding the traditional barriers against killing people outside of wartime.

Second, this book argues that drones accelerate the trend toward information-rich warfare and place enormous pressure on the military to learn ever more about the battlefields that it may face. For the Pentagon, the collection and delivery of information to soldiers on the ground has long been essential to controlling the "battlespace" and ensuring that its operations are precise and effective. Over twenty years ago, the US military was openly declaring its desire to develop "dominant battlespace knowledge" integrated into a "system of systems" that will give it a decisive advantage over its opponents.[102] Today, aided by drones, that desire is accelerating as the Pentagon increasingly conceives of war as a contest for information. This has had an organizational effect: the ability for the United States to know more through drone imagery has turned into a necessity to know more. As chapter 4 shows, the dangers of overreliance on images and data from drones are multiple and clear: militaries may underestimate the risks that they face on the battlefield due to overconfidence based on superior information coming from drone imagery. Alternatively, they may become so overloaded with information that they cannot convert imagery into intelligence, thus slowing the tempo of kinetic action so they can learn more. This might work with targeted killings, which operate on the logic of execution rather than combat, but it will not work if future wars are as lethal and fast as predicted.[103] The US military is becoming so enamored of its ability to know more through drone surveillance that it is overlooking the operational and organizational costs of "collecting the whole haystack."[104] Using drones for a vast surveillance apparatus, as the United States and other countries have been doing, has underappreciated implications for the workload, organizational structures, and culture of the military itself.

Third, as chapter 5 shows, drones allow non-state actors like rebel and terrorist groups to level the playing field with powerful governments and to expand their goals as a result of being able to take to the skies. Traditionally, conflicts between governments and their non-state challengers are marked by asymmetry of power, with governments far better equipped with resources to defeat their non-state opponents. This pattern will remain intact, but the low marginal cost of drones will be a boon to non-state actors who will use them to blunt the advantages that governments have. In conflict zones, drones enable groups like Hezbollah and the Islamic State to watch the battlefield in a way that they never could before and to strike at their enemies in surprising ways. As the example of the Islamic State shows in its campaign in Iraq and Syria, drones may even help some of these groups to hold on to territory and make more powerful militaries pay a cost in lives for fighting in the open. At home, drones will enable terrorist organizations to strike civilian targets and perhaps even to assassinate world leaders. While the United States and a few other powerful countries had a virtual monopoly on drone strikes for many years, the practice is now coming home and distributing new risks to civilian populations in North America, Europe, and elsewhere who previously had little reason to fear terror attacks from above.

Fourth, drones will transform the dynamics of protest and surveillance in democratic and non-democratic states, as chapter 6 will illustrate. It is well acknowledged that the use of drones by law enforcement and private companies poses a serious challenge to the protection of privacy and could contribute to a creeping surveillance state. What is less noticed is how drones have empowered new groups—civil liberties groups and activists—to identify and publicize human rights abuses and to press democratic governments for action. In democracies, we are moving to a world in which both police and protesters will be empowered by drones in contrasting, perhaps offsetting, ways. But the stakes around drones are even higher in authoritarian states. The embrace of drone technology by authoritarian states raises the possibility that anonymous dissent may eventually become difficult if not impossible. Surveillance drones may allow authoritarian governments like Russia to monitor protesters and to punish them for opposing the government.[105] If that happens, drone technology may actually diminish the chances of democratic reform. Drones may also provide another tool for governments to engage in surveillance and repression of their secessionist regions.[106] As they have for China in Xinjiang, drones may also enable powerful governments and some non-state actors to experiment with new forms of political and social control in less governed territories, discarding the legitimate grievances of repressed populations and managing them with ever-increasing levels of drone surveillance.[107]

Fifth, drones will enable international organizations, NGOs, and advocacy groups to monitor human rights abuses, deliver relief, and pressure governments for change. As chapter 7 discusses, small surveillance drones are ideally suited for taking on the "dull, dirty, and dangerous" jobs that are needed in these situations.[108] In the future, drones will be able to transport and drop food and medicine in crises where humanitarian organizations are reluctant to send their own personnel. Drones will ultimately give these actors another tool by which they can monitor events on the ground and possibly shame governments into stronger action. But there are risks on the horizon too. More than just another tool, drones may also increase the ambitions of international organizations (IOs) and NGOs to intensify the pace of humanitarian relief and social change, even if doing so is unsustainable. While some international organizations and non-state actors feel considerable unease over adopting drones because these so-called "bots without borders" bring with them a host of logistical, political, and ethical obstacles, an activist community is taking matters into its own hands, launching bold but sometimes unsustainable interventions in an increasing number of locales worldwide.

Finally, drones will amplify the competition for power and influence between states in conflict zones and produce new risks of deterrence breakdown and crisis escalation. Chapter 8 shows how states are already using drone technology to test the nerves and strategic commitments of their rivals. This is because risk calculations with drones have changed: what was once too dangerous to do with a manned aircraft is now possible with a drone. Today, we live in a world where India and Pakistan are flying drones over the disputed territory of Kashmir and where North Korea flies rudimentary drones across the demilitarized zone (DMZ) to provoke South Korea. These strategic gambits are now possible because of their low financial and human cost, as well as the illusion that drones can be used without the risk of conflict escalation. As drones are used in more conflict zones around the world, they will begin to quietly reorder the risk calculations behind deterrence and coercion and produce greater chances of miscalculation, error, and accident.

The Way Ahead

The six consequences of the diffusion of drone technology will be explored in the main chapters of this book. Throughout, we will see how drone technology altered strategic choices of its users through changes in risk calculations and goal displacement. It was these changes that enabled President Barack Obama to kill Anwar Awlaki on that bright September morning. To understand how drones led to this, it is important to return to the beginning of the story, when drones were an impractical hope of aviation enthusiasts. Chapter 2 traces the history of drones, with all of its fits and starts, and shows how the dream of unmanned flight was turned into a reality by the necessities of war.

2

Automated Warfare

BY LATE 1944, THE tide of World War II had turned against Nazi Germany. Under the leadership of General Dwight D. Eisenhower, Allied forces were preparing for an invasion of Europe to liberate occupied countries and destroy the government of Adolph Hitler. Nazi military forces were stalemated or losing on both fronts, while their leaders were getting glimpses of the ruin that awaited the German people once the invasion began. This dark future left the Nazi leadership desperate for ways that they could forestall a disaster and, if not, punish their enemies with a last, desperate strike. German scientists were put to work by Hitler to create what he referred to as "vengeance weapons" which could strike at civilians in the United Kingdom and other Allied countries.[1] Among these were the V-1 rocket, otherwise known as a "buzz bomb" or the "doodlebug" for its buzzing sound as it slowly flew over British cities, and the terrifying V-2 rocket. Fueled by liquid propellant and equipped with a sophisticated gyroscope for direction finding, V-2 rockets were lightning fast—they could reach their target in under five minutes from launch without warning—and extremely lethal.[2] One estimate suggests that V-1 and V-2 rockets killed more than 8,000 people and injured another 22,000 in Britain alone.[3] In some ways, the psychological effect was almost as bad as the physical impact. In language that echoes the contemporary criticisms of drones, the writer Evelyn Waugh described the V-1 rocket as "impersonal as a plague, as though the city were infested with enormous venomous insects."[4] Beyond their impact on their victims, the V-1 and especially the V-2 rockets were an alarming symbol of the technological sophistication of the Nazi regime. Under the leadership of Werner Von Braun, German scientists had created nearly space-age technology with a V-2 rocket that could fly hundreds of miles to a preprogrammed destination according to an automatic guidance system. Aware that Germany had plans for even more lethal rockets and nuclear weapons on the drawing board, Allied leaders began to consider ways to destroy V-1 and V-2 launch sites to slow their progress. The problem was that V-1 and V-2 launching pads located on the coast of northern France were difficult to hit. Most of the contemporary Allied missiles were not precise enough to strike them directly or powerful enough to shatter their steel and concrete frames.[5]

To tackle this problem, US Air Force officials began to consider ways to turn nearly out-of-service aircraft into missiles that could strike hardened German bunkers along the coast. These explosive-laden planes would fly low, at around 2,000 feet, toward the target but be followed by a larger plane, dubbed a mothership, which would adjust its auto-pilot functions and help steer it to its final outcome.[6] To hide their approach in daytime, they were painted yellow or white. At the time, these "aircraft turned cruise missiles" were called drones because they were controlled by radio from another aircraft.[7] According to US Air Force Chief General Henry "Hap" Arnold, these drones were desirable because "if you can get mechanical machines to do this, you are saving lives at the outset."[8] Working with General Carl A. "Tooey" Spaatz, Arnold ordered the Air Force to retrofit B-17 planes with the newest military autopilots and radio systems and to pack them full of explosives. Their secret plan, called Operation Aphrodite, began test flights in Florida and Britain to see if strikes against German missile bunkers would even be possible.

As the plans moved forward, Air Force officials gradually hollowed out both B-17 and B-24 planes and transformed them into remotely controllable cruise missiles. They packed the planes, newly dubbed BQ-7s, with 30,000 pounds of RDX explosives by stripping out every piece of payload not necessary for converting the plane into a flying missile.[9] These planes were heavy, weighing over 64,000 pounds, but still airworthy. The biggest technological problem that they faced was the remote control of the aircraft. While the mothership flying behind the explosive-laden aircraft could make adjustments to the trajectory of a BQ-7 in flight and even convey television pictures of the cockpit and control panel, Operation Aphrodite required pilots to be in the aircraft to get it airborne and to point it in the right direction of the target. The control of the plane from the mothership was not sophisticated enough to allow for the plane to take off on its own as a truly unmanned vehicle. It was also not good enough to ensure that it was going toward the right target. The remote control system, originally designed for directing weapons to narrow railroad bridges, did not allow for the precise targeting of its destructive payload across significant distances.[10] As a result, two airmen—a pilot and an autopilot technician—would need to be on board for the initial phase of the flight. The pilots would get the plane into the air and make sure it was headed toward the intended target. If all went to plan, the pilots would bail out with parachutes over the English Channel and later be collected by boat.

The problem was that these missions rarely went to plan. The dry-run tests were beset by technical problems and difficulties reaching their targets. Some planes—nicknamed "baby ships" or "topless roadsters" by the pilots—strayed far off course. Others were shot down by friendly fire.[11] Although the English antiaircraft artillery crews were warned not to fire on the flights, they often did, causing the pilots to frantically bail out.[12] Some pilots found it difficult to get out of the plane in time, with one pilot winding up in the English countryside trying to explain to a puzzled crowd how he got there.[13] On August 4, 1944, during a particularly disastrous test run of four BQ-7s, one aircraft exploded and the other three missed their target.[14] Two airmen—a pilot and an engineer—were killed in the disaster, and two acres of English countryside were scorched in the fireball that followed one of the crashes.[15] A few days later, additional tests failed, resulting in the death of another pilot when his parachute did not open.[16] By one estimate, none of the eleven test flights of Operation Aphrodite were successful.[17] Yet the test flights continued, and the US Navy even joined in the exercise, retrofitting PB4Y-1 bombers (equivalent to B-24s, also called Liberators) to be flown at the German submarine pens in Heligoland.[18]

In September 1944, the US Navy set its sights on a massive underground complex in France, called the Fortress of Mimoyecques, where, it was rumored, Germans were developing next-generation V-3 rockets.

The target was even more difficult than usual: the navy would need to guide the plane into a doorway of the fortress, no more than 15 by 16 feet, which was the most vulnerable part of the structure.[19] To hit such a precise target, they needed skilled pilots to fly a B-24 Liberator (renamed by the navy a B-8) packed with Torpex explosives to the target. Among the volunteers was Joseph P. Kennedy Jr., the older brother of future president John F. Kennedy. The oldest Kennedy brother was already being groomed for high political office by his ambitious father and had served for two tours of duty with the US Navy flying B-24 Liberator planes. He volunteered for the mission despite being eligible for state-side duty. He was publicly optimistic about its outcome, but privately he was aware of the risks involved. Before the mission commenced, Kennedy passed a message to a family friend: "if I don't come back, tell my dad—despite our differences—that I love him very much."[20] Paired with co-pilot Wilford "Bud" Willy from Texas, Kennedy took off on August 12 from an airfield near Norwich.[21] Shortly after the handoff of remote control to the mothership, less than a half an hour into the flight, the plane exploded in an enormous fireball, raining down debris on the English countryside.[22] No trace of Kennedy or Willy was ever found. The Kennedy family lost one of their favorite sons, one tipped for high political office, in what the navy only described as "an exceptionally hazardous and special operational mission."[23]

The disaster was ultimately investigated by an Air Force top secret panel which identified a number of technical reasons for the disaster, but could not agree on a single cause. Some believed that it was due to a technical fault; others speculated that signal jamming may have detonated the charge and produced the explosion.[24] The navy concluded that the Torpex explosive was too unstable to be reliably used and soon replaced it with TNT. As a program, the conversion of these old air-craft into flying bombs was deemed an abject failure. Almost no Nazi targets were ever hit, and the program was canceled as retreating Nazi forces abandoned the missile launch sites along the French coast.[25] For the Kennedy family, the disaster was compounded by the navy's de-cision to keep the details of the failed mission a secret.[26] This disaster also subtly altered the course of history. It transferred the ambitions of the Kennedy family to the younger son, Jack, who was also serving in the navy at the time. The persistence of the US military with this dangerous project showed how far it would go to achieve something it had long dreamed of: the ability to fly and to strike long-distance enemy targets precisely and by remote control. This dream, long in the making, would ultimately not be realized for nearly another fifty years.

Early Drone History

The early history of drones is not a linear or clean one. It is a story of mistakes, false starts, disasters, and failed prototypes.[27] There were dozens of prototypes and failed experiments that contributed in varying ways toward the drone models that we see today. These prototypes were also not often seen, or described, as "drones" in the modern sense, but they did help to clarify and resolve some of the technical issues that afflicted early drone development. The evolution of the drone was also inextricably tied to the development of satellites and cruise missiles; indeed, in many respects, the technology behind Kennedy's failed mission was closer to modern cruise missiles than to today's drones. The early history of drone technology is also not a story that could have been told without the financial support of the military. Like the Internet itself, drone technology was incubated over decades by the military-industrial establishment and was used for secret military operations for years before appearing in the public domain and broadened their capabilities. Drones also might have also remained an obscure, military-only technology in the absence of significant technological changes—for example, the rise of Global Positioning Systems (GPS) and satellite communications—that spilled from the military into the commercial domain.

The earliest evidence of the thinking behind drones lies in efforts to create remote-controlled cruise missiles at the beginning of the aviation age. In some respects, the effort to create unmanned aircraft predates the revolution that the Wright brothers kicked off with manned aircraft. The eccentric but brilliant inventor Nikola Tesla was among the first to propose a remote-controlled cruise missile in 1898, creating a prototype called the "teleautomaton" which he later tried to mass-produce.[28] One of his colleagues, Elmer Sperry, was the first to solve several of the technical problems that Tesla encountered—for example, stabilization of the plane in mid-air and a crude version of remote control by radio to allow the plane to be steered in mid-air. The steering was limited, however, as these aircraft were generally able to fly only toward a pre-determined location. Funded by the US Navy, Sperry's experiments were designed to develop an "aerial torpedo" to be launched into German U-boats and other targets, much in the same way that Kennedy's craft would later aim to do.[29] He had to experiment with different ways to get these planes into the air, including most notably attaching the aircraft to the top of his car and driving it at 80 miles per hour down the Long Island parkway to create a wind tunnel. The airplane nearly lifted the car off the track once it hit flying speed.[30] By March 1918, he had devised a way to launch the bomb-laden unmanned aircraft via a catapult system that would allow it to fly approximately 1,000 yards and still be reused.[31]

This issue—reusability—would later become central to the distinction between drones and cruise missiles. Today, the typical assumption is that drones are reusable but cruise missiles are not. But in the early twentieth century, the distinction was not as clear. Many of the forerunners of drones were explicitly designed to be flying bombs or cruise missiles. Some advocates of unmanned aircraft experimented with reusable prototypes; others worked with crafts that were deliberately expendable. Many of the technical issues—remote control, precision, and payload capacity—were the same. As usual, wartime necessity forced inventers to confront these problems and develop solutions. In World War I, British military officials commissioned an eccentric engineer, Archibald Low, to create a radio-controlled aircraft that could be used as a flying bomb against the Germans.[32] The prototype worked and showed in-flight navigation was possible over short distances, but it was hardly a rousing success. After watching one unsuccessful test flight, one British military observer snapped, "I could throw my bloody umbrella farther than that!"[33] Across the Atlantic, the US Army turned to Charles F. Kettering, founder of the Dayton Engineering Laboratories Company, and to Orville Wright.[34] Together, these men—sometimes rivals but here joined in the war effort—produced the Kettering Bug, one of the first mass-produced, widely used unmanned aircraft. The Kettering Bug was relatively small—only 12 feet long and made of wood—but it could carry a 180-pound explosive.[35] The underlying idea was that the Kettering Bug would be able to fly approximately 75 miles to a predetermined target so long as the engineers had preprogrammed the engine to do enough revolutions to reach the target.[36] The Kettering Bug was the first airplane that could be both stabilized and navigated remotely through the use of pneumatic sensors and gyroscopes.[37] The army considered developing between 10,000 and 100,000 Kettering Bugs, but in the end fewer than fifty were actually made.[38] The British Royal Navy also experimented along these lines, creating aircraft that blurred the lines between drones and cruise missiles. One of their experiments was the Larynx, a fixed-wing, unmanned aircraft which could fly as fast as 200 miles per hour. The early trials of the Larynx showed promise as it was able to fly 112 miles toward its target, even if it crashed 5 miles short of ultimate destination. Just as the US military would later do, British military officials flew the Larynx over the deserts of Iraq in 1929.[39] But only a few prototypes were made, and research on drones as cruise missiles halted for a while.

In the 1930s, government attention shifted from turning planes into bombs toward forms of remote control of a craft by radio. Both the United States and United Kingdom started looking for small un-manned aircraft that could be used for target practice for anti-artillery weapons. The idea was to use inexpensive drones to improve the accu-racy of gunners in shooting down moving targets. In 1932, the Royal Navy introduced an early target drone prototype, called the Fairey Queen, which was then used for a series of test runs but was eventually destroyed.[40] Royal Navy officials then took parts from two manned aircraft—the de Haviland Tiger Moth and the Moth Major—and created a radio-controlled aerial target aircraft. The Queen Bees, as they became known, were controlled remotely by a radio signal to circle boats in the water for anti-aircraft artillery practice.[41] They were considered so successful that eventually 400 were constructed by the Royal Navy, a fact which drew attention and investment from the US Navy.[42] By the mid-1930s, the Northrup company had a division wholly devoted to UAVs and planned to develop a thousand of their own US version of the Queen Bee, called the OQ-2A target drone.[43] In the end, the Queen Bee was influential in establishing the contours of what a drone is supposed to be—small, mobile, and reusable—and it also arguably gave its descendants the name "drone."[44]

By the mid-1930s, the growing sophistication of radio tech-nology had led to the emergence of a hobbyist community devoted to radio-controlled tiny planes. Among these hobbyists was Reginald Denny, a British émigré to the United States. Denny began his career as B-list Hollywood actor in silent films during the 1920s and continued to act sporadically until the 1960s.[45] After founding a company in Van Nuys, California, in 1934, Denny and his collaborators came up with a prototype radio-controlled plane—nicknamed the Dennymite—which attracted the interest of the US Air Corps (later the US Air Force).[46] In 1940, he successfully pitched the aircraft to the US Army, which agreed to buy fifty for target practice. After the Japanese attack at Pearl Harbor, the air force and navy also expressed an interest and that number increased to a planned 15,000 drones.[47] The Radioplane Models had a wingspan of nearly 12 feet, weighed 130 pounds, and later models could fly as much as 140 miles per hour.[48] Because of the insatiable needs of the military service during World War II, thousands of Radioplane models were quickly constructed and put into service, but never anything close to the numbers the military originally planned.[49] This effort drew the interest of Captain Ronald Reagan of the US Army's First Motion Picture unit based in Culver City, California. Recognizing a good story that would aid the war effort, Reagan assigned Private David Conover to take pictures of workers on the shop floor.[50] Conover spotted a beautiful, brown-haired woman named Norma Jean Dougherty whom he thought had potential as a model. Dougherty—who dyed her hair blond and later became known to the world as Marilyn Monroe—was first photographed as a young model smiling with a Radioplane propeller (fig. 2.1).[51]

Among the most important technological innovations was the adaptation of televisions for use in unmanned aircraft in order to provide a view of target beyond the line of sight of the operator. The US Navy began experimenting with placing TV transmitters on aircraft to allow operators located in a nearby aircraft to see out its cockpit and steer it to the proper destination. By 1941, the navy expanded this to a full-fledged, 70-pound RCA television in the drone that could transmit a picture up to 30 miles away.[52] That picture was sent to an accompanying aircraft that could make decisions on targets. Initial tests showed that these drones could hit targets with torpedoes with remarkable accuracy even when the drone was beyond the operator's line of sight.[53] Navy officials soon began to work on radar technology to allow

FIGURE 2.1 Norma Jean Dougherty—later known to the world as Marilyn Monroe—photographed in June 1945 with one of Reginald Denny's OQ-2 Radioplanes.

transmissions of images to the accompanying manned aircraft at night and in poor weather.[54]

During World War II, US military production of drones involved an array of other target drones, including both biplanes and monoplanes, the efforts to convert B-17 and B-24 bombers into radio-controlled missiles, and the construction of Katydid drones which could compete in lethal accuracy with the German V-1 rockets.[55] There was also a program for a US version of the V-2 rockets, but this was beset by logistical problems.[56] Under the leadership of Lt. Commander (and later Rear Admiral) Delmar Fahrney, the US Navy developed aerial assault drones, and even contemplated using drones to ram enemy fighters.[57] The navy converted a TG-2 biplane torpedo bomber into a remotely controlled drone and conducted a demonstration attack on a US destroyer. As Ann Rogers and John Hill have noted, "this can be seen as the first use of an unmanned combat air vehicle (UCAV)."[58]

By late 1944, the US Navy had developed drones that were capable of ramming targets and also dropping bombs and returning home. For the first time, drones were used in combat in the Pacific theater with several TDR-1 drones dropping bombs on and then later diving into stationary Japanese cargo vessels.[59] The TDR-1 drones were built with a wood and steel frame using parts supplied by famous US companies such as Wurlitzer and the Schwinn bicycle company.[60] It was controlled in flight by a manned aircraft flying alongside it and beamed its crude television images back to the pilots in that plane. Although hundreds of TDR-1 drones were ordered and constructed, the program was canceled in October 1944. Admiral Fahrney bitterly remarked that "the great broom of victory swept all new projects into the ashcan of forgotten dreams."[61] Yet the World War II investment in drones also deepened the commercial and industrial base for drone technology, with companies like Northrup, McDonnell, and Lockheed, the Glenn Martin Company, and others—all later big players in the modern military drone industry—getting their first experience with drones during this period.[62]

Cold War Drones

While the use of drones during World War II was more extensive than was widely known, the conduct of the war also led to a series of changes in the way that drones were conceived. Until World War I, drones were seen primarily as flying bombs. By World War II, military planners had concluded that they could also be used as target practice. Perhaps more importantly, they also saw the demand for intelligence on the activities of the enemy was increasingly essential, especially if the United States wanted to avoid the large-scale bloodshed that marked so many engagements during the world wars. The problem was that unmanned aircraft lacked the technological sophistication to be useful for this purpose.[63] Once the war ended, drone production was significantly scaled back, and the successes of many prototypes were hidden from the public due to government secrecy. Yet privately the US military never wholly surrendered its desire to use drones, especially for surveillance and reconnaissance, and found new, often secret, uses for drones in the early days of the Cold War.

One of the first uses of drones for reconnaissance occurred at the birth of the atomic age. Shortly after the nuclear blasts at Hiroshima and Nagasaki, Japan, the US military realized that drones would be suitable for measuring the lethal radiation in the cloud that emerged after these blasts.[64] In July 1946, the US Air Force flew two B-17s into the radioactive cloud that emerged after a nuclear test over the Bikini Atoll in the Pacific Ocean. These drones—called "babes"—were controlled by motherships 25 miles away, well out of the range of the radiation.[65] Although the drones were damaged and buffered by the force of the explosion, they were still functional and managed to fly from Hawaii to California, a 2,174-mile journey, after capturing data on the radiation. In an August 1946 edition of the *New York Times*, the military expert Hanson W. Baldwin hailed the use of the drones as a rehearsal for a "push button war" and speculated "that what this can mean for the future of warfare—for that matter the future of air transport—requires little imagination to conceive."[66] He even speculated about the use of infrared that drones can see and how swarms of drones might be used in the future. Yet despite the advances in radio control, stability and the ability to deliver explosives, and the ability to "see" a target from a drone, drone technology remained too unstable to have the widespread impact that many assumed it would at the end of World War II.

Instead, drones were initially used for novel experiments to cope with some of the problems produced by the atomic age. For example, US military leaders began to consider how they could use drones to deliver nuclear weapons to targets. The first problem with thermonuclear bombs was that they were heavy (10,000 pounds) and would need to be hauled distances up to 10,000 nautical miles.[67] The second, and far more serious, problem with thermonuclear bombs was that they "would produce a lethal area so great that, were it released in a normal manner, the carrier would not survive."[68] Thermonuclear weapons were so powerful that they would create an "inferno capable of charring wood at a distance of 20 miles" and a small-size hurricane with gusting winds that could destroy a manned aircraft.[69] In 1949, in Operation Brass Ring, the US Air Force secretly examined delivering a thermonuclear bomb via a retrofitted B-47 bomber flown directly into the target. The idea was very similar to the mission that killed Joseph Kennedy Jr.: that B-47, equipped now with a nuclear payload, would be rammed into a target almost as a cruise missile. The air force experimented with deploying the aircraft from a mothership and using a ground control station, but its efforts were beset by technical problems. Ultimately, bureaucratic infighting, an increase in the size of the nuclear payloads, and technological innovations that would allow a manned fighter to survive this operation led the US Air Force to abandon plans for drone delivery of thermonuclear weapons.[70]

By the early 1950s, the US military began to experiment with drones that could conduct reconnaissance of Soviet battle positions. For example, one US Army experiment involved drones laying communication wire across a battle front; others were designed to be more flexible and to allow army units to cast the drones skyward for a picture of the battlefield, much in the same way that Raven and Black Hornet drones are used by the military today.[71] One all-weather reconnaissance model, later known as the Falconer, was a precursor of the small drones like the RQ-11 Raven now seen on battlefields. A number of successors followed, and by 1964 the army had drones that were capable of real-time transmission of "over the hill" imagery.[72] The US Marine Corps developed its own reconnaissance drone—known as the Bikini because, in the words of its Commandant, it was "a small item that covers large areas of interest"—which was able to produce photographs of the battlefield within minutes of landing.[73] The US Navy also developed a number of prototypes, including anti-submarine drone helicopters. One series of models, called DASH drones, was designed to be launched off an aircraft carrier in order to spot enemy drones and even launch missiles at approaching submarines. All of these models were beset by technical difficulties and cost overruns, making it difficult to put them into service.

Within the US Air Force, research also continued on drones that could be used as targets to test pilot skills in aerial combat and the accuracy of air to air missiles. By 1947, President Truman became the first president to observe the operation of a drone from aboard a US aircraft carrier. In 1951, the air force introduced the jet-powered Firebee target drone, built by the Ryan Aeronautical Company.[74] This fixed-wing drone resembled a missile and could be launched on the ground or in the air. It was eventually purchased by the other military services and became one of the most popular and widely used drones of all time. The

US also sought to develop fast aerial target drones which could be used to simulate the movement of a missile or manned aircraft. One example was the MQM-74 subsonic drone, also known as the Chukar, which was developed by Northrup and was put into regular use in the United States, United Kingdom, and Italy in the 1960s.[75] By 1975, an even faster stratojet aerial target drone, the MQM-107 or Streaker drone, emerged. It could simulate the heat and radar signals of real aircraft and maneuver accordingly, which was valuable for training operators of surface-to-air missile. Today, there are over 2,000 Streaker drones in a dozen countries.[76] Many of the technical issues surrounding the use of target drones were largely resolved by the early 1960s, but conceptually they remained closer in function and form to cruise missiles than to the drones we know of today. The United States was also not alone in its pursuit of target drones. Russia and China had programs for target drones by the mid-1950s, with both converting manned aircraft into drones for naval gunnery practice.[77]

The real technological advances seen in modern drones came about through the use of drones for high-level Cold War surveillance from the early 1960s onward. Many of the novel developments with drones occurred in the shadows, as their development was ever more entwined with the priorities of the intelligence agencies during the Cold War. But as Lawrence Newcome has observed, the problem was that the Cold War often ran hot when it came to high-level surveillance. Between 1946 and 1990, 179 airmen were lost due to routine high-level surveillance of sites of interest.[78] Crashes of military aircraft could result in international incidents or even crew being held as hostages; for this reason, unmanned aircraft were seen as preferable, assuming that the technology could be made to work. The necessity of developing unmanned aerial surveillance planes was made apparent in May 1960, when Francis Gary Powers's U-2 spy plane was shot down by an SA-2 missile over the Soviet Union. This has been called the "genesis event" of UAV systems.[79] Almost immediately, both the CIA and US Air Force announced bids for contracts for unmanned surveillance aircraft.[80] The air force secretly awarded its contract to the Ryan Aeronautical Company to adapt the Q2 Firebee target drone.[81] Yet the entire effort to build surveillance drones ran into bureaucratic headwinds. When the air force sought $1 million to invest in drones, they were flatly refused by Secretary of Defense Thomas S. Gates Jr., who responded, "I thought we weren't going in this direction."[82] The Kennedy administration also toyed with drones but shied away from sustained investment. Instead, the immediate effect of the U-2 crisis was to accelerate efforts to invest in other surveillance platforms, such as satellites and manned aircraft. Thomas Ehrhard, who wrote one of the most detailed histories of US Air Force investment in drones, concluded that in the three-way competition between drones, satellites, and manned aircraft, "drones lost all the early battles."[83] Drones particularly lost the battle compared to the CIA's manned spy plane, the SR-71, which was capable of flying at three times the speed of sound and could operate at altitudes that no Soviet missile could reach.[84] Many of the problems usually associated with drones—control, stability, and the risk of accident—made them pale in comparison to manned aircraft like the SR-71 and even early Cold War satellites.

With the US military divided on the utility of drones, research and development into many drone prototypes only survived due to sustained investment by branches of the military intelligence agencies. The geopolitical competition with the Soviet Union meant that these agencies were rich with funds and could take risks by funding prototype drones, even if the models never paid off. The National Reconnaissance Office (NRO)—an agency so secretive that its name was once classified—poured millions into prototype research for drones that could perform the functions of U-2 aircraft.[85] Created in 1961, this agency became a crucial vehicle for investment in drones by both the US Air Force and the CIA. Since much of the funds for drones were drawn from secret intelligence budgets, developers could experiment without drawing the Congressional attention and public scrutiny that they would with normal military procurement. One way that they did this was to work through the auspices of the US Air Force's Big Safari program, which was designed to streamline the acquisition and development process for experimental aircraft and enable the building of riskier, cutting-edge prototypes.[86] This secret infrastructure behind drone development meant that the US public remained largely unaware of the scale of drone development in the Cold War, even when drones became widely used in the military and began to have real consequences.

One moment when spy drones mattered was during the Cuban Missile Crisis. When the Soviet Union placed missiles in Cuba in October 1962, the Kennedy administration had to somehow get aerial reconnaissance photos without risking an accident that might spiral into nuclear war. The SR-71 was an obvious option, but it was high cost and had only limited availability.[87] It also carried with it the risk that it would be shot down by a Soviet SA-2 missile, thus escalating the crisis. To address this, Undersecretary of the Air Force Joseph V. Charyk sought to use a new Firefly drone—essentially a Firebee drone modified for reconnaissance—and had these drones loaded on a C-130 mothership at a Florida air force base in the middle of the crisis. Air Force Chief of Staff General Curtis E. LeMay reacted badly to this plan. In the middle of the crisis, he killed the potential flight for fear that the overflights would reveal the capabilities of the US drone program to the Soviet Union and jeopardize their potential as a constant asset for strategic surveillance of Soviet military facilities. LeMay, known for his aggressive style and love of cigars, stormed into a briefing on the potential drone flights, threw out the participants, and told Charyk "hell no."[88] The need for unmanned aircraft became apparent soon enough. On October 27, 1962, only days before the end of the crisis, a U-2 plane was shot down over Cuba and its air force pilot, Major Rudolph Anderson Jr., was killed.[89]

This incident did not lead the air force to give up on drone surveillance. Rather, it began to institutionalize its drone operations. In July 1963, the US Air Force developed the first drone reconnaissance unit as part of the 4080th Strategic Reconnaissance Wing.[90] Even more importantly, the NRO authorized the creation of a new and expanded Firefly model, dubbed the Lightening Bug (147B), to conduct surveillance at an altitude of up to 62,500 feet.[91] The Lightening Bug could operate at high and low altitude, conduct surveillance of territory, and provide tactical benefits to US units on the ground.[92] It was expensive, but it was so capable that it became among the most important of the Cold War's drones. The Lightening Bug could pick up electronic intelligence on the Soviet Union's SA-2 missile system;[93] it could be deployed underwing from larger aircrafts like DC-130s and could travel a preprogramed range of up to 1,300 miles;[94] it was capable of avoiding radar lock while in flight; and it could reach speeds of 600 mph.[95] This small drone, with a wing span between 13 and 32 feet, could fly as low as 300 feet off the ground.[96] Even though its initial performance left something to be desired, due to poor navigational accuracy and the risks of damage in flight, it nevertheless became the first drone to be put to widespread use by the US military for a range of surveillance purposes.

The first major use of the Lightening Bug was for surveillance over China's key military and nuclear sites following its intervention in the Korean War. Under the code name "Blue Springs," the United States began deploying Lightening Bug drones over Chinese military locations starting in 1964. The pace of drone surveillance picked up as the operational demands of the Vietnam War became more pressing. China's decision to supply arms and other material to North Vietnam and its first nuclear test in October 1964 made US officials eager for greater surveillance of its activities. By 1965–1966, the United States had flown 160 Lightening Bug sorties over Chinese targets. At the same time, US officials were wary of another U-2 incident and wished to ensure that they kept plausible deniability in case one of the Lightening Bug drones were shot down. One way that they did this was to cover the drones with Nationalist Chinese markings to make it look as if it came from nationalist forces based in Taiwan.[97] This ruse was unlikely to have convinced Beijing, but it did provide nominal political cover if drones were shot down.

The United States had good reason to be worried about drones being shot down. On November 15, 1964, only months after the drone overflights began, China shot down a US drone over its territory. This Lightening Bug was designed to trigger Chinese air defenses so the United States could see how they reacted. Although both sides knew the truth, the United States claimed to be baffled by the claims of intrusion and news of the incident gradually died down. This incident demonstrated to the United States that the Chinese and Soviets would implicitly accept that drone overflights were part of normal intelligence tradecraft, rather than a military incursion demanding a response.[98] By 1965, the Chinese military had shot down eight Lightening Bugs. The United States was not deterred by losing the drones, and the intelligence collection via drone continued unabated.[99] Unlike the Francis

Gary Powers U-2 incident, no international crisis followed from any of the Lightening Bug interceptions. It did, however, spur the NRO to fund new experimental prototypes that could operate at high speed and in stealth in order to reach Chinese nuclear sites that were out of reach of existing drones and too risky for U-2 flights.[100] Among these was the D-21 Tagboard drone, which had impressive speed and range but also high rates of accident and loss. Only the Lightening Bug remained in common use, and its models became adapted to a wide variety of tasks, including signals intelligence. By 1971, a modified version of the Lightening Bug, called the Buffalo Hunter, was even capable of flying amid cloud cover and poor weather conditions, two factors that had made flying with manned aircraft difficult if not impossible.

Even though they were commonly used, the Lightening Bugs had a capacity for accuracy that left something to be desired: they were not always able to direct themselves exactly to the target and could be as much as 3 or more miles off.[101] One estimate even suggested that the Lightening Bug could drift as much as 6–9 miles off target during a 100-mile photo run.[102] Course corrections could be done from the mothership, but this was difficult given the telemetry to the C-130 ship. Early forays with the Lightening Bugs produced little in the way of usable film: the first seven missions in 1964 produced only two reels.[103] For tactical operations, Lightening Bugs needed to be extremely close to the target to be effective in producing photos. The film would also have to be flown to a location outside the flight path to be developed and used, which took time.[104] Launch and recovery was difficult, even after the air force developed a sophisticated mid-air retrieval system to "catch" the drone in flight. Drone crashes were common, and recovery itself tended to damage the drone. This would not have been a problem if they were as cheap as typical target drones, but they were relatively expensive for the time.[105] As a result, advocates of the Lightening Bug often had to make the case for a faulty technology against superior, manned alternatives, like the SR-71. These arguments became easier to make when the Soviet SA-2 missiles became much more effective and capable of knocking manned aircraft out of the sky, because pilots became more eager to hand off dangerous missions to their remotely operated cousins.[106]

One less-discussed obstacle to the adoption of the Lightening Bug was the organizational culture of the US Air Force during the 1960s. The air force often struggled to find commanders willing to support drones, in part because flying unmanned aircraft was seen as a poor career choice for pilots, as it was "drab and unexciting" and not professionally rewarded by the air force itself.[107] It also ran against the organizational culture of the air force which promoted becoming a pilot, with all of the prestige and glamor that this implied, as the pinnacle of a service member's career. This organizational culture produced resistance throughout the air force, in some cases because pilots feared being replaced and in others due to a lack of confidence in the technology.[108] The command was not any more enthusiastic; in fact, some were so hostile that they did not even want to look at the drones. For example, in 1964, one former vice chief of staff for the air force, John Dale Ryan, was ordered by his boss to inspect the drones personally to reduce his skepticism. He flew to Eglin Air Force Base, got off his plane, touched one of the Lightening Bugs in its hanger, and said "There, I touched that little son of a bitch, now I can go home."[109] This strong organizational preference for manned aircraft could have led to drones being wholly abandoned during the Cold War, but the programs survived due to a combination of factors, including some motivated leaders within the air force, generous Cold War budgets that permitted experimental work through Big Safari and Lockheed's famous Skunk Works lab, and the relentless demands of the intelligence agencies to know more about the battlefields that they might face if the Cold War ever turned hot.[110]

This last factor is what ultimately led to the expansion of the drone fleet and routine use of Lightening Bug drones during the Vietnam War for a wide variety of tasks, including gathering electronic signals, providing targeting data for missiles and ground units, and producing photographic and infrared imagery for pre- and post-battle assessments. Some uses were less straightforward. A few Lightening Bug drones, modified to drop propaganda leaflets on North Vietnam, were affectionately renamed "bullshit bombers" by their air force operators.[111] Others were used as decoys to deliberately draw fire from Soviet SA-2 missiles stationed in North Vietnam, which allowed the CIA to find new ways to track and avoid those missiles.[112] By the early 1970s, a few

Lightening Bugs were equipped with their own Maverick missiles to directly fire at targets, making them capable of crude versions of the targeted killing missions that modern armed drones perform today.[113] Initially used for reconnaissance of military sites in the Chinese mainland, Lightening Bug drones were flown in Vietnam in 1964 and were eventually relocated and flown directly out of Da Nang in South Vietnam by 1966. Although not widely reported at the time, their missions became routine. The 350th Strategic Reconnaissance Squadron reported flying 3,466 sorties in Vietnam with Lightening Bugs between 1964 and 1975. Perhaps more surprising was their survival rate: the air force managed to recover them in mid-air nearly 97% of the time, and only 578 drones were destroyed by Chinese or Vietnamese air defenses.[114] One Lightening Bug drone, nicknamed the "Tom Cat," survived sixty-eight missions before being shot down in 1974. Yet these drone operations, now routine and survivable, were not always effective: by one estimate, only 40% returned reconnaissance images, with a cost five times greater than that of manned aircraft.[115]

Once the Vietnam War ended in 1975, the use of reconnaissance drones was scaled back and US interest in developing the technology began to wane. The economic recession that afflicted the Carter administration made the Pentagon tighten its belt and cut back on programs that were not needed. In this environment, drones would naturally suffer: they remained more expensive and less reliable than satellites, which were showing tremendous gains in their capabilities as technology developed. They were also ill-suited to use within the European theater of war because of a congested flying environment with more manned military and civilian aircraft; moreover, their use was governed by some arms control regimes put in place by agreement with the Soviet Union.[116] In this respect, the shared lineage of the cruise missile and drones proved fateful to the latter, as UAV systems were defined as restricted military technology and subject to limitations on their use within the European theater of war. Perhaps the most consequential development was the transfer of control over drones from the NRO to the US Air Force's Tactical Air Command (TAC), which sought control of the drone fleet but was ultimately opposed to the huge cost overruns that accompanied drone development. The result was that each tightening of the defense budget led to a reduction in drones as many models were seen as wasteful and unnecessary compared to manned aircraft.

In the 1970s, research continued on a number of high-altitude surveillance drones, with mixed results. Some drones showed remarkable capabilities. For example, the HALE Compass Cope drone could fly for up to twenty-eight hours, but they were astonishingly expensive and could not be put into regular production.[117] The Air Force wanted to have a Compass drone model provide nearly continuous coverage of military units in the Warsaw Pact and experimented with a number of Compass models that could achieve that capability. It also experimented with solar-powered drones to allow for greater endurance and intelligence collection capacity.[118] In the late 1980s, the Defense Advanced Research Projects Agency (DARPA) funded a number of drones, such as the Boeing Condor and the RQ-3 Dark Star, all of which were designed to fly at high altitudes and at high speeds. Both models were costly failures, yet they yielded some design insights. The Condor drone proved that drones with longer wingspans could conduct long-range surveillance while the Dark Star showed that autonomous flight based on GPS data was indeed possible.[119]

The US Army wanted to know more about the battlefield as well and devoted its resources to developing drones for tactical use, including the SkyEye and Aquila, both of which were small drones that could be deployed in support of ground units with reconnaissance. Both of these drones reflected the hope that drones could be used for attacking targets on the ground: one prototype of the SkyEye could be equipped with tiny Viper missiles; the Aquila could use lasers on targets up to 7 miles away.[120] The problem is that they were either unreliable (SkyEye) or too costly (Aquila) to be put into widespread use. Even the few times when the SkyEye was used, such as in El Salvador in the mid-1980s, it was derided as a "toy plane" by the rebels and had minimal real-world impact.[121] The Aquila was a particularly devastating failure, as it cost $750 million over twelve years and only succeeded in seven of 105 test flights.[122] Having already incurred this cost, the Pentagon was reluctant to spend another $1.1 billion for 376 Aquila drones.[123] Throughout the 1980s, drones were increasingly seen as a boutique technology: an expensive novelty that was unnecessary given the information collection capabilities of satellites and manned aircraft.

Birth of the Predator

Although US drone development stagnated in the 1980s, the desire to know what was happening on battlefields—in effect, to see around a corner or to anticipate danger—was increasing rather than diminishing. In World War II, the emphasis with drones had been on inflicting damage, as evidenced by the retrofitting of drones as missiles or even nuclear weapons delivery systems and the efforts to sharpen the skills of those manning antiaircraft weapons with target drones. The Cold War subtly changed some elements of the way drones were conceptualized; most Cold War drones were seen as a way of learning more about real and potential battlefields rather than directly inflicting harm. Despite the fact that a few prototypes of drones had been equipped with missiles, the majority of Cold War drones were unarmed and devoted to collecting information that could be converted to intelligence and put to use in the superpower competition.

The problem with US Cold War drones were that they were directed mainly at the superpower competition and were both highly sophisticated and expensive. The predominant focus of the United States was on conducting surveillance of the Soviet Union and to a lesser extent China. This meant a natural emphasis on high-altitude surveillance drones and satellites, as opposed to medium-altitude or other tactical drones. By contrast, other countries were beginning to look at the information-collection capabilities of smaller drones differently, seeing them as a low-cost, disposable mechanism to be deployed in great numbers to get information on more immediate threats.[124] Israel was among the first to demonstrate the value of this alternative approach. After Egypt shot down two Israeli F-4 Phantom reconnaissance jets in early 1970, Israel started looking into drone technology for its ongoing battle with Egypt over control of the Sinai. Israel came first to the United States to investigate buying the Compass Arrow drone, but President Richard Nixon ordered it destroyed rather than sold for export.[125] Israel later purchased Lightening Bug–style drones from the United States which it then used for surveillance and to patrol the Suez Canal zone.[126] By 1978, Israel had created its first drone, the Mastiff, a small reconnaissance drone with a 13-foot wingspan that could be flown over battlefields and also work as a decoy to draw fire and identify weaknesses in another country's defenses.[127] By 1982, they had managed to get Syrian air units in the Beka'a Valley to fire on Israeli drones in mid-air and exhaust most of their SA-2 missiles, thus allowing Israel to strike against those units decisively. When Israeli forces invaded Lebanon in 1982, they also deployed drones in advance to spot Hezbollah forces hiding in civilian population centers.[128] Notably, Israeli drones were pitched at a tactical level, and they were successful in drawing out information about the signals and emissions of particular missiles, which in turn allowed the Israelis to develop countermeasures.[129]

The US military took notice of Israel's use of drones to adjust artillery fire after the bombing of the US Marines Corps' barracks in Lebanon in 1983.[130] Two US Navy captains were so impressed by the Mastiff's ability to support artillery fire that they recommended that the United States buy Israeli drones. It was clear that Israel had come up with tactical drones that were more nimble and effective than the costly, high-altitude drones that the United States spent more than a decade developing.[131] Although the United States has a strong preference against purchasing drones from foreign suppliers, the army and navy made an exception in this case and purchased a descendent of the Mastiff, the Pioneer drone, in 1986. The Pioneer is a medium-altitude (15,000 feet maximum) sleek gray drone with a 16-foot wingspan.[132] Equipped with infrared cameras, the Pioneer was used for reconnaissance and aerial spotting of targets to be hit with conventional fire. This model not only showed the value of tactical drones against insurgents, terrorists, and other forces, it also revived interest in the technology at a point when it was collapsing due to failures with high-end drone prototypes.

The Pioneer drone became the first unmanned aircraft to be used in war since the Vietnam War. In Operation Desert Storm in 1991, some Pioneer drones helped to track Iraqi armored vehicles and provided artillery spotting for a US Marine artillery assault on a small Kuwaiti island held by Iraqi forces.[133] In the latter instance, Iraqi soldiers surrendered to the drone, appearing on video at the Pentagon waving white flags in order to avoid yet another artillery bombing.[134] Across the war, the Pioneer drones flew 330 missions, spending about 1,000 hours in the air.[135] While they mainly flew in uncontested airspace, the Pioneer drones were able to provide limited, real-time video of events on the ground, leading many US military officials to declare them an unalloyed success. Pioneers were also used to confuse and disable Iraqi air defense systems at the beginning of the air campaign, but their actual tactical value was limited in most other respects.[136]

One of the reasons why drones so often failed is that they had low endurance and tended to crash at a high rate.[137] Pioneer drones crashed regularly, with more than a dozen lost to mechanical failure or operator error.[138] While the US military was experimenting with the Pioneer drone, an effort was underway elsewhere in government to build drones that could address these problems. Abraham Karem, an Israeli engineer and aviation enthusiast, was fascinated by the possibility of drones used as a decoy to reveal the capabilities of air defenses but also to defend land borders against invading armies.[139] Having experimented with drones during his career at Israel Aerospace Industries (IAI), Karem left the company to start his own company, Leading Systems Incorporated, in 1974.[140] After a few years of tinkering with drone models, and building rough prototypes in his garage, Karem proposed building a low-endurance drone to DARPA. While DARPA had been involved in drones in the 1960s, it stopped research and development for most unmanned aircraft for years, experimenting only with HALE drones that were kept aloft with cutting-edge motors driven by nuclear or solar power.[141] Convinced of the value of Karem's designs, DARPA funded his research between 1984 and 1990. The result was the Amber drone, which was medium-sized—15 feet long, with a wingspan of 28 feet—and had a liquid-powered piston engine. Made of lightweight, composite materials and equipped with a sophisticated hardware and software package,[142] it was capable of flying for 38 hours or more—most of its competitors could manage no more than 12 hours.[143] It was, in the words of a senior defense, official "a marvelous aircraft."[144]

Yet the fact that the Amber drone had fans was not enough. It was an important technological advance on drone technology but it was operating in a political climate hostile to drones. In 1988, the Pentagon reorganized its drone research and development under a new bureaucratic entity, canceled the Aquila drone, and halved the research budget for drone technology.[145] Recognizing that US funding was drying up, Karem developed a variant of the Amber, the Gnat 750, for foreign sales. Facing declining US interest, the Amber program was canceled in late 1988, and Leading Systems International drifted toward bank-ruptcy. Sensing an opportunity, General Atomics, under the leadership of Linden and Neil Blue, purchased Karem's company and its surviving arsenal of Gnat 750 drones in 1991. The Gnat drones languished until 1993, when the CIA purchased a number of them for battlefield recon-naissance in Bosnia. During that brutal civil war, Serbia's military forces laid siege to Sarajevo and shelled its citizens on a regular basis despite a "no fly zone" and UN presence. Frustrated with the inability of U-2s and satellites to get reliable imagery of the territory and position of forces due to cloud cover, the CIA deployed Gnat 750s to collect im-agery of tanks, surface to air missile sites, and other targets of interest in Bosnia in 1994.[146] The Gnats' record was not wholly successful, as the video feed was shaky and subject to breaks and interruptions, and many planned missions had to be aborted because of bad weather, but it was sufficient for the military and CIA to be interested in an upgraded GNAT 750. This upgraded model was called the MQ-1 or, more pop-ularly, the Predator.[147]

In many respects, the birth of the Predator can be thought of the birth of the drone age itself. Although the United States was also pouring money into building a high-altitude surveillance drone, the Global Hawk, it was the Predator that captured the attention of policymakers. More than just the most successful model since the Lightening Bug, the Predator functioned almost like a proof of concept of the capabilities and promise of drones. Even on its own terms, the Predator drone was a significant improvement over the Gnat 750 model. It was more ca-pable of carrying a payload due to its longer wing span and fuselage; it also had a fuel capacity allowing for 24 hours of flying endurance.[148] While this did not break the upper limit of the Gnat 750's flight time, it was far more reliable and stable. It was also significantly quieter than the Gnat—once described by a journalist as a "lawnmower in the sky"—which had significant advantages for reconnaissance and support of combat troops.[149] Perhaps more crucial was what was "under the hood": the Predator had a large snub nose which contained integrated satellite communications that allowed communication via voice, dig-ital imagery, and text messages with a ground station as opposed to a mothership.[150] This allowed the aircraft to be flown and commanded from 400 miles away.[151] Since it was connected to satellites, Predator drones could also now beam back images to ground stations and to the desks of officials in the Pentagon and intelligence agencies. CIA Director James Woolsey was even able to watch foot traffic over the damaged bridge in Mostar from his office in Langley, Virginia.[152] The Predator functioned, for the first time, like a "television camera in the sky."[153]

The Predator could only be used in this way because of a series of cascading technological innovations in the late 1990s. The information revolution had resulted in vast increases in processing power for computers and data imagery as well as satellite and other forms of electronic communication. In many respects, this was a natural result of the famous observation by Intel founder Gordon Moore that the number of transistors on a computer chip—and hence its processing power—doubles every two years while the cost of making the chip plummets. In other words, the growth of computer processing power is exponential. As Peter W. Singer has argued, this speed of change has been replicated in other forms of technology, including wireless capacity, optical storage, and internet bandwith.[154] Although the speed of this change has recently slowed, leading some to wonder whether Moore's law will continue to apply, the change in data collection and transmission capabilities transformed in a short period of time what drones like the Predator were capable of doing.[155] The early evolution of the Predator capitalized on existing investments and encouraged the Pentagon to go even further with research and development into the transmission of data and imagery. It also encouraged other crucial adaptations, such as affixing a laser designator to spot and transmit the coordinates of a target to the nose of the drone.[156]

Among the most important developments was the rise of GPS as a navigation software. The earliest drones had to be within the line of sight of the operator, either on the ground or on an accompanying mothership, for the remote control to work. This almost always exposed the operator to some degree of risk, as the discussions of Operation Brass Ring showed clearly, and it imposed some natural limits on how far drones could fly. In 1983, the possibilities for drones expanded through a crisis and a little-heralded announcement by President Ronald Reagan. On September 1, 1983, a Soviet fighter plane shot down a Korean Airlines Flight 007 over the Sea of Japan. The plane had been traveling from

New York to Seoul and among the 269 passengers and crew killed were 62 US citizens, including a member of the House of Representatives from Georgia. President Ronald Reagan was infuriated and weeks later announced that the US government would make the Navigation Signal Timing and Ranging Global Positioning System (NAVSTAR), later known simply as GPS, freely available to help civilian airliners avoid Soviet airspace.[157] One of the earliest to notice the benefits of GPS for drone technology were Neil and Linden Blue. They came to the conclusion that GPS technology could be used to enable planes to fly unmanned as cruise missiles toward targets either against the Contras in Nicaragua or even potentially in a war with the Soviet Union in Europe. As Neil Blue remarked, "you could launch them from behind the line of sight so you would have total deniability."[158] Ironically, this conceptualization of drones echoed the ambitions of Elmer Sperry and others that unmanned aircraft could be guided along a set path to destroy a target. GPS made it possible to extend the distances involved and to envision a world in which drones run along waypoints set on routes hundreds, even thousands, of miles long. Although this was not well understood at the time, the linkage between drones and GPS gradually made it possible for unmanned aircraft to pinpoint and convey target locations around the world with remarkable accuracy.

Building off these technological developments, the Predator represented an important but imperfect advance on previous drone models. Its early record was mixed. Early in the mission in 1995, at least one Predator crashed and was recovered by Serbian forces. More Predators crashed due to weather and problems with de-icing. Early Predator drones had problems conveying images due to limitations of the UHF antenna or to the fact that the images were compressed for transmission, rendering them grainy.[159] In Kosovo in 1999, Predator and US Army Hunter surveillance drones were deployed to collect imagery of Serb forces on the ground, to detect sound, and to collect electronic signals.[160] One estimate suggests that between 1996 and 1999, only half of Predator operations in the Balkans were actually completed due to enemy fire or operator error.[161] Even when they found the targets, Predator drones tended to struggle to destroy them because there was a costly time lag between spotting a target with a laser designator, conveying the information to an armed, manned aircraft, and firing a missile. One study found that only three of the nearly 300 Serb vehicles located by Predator drones had been destroyed due to this time lag.[162] In a scathing internal report in 2001, the Pentagon's director of operational tests and evaluation Thomas P. Christie wrote that the Predator was not "operationally effective" because of poor target location, accuracy, and limits to communication imposed by relatively benign weather such as rain.[163]

By early 2001, the Predator was at a crossroads. It was still criticized as ineffective, with some in the Pentagon convinced it was expensive and wasteful, especially given that manned aircraft and satellites could perform many of the same functions. But unlike previous drone models it had built an ardent constituency of military, intelligence, and other government officials who were addicted to the imagery coming from its operations. Enticed by the possibility of learning more about their enemy and the contours of the battlefield, high-ranking military and government officials began to push for even more Predator flights over war zones, and in some cases began to call pilots to make special requests for obtaining specific images.[164] Some inside the military dismissed this as merely "Predator porn,"[165] but others began imagining new possibilities for these drones. In 2000, General John Jumper began to lead an effort to arm Predators with Hellfire missiles.[166] By mid-2001, the United States had conducted its first successful tests of an armed Predator drone in the desert in China Lake, California.[167] As the September 11 attacks loomed, the United States was almost ready to begin arming Predator drones for the hunt for Osama bin Laden. It took only that attack for drones to be quickly converted into a weapon against terrorist organizations.

Conclusion

For decades, advocates of drones experimented with the technology and attached a number of purposes to them. Early drones were viewed as crude cruise missiles, target decoys, and reconnaissance crafts, sometimes at the same time. The goals for the technology grew larger over time; as capabilities improved, the calculation of the risks involved in deploying drone technology changed as well. By the end of the twentieth century, the technology behind unmanned aircraft had matured enough to realize the dream that US planners had at the time of Joseph P. Kennedy Jr.'s death. The Predator drone finally allowed the United States to fly and to strike long-distance enemy targets precisely and by remote control. What the planners could not anticipate at the time was how the remote control of unarmed and armed drones would dramatically transform warfare—and indeed the military itself—in the twenty-first century.

3

Death from Above

ON MARCH 16, 2004, the Pakistani army surrounded a large, fortress-like house in South Waziristan along the Pakistan-Afghanistan border.[1] The house belonged to a twenty-nine-year-old charismatic tribal leader, Nek Muhammad Wazir, and two other more senior militants, Haji Muhammed and Nar-ul-Islam.[2] In this tribal region, long known for its violent revolts against the authority of the central government in Islamabad, Nek Muhammad was an unusual figure. Aside from being a young leader in a society where age confers political power, he looked to some observers like a young Che Guevara, with a scruffy beard and long, flowing, black hair.[3] Like many revolutionaries before him, he was not shy of journalists or television cameras. Among his followers, he was reputed to be so fearless in battle that he earned the Pashto nickname "Bogoday" (the stubborn one).[4] His death—the first targeted killing by a Predator drone in Pakistan—would set in motion a chain of events that would lead to international condemnation of the use of drones for targeted killings and a crisis in US-Pakistani relations.

Born near the market town of Wana, Nek Muhammad was a shopkeeper in the local bazaar and a petty car thief before joining the Taliban at the age of 18.[5] Once the US invasion of Afghanistan began, a number of Arab and Chechen militants affiliated with al Qaeda and the Taliban sought refuge across the border in Pakistan. Nek Muhammad offered them sanctuary at inflated rents, making him a wealthy powerbroker in the region.[6] Aided by Uzbek militants from the Islamic Militant Union (IMU) and local tribes, he launched a series of attacks on US military outposts across the border in Afghanistan.[7] This drew the attention of policymakers in Washington, who began to press Pakistani president Pervez Musharraf to hand over the al Qaeda–linked militants and to pacify the unstable border regions threatening America's occupation of Afghanistan.

Musharraf was initially reluctant to confront Nek Muhammad and his allies because he knew that a military operation in South Waziristan would enrage the local tribes. The region had long been out of Islamabad's formal control, and the modus vivendi that existed between the government and the tribes relied on neither overly interfering in each other's affairs. Yet two assassination attempts and a fatwa calling for his death from al Qaeda's second in command, Ayman al-Zawahiri, changed his mind.[8] In late 2003, Musharraf ordered 80,000 troops into South Waziristan and demanded that Nek Muhammad and his allies turn over all of the al Qaeda–linked militants that he was sheltering, especially Tahir Yuldashev, the powerful head of the Islamic Movement of Uzbekistan (IMU).[9] Weeks of intense fighting and bombardment from helicopter gunships did little to break the resistance of Nek Muhammad and his allies. Instead, they launched scattered attacks on the Pakistani Frontier Corps troops in a grinding battle for control over South Waziristan, a campaign that eventually brought the Frontier Corps to Nek Muhammad's compound in March 2004.

The tense standoff quickly turned into a disaster for the Pakistani Frontier Corps: Nek Muhammad's allies in the Ahmedzai Wazir tribe had set a trap for them. In a sudden attack, fifteen Frontier Corps troops and one Pakistani army regular were killed and another fourteen soldiers were taken hostage.[10] Dozens of trucks, armored personnel carriers, and pieces of artillery were burned.[11] Across the region, fighting raged, with ambushes of military vehicles and house-to-house fighting in Wana.[12] Recovering from their setback at Nek Muhammad's compound, the Pakistani Frontier Corps drew back, but continued their cordon operation elsewhere, while, undaunted, Nek Muhammad continued to launch periodic rocket attacks on their outposts. Amid growing concerns over civil casualties and calls for the people of South Waziristan to resist Islamabad's imposition of direct rule, Musharraf ordered his troops to pull back in April 2004. In an embarrassing turn of events for the Pakistani military, General Safdar Hussain, the commander of the operation, was forced to meet Nek Muhammad and his allies to personally negotiate a ceasefire. General Hussain was photographed presenting flowers to the young firebrand and drinking tea with him, effectively enshrining Nek Muhammad as the most powerful man in South Waziristan.[13] The Shakai peace treaty signed between the Pakistani military and Nek Muhammad even referred to him as a "mujahid" or holy warrior.[14] Nek Muhammad was keenly aware of how he had turned the tables on the military, noting that Hussain was forced to come to his madrassa so "that should make it clear who surrendered to whom."[15]

Only moments after the peace ceremony, Nek Muhammad told the assembled journalists that he would continue to wage jihad and swore his allegiance to Mullah Omar, the leader of the Afghan Taliban. He also immediately backtracked on his promise to turn over all foreign fighters to Islamabad.[16] Although the government had offered generous terms—including amnesty for all foreign fighters and reparations for civilians in South Waziristan—Muhammad resumed attacks on US positions in Afghanistan.[17] He also began a brutal campaign killing all of the local leaders (maliks) who had supported the government's assault.[18] Soon, Musharraf ordered his troops back into South Waziristan and fighting resumed with tribes and foreign militants. For Islamabad, the situation was growing more dangerous. Widespread resentment of the Pakistani Army's attacks on Waziristan pushed many tribes toward Nek Muhammad and his allies. Opposition to Islamabad's rule was beginning to coalesce into what was called the Tekrik-i-Taliban (Pakistani Taliban) movement, which would later launch a string of terrorist attacks and bring the violence of the frontier closer to the capital.

Furious at the betrayal and humiliation, Islamabad was willing to consider previously unacceptable options to eliminate Nek Muhammad. The CIA seized the opportunity: it proposed regular overflights by armed Predator drones over South Waziristan. This offer was partially due to necessity—export restrictions forbade the Bush administration from giving drones directly to Pakistan—but it was also a long-sought goal by some within the CIA to have the ability to directly strike militants in ungoverned territories, especially along the border within Afghanistan.[19] For many in the CIA, it was clear that the militant movements in Pakistan were a vital source of arms and fighters for the Afghan Taliban. Pakistan's intelligence service, the powerful Inter-Services Intelligence (ISI), was willing to grant the United States the right to fly armed drones over the region, but it imposed strict conditions in a secret deal with the CIA. The United States could only fly drones over approved spaces and at approved altitudes—designated "flightboxes"—and the ISI had to approve each drone strike before it happened.[20] These attacks would never be acknowledged by the Pakistani government and would be conducted solely under the CIA's authority. This granted plausible deniability to the Pakistani government for, as President Musharraf allegedly remarked, "in Pakistan, things fall out of the sky all the time."[21]

On June 17, 2004, Nek Muhammad was on his mobile phone in the outside courtyard of the house of his friend Sher Zaman Asrafkhel, surrounded by a number of men and boys having dinner.[22] Above them came a buzzing sound and then a blinding flash.[23] A drone strike killed five of his companions, including two children, and severed the left hand and leg of Nek Muhammad.[24] In shock, he was rushed to the hospital in Wana, where his last words allegedly were, "why aren't you putting a bandage on my arm?"[25] The Pakistani army claimed credit for killing Nek Muhammad, crediting night vision helicopters with tracking him down, but within days it was reported that the United States was behind the strike. Pakistan never held the US government to account for civilians killed in the strike, including the children.[26] Nek Muhammad's grave in Waziristan was accompanied by a handmade sign which said only "he lived and died like a true Pashtun."[27]

The killing of Nek Muhammad was the first targeted killing in Pakistan, but it was not the first time the United States had conducted a targeted killing with a drone outside a declared war zone. In 2002, the United States had struck a convoy in Yemen and killed Qa'id Salim Sinan al-Harithi, one of the planners of the USS *Cole* bombing in 1998; Kamal Darwish, a naturalized US citizen; and four others. At the time, that targeted killing—initially denied, then confirmed by senior US officials—was an exception. In Pakistan, targeted killings by drones soon became the rule. Between 2004 and 2015, the United States conducted 400 drone strikes in Pakistan, killing between 2,276 and 3,614 people, according to data collected by the New America Foundation.[28] The use of drones to kill militants had become an almost routine event;

by 2010, the Obama administration was conducting a drone strike in Pakistan almost every three days.[29]

The expansion of the targeted killing program in Pakistan was gradual but steady. At first, the bargain that the United States struck with the ISI held, and Pakistan retained final approval over any prospective drone strike target. Throughout 2004–2005, US officials shared detailed images of training camps with their ISI counterparts and, in the words of the one senior CIA counterterrorism official, "every one of those shots was with Pakistani approval."[30] Yet as evidence mounted of Pakistani's complicity in hosting the Taliban and other militant groups targeting US troops in Afghanistan, Washington's patience wore thin.[31] The United States had evidence that the ISI was working with a variety of militant networks, including the Taliban, the Haqqani network, and even al Qaeda, and was actively resupplying Taliban forces battling US troops in Afghanistan.[32] In January 2008, the Bush administration decided to abrogate the secret deal of 2004, that had led to the death of Nek Muhammad, and began launching drone strikes without Islamabad's formal consent—at will and with changed targeting practices.[33] The CIA moved from personality strikes—in which the identity of the victim is known with a high degree of certainty—to signature strikes—based on the patterns of activity present among a target population. It was rumored that the threshold for likely success of a strike was lowered from 90% to 50%, although this has been disputed by some US officials.[34]

For the most part, Pakistan stood by as the US drone campaign ramped up throughout the Federally Administered Tribal Area (FATA), a semi-autonomous tribal region in its northwest that was home to some of the most dangerous Islamist groups. It did not contest the use of its airspace by shooting down US drones. In reality, they were playing, in the words of US officials, a "double game."[35] The Pakistani military and ISI secretly allowed the CIA to launch drones from some of its airbases and occasionally even suggested targets. While Islamabad publicly pretended that it condemned drone strikes, it was more than happy to see Washington take out its most bitter enemies. Prime Minister Gilani privately told US officials "I don't care if they do [drone strikes] as long as they get the right people. We'll protest in the National Assembly and then ignore it."[36] The killing of Baitullah Mehsud, the leader of the Pakistani Taliban and a spiritual successor to Nek Muhammad, was widely applauded within the security and intelligence establishment in Pakistan.[37] Privately, some Pakistani officials supported expanding drone strikes even if it caused civilian casualties. President Zadari told US officials to "kill the seniors. Collateral damage worries you Americans. It does not worry me."[38] But when drone strikes killed large numbers of civilians, they were publicly condemned by the Pakistani government and blamed on the CIA. The wary, often hypocritical relationship between Washington and Islamabad was built upon a tacit understanding that targeted killings by drones were useful against their joint enemies, even if they did not have the same list of enemies.

Although there were expectations that President Barack Obama would scale back the targeted killing program once he came into office, he instead increased the tempo and scale of the attacks. Only three days after his inauguration, he authorized two drone strikes in Waziristan.[39] The strikes killed fourteen people, including some civilians, but reportedly missed their high-value targets.[40] Although President Obama was reportedly angry about the civilian casualties, he nevertheless authorized the drone program to continue.[41] Senior administration officials, including chief counterterrorism advisor John Brennan and CIA Director Leon Panetta, backed the drone strikes as the only way to keep pressure on the dangerous militant networks and to enable the Obama administration to end the war in Afghanistan.[42] As the drone strikes ramped up in 2010–2011, the Obama administration remained silent on the legal rationale for the program, even refusing to acknowledge in the courts that the drones program existed.

Behind closed doors, it was an entirely different story. The Obama administration undertook a program to institutionalize the targeted killing program, even producing an internal "kill list" with clearly specified criteria about who could be added to it.[43] Soon, this process had settled into a bureaucratic routine. The Obama administration compiled dossiers on possible drone strike targets and debated the merits of striking each person in a weekly meeting dubbed "Terror Tuesday."[44] Over one hundred people from the Pentagon, CIA, State Department, and other government agencies would participate in these secret teleconferences. Led by President Obama's chief counterterrorism adviser John Brennan, participants debated whether the target was considered "high value," whether the threat that they posed was "imminent," and how connected they were to al Qaeda and the Taliban.[45] President Obama presided over the discussions of potential targets, signing off on approximately one third of all drone strikes before they happened.[46] In Pakistan, because there were so many drone strikes, President Obama approved only the riskiest strikes.[47] The CIA-led targeted killing in Pakistan—once considered an exception when Nek Muhammad was in the crosshairs—had now become routine. Targeted killings were, in the words of Panetta, "the only game in town in terms of trying to disrupt the al Qaeda leadership."[48]

Yet the truth was that drones were increasingly targeting neither al Qaeda nor its leadership. By 2012, the targeted killing program had expanded beyond al Qaeda to take on an array of different militant networks inside Pakistan, including the Afghan Taliban, Pakistani Taliban, and the Haqqani network. Moreover, it was not only targeting leaders. The expansion of drone strikes from personality strikes to signature strikes had left Predator and later Reaper drones targeting low-level operatives or "foot soldiers" in the words of US officials. An estimate by Peter Bergen suggested that only 2% of all drone strikes were directed against leaders of militant organizations.[49] As he later put it, the drones program had "increasingly evolved into a counterinsurgency air platform, whose victims are mostly lower-ranking members of the Taliban (Pakistan) and lower-level members of al Qaeda and associated groups (Yemen)."[50] One mid-ranking Haqqani network member remarked in 2010 that "it seems that they want to kill everyone, not just the leaders."[51] The definition of an "imminent threat" had begun to slip as well. The US program was increasingly targeting individuals who posed a speculative rather than real or evident risk to US personnel or interests.[52] The United States also expanded the geographic reach of the targeted killing beyond Pakistan to include new targets in Yemen, Somalia, and Libya, with unconfirmed reports that additional strikes had made in Mali and the Philippines. The dramatic expansion of the program suggested to UN investigator Ben Emmerson that the CIA's drone program had slipped beyond the control of the White House.[53]

By 2012, the consequences of this expansion of targeted killings had begun to show. There was a growing international backlash against the United States for engaging in targeted killings with drones outside declared war zones. UN Special Rapporteur Christop Heyns attacked the US targeted killing program as a form of "global policing" and argued that the frequency of targeted killings risked loosening the restraints on the use of force by other actors.[54] Human rights groups such as Amnesty International, Human Rights Watch, and others condemned the US drone strikes for producing excessive civilian casualties. Moreover, despite the complicity between the ISI and CIA over drone strikes, the US relationship with Pakistan began to break down. Islamabad began to worry over growing popular protests over the drones program and unrest in the tribal regions as militants capitalized on public discontent over drones to recruit more operatives. Enraged by the murder of two Pakistanis by CIA contractor Raymond Davis and the killing of Osama bin Laden by US Special Forces in Abbotabad, Pakistan expelled the US personnel responsible for the drones program from their airbase in Shamsi. The Pakistani government publicly blasted the United States for its "illegal and counter-productive" use of drones and argued that it was fueling the insurgency against the central government.[55] By June 2012, the Pew Foundation found in a poll that 74% of Pakistanis considered the United States an enemy.[56]

In May 2013, at a speech at the National Defense University in Washington, DC, President Obama addressed the growing controversy over the drones program and defended the practice of targeted killing. Acknowledging the moral dilemmas that drone technology produced, he insisted that the program was legal and effective, but nevertheless promised to cease signature strikes, to raise the standard for targeting individuals, and to provide more oversight over the program.[57] While US drone strikes in Pakistan declined in tempo, the targeted killing program never stopped there or elsewhere. Since assuming office in 2017, President Donald Trump has continued many of President Obama's targeted killings policies and even ratcheted up the strikes, launching 238 between January 2017 and November 2018.[58] He also swept aside many of the self-imposed limits on the strikes designed to reduce civilian casualties.[59] Other countries—such as the United Kingdom, Nigeria, Iraq, and Pakistan—began to follow the US example in using targeted killing. What had begun as a way to strike a dangerous enemy in Pakistan and to control risks for US pilots had become a new and possibly dangerous global practice at odds with much of international law. Was it something about drone technology that made targeted killing not only permissible but actually a normal, everyday occurrence? In other words, did the ability to kill at a distance make killing more likely?

A Short History of Targeted Killing

The term "targeted killing" is a relatively new one, having been brought into common usage over the last twenty years. The practice of targeted killing began with Israel's response to the al Aqsa intifada in September 2000. The intifada began as a series of violent protests over Israeli prime minister Ariel Sharon's visit to the al Asqa mosque, but escalated into a general protest over Israel's occupation of the West Bank and Gaza. Facing growing unrest and suicide attacks, Israel began to use helicopters and manned aircraft to launch attacks against militant Palestinian groups in the Occupied Territories. Over the period 2000–2005, Israeli airstrikes killed over 203 Palestinian militants and 114 civilians.[60] Although many Israeli airstrikes targeted specific individuals, Israeli officials maintained that these killings did not constitute assassinations, but were rather "targeted killings."[61] According to their legal arguments, targeted killings were different from assassinations in three ways. First, they were directed by government-controlled military forces in an active armed conflict. Unlike assassinations, which were typically conducted by spies or hired guns and involved subterfuge or deception, these killings were authorized and conducted through a military chain of command. Second, Israel was in a state of war against Palestinian militants, so the rules prohibiting assassinations in peacetime did not apply. If killing was legally and morally permissible in combat, there was no reason not to engage in targeted, and presumably precise, forms of killing in wartime.[62] Third, Israeli authorities argued that militants were combatants, not political leaders, so their killing did not constitute assassination in the formal sense.[63] In a war against terrorists who embed themselves in the civilian population and conduct suicide attacks, Israeli officials argued, governments had the right to conduct targeted killings as a form of active self-defense. Israeli officials later admitted that they made a concerted effort to legitimize the concept of targeted killings to permit the elimination of militants in the Occupied Territories.[64]

Initially, few states backed Israel's position on targeted killings. Even sympathetic US officials claimed that they remained opposed to the practice and did not accept the distinction between targeted killings and assassinations.[65] US Ambassador to Israel Martin Indyk remarked in July 2001 that the "United States government is very clearly on the record as against targeted assassinations. They are extrajudicial killings and we do not support that."[66] European governments were similarly opposed, arguing that these attacks were both illegal and counterproductive. At the same time, the US government was privately wrestling with the possibility of killing Osama bin Laden after he issued his fatwa urging all Muslims to kill US citizens and their allies. The Clinton administration quietly commissioned a series of internal legal findings that attempted to get around the ban on US officials engaging in assassinations, which had been in place since 1976.[67] Aside from political indecision, there was resistance against targeted killing from the intelligence community and the military. CIA director George Tenet did not believe that he had the authority to "pull the trigger," in part because the order to fire weapons should only come through a military chain of command.[68] Military officials were equally concerned about using lethal force in a region where the United States was not formally at war.[69] No one wanted to pay the political, and potentially legal, price if a targeted killing resulted in civilian casualties.[70]

The rapid advance of drone technology had begun to make targeted killings in places like Afghanistan feasible, if not easy, for the United States. The Israeli targeted killings had been conducted by manned aircraft, supported by drones, and launched at targets that were only a few miles away in the Occupied Territories. The United States had to be able to strike at a much greater distance at targets half a world away in Afghanistan. This produced an array of technical and logistical obstacles, including how the United States would pre-position drones for strikes in Afghanistan, how they would maintain communication with the drones, and how they would be sure at that distance that they were getting the right target. The chief technical problem was that there was a time lag between identifying a target and launching a missile. Under a program called "Afghan Eyes," the United States had deployed Predator drones to find Osama bin Laden and other al Qaeda operatives. Their efforts had succeeded. Unarmed Predator drones had detected bin Laden a number of times and recorded video footage of a tall man in white flowing robes surrounded by a security detail in an al Qaeda compound, which some CIA analysts were convinced was Osama bin Laden.[71] But the United States was not ready to strike him. Since the Predator drones were not affixed with Hellfire missiles, US officials would have to call in airstrikes from manned aircraft or use cruise missiles, both of which would not arrive for some time after the target was spotted. This meant that the bin Laden could escape and, even worse, that civilians could be killed if they moved into his place in the meantime.

By January 2001, the United States had managed to get around the technical difficulties of affixing Hellfire missiles to Predator drones and began to test them on dummy targets in the Nevada desert.[72] Neither the Pentagon nor the CIA was convinced that the Predator was sufficiently precise or reliable to be used in a targeted killing on a real battlefield.[73] In its first year in office, the Bush administration was equally unwilling to cross the line against assassinations.[74] The Bush administration slowed down the decision-making process, leaving no consensus over whether the United States had the right to take the shot if it had bin Laden in its sights.[75] The administration was divided over who had the right to authorize the killing, who pulled the trigger, and who paid for the missiles.[76] The technology was ready, but the policy for using Predator drones for targeted killing remained confused and uncertain.

The September 11 attacks produced a sea change in US policy as the Bush administration authorized previously unacceptable lethal actions to destroy al Qaeda.[77] While the US government did not formally abandon its legal ban on assassination, it embraced Israel's arguments about the distinction between assassinations and targeted killing, although the extent of explicit coordination between US and Israeli policymakers on this point remains unclear.[78] Secretary of Defense Donald Rumsfeld became convinced that "the techniques used by the Israelis against the Palestinians could quite simply be deployed on a larger scale."[79] The Bush administration won Congressional support for an Authorization to Use Military Force (AUMF) against those responsible for the September 11 attacks, which it subsequently interpreted as blanket approval for worldwide operations against militant groups even loosely affiliated with al Qaeda. The AUMF became a core part of the expansive legal rationale for drone strikes under the Bush, Obama, and Trump administrations, essentially greenlighting military action world-wide against any force presumably related to al Qaeda.[80] This would be interpreted later to include the Taliban, Islamic State, and other Islamist groups less directly related to al Qaeda. A series of executive orders signed by President Bush authorized new techniques and expanded the capacity of the CIA and branches of the military to conduct manhunts, interrogations, and targeted killings. On the grounds of self-defense, the United States would have the right to strike at al Qaeda and asso-ciated forces worldwide. The Bush administration granted the CIA an almost unlimited authority to kill terrorists, using a combination of manned aircraft, drones, special forces, and other means.[81]

Initially, a relatively small number of high-ranking al Qaeda operatives—estimated between seven and two dozen—were pre-approved for targeted killing if the CIA, military forces, and even some contractors could locate them.[82] According to former Deputy Secretary of State Richard Armitage, the Bush administration also issued a legal finding that killing terrorists worldwide did not constitute assassina-tion and hence did not violate the long-standing legal ban on this ac-tivity.[83] The United States did not accept that its embrace of targeted killing produced a generalizable right of targeted killing for all states, as it emphasized its unique situation in a global war against al Qaeda.[84] This assertion of a unique right to worldwide targeted killing for the United States was not widely accepted. In 2012, former CIA director Michael Hayden remarked that "right now, there isn't a government on the planet that agrees with our legal rationale for these operations, except for Afghanistan and maybe Israel."[85]

Yet the law was not a real obstacle to the creation of a targeted killing program with drones. In the post–September 11 period, both the CIA and Pentagon overcame their squeamishness about targeted killing and ordered that missiles be attached to many Predator drones already in service. The technology was now ready to be used. With the technical issues resolved, Predator pilots were now circulating over Afghanistan with Hellfire missiles, ready to either strike targets directly or call in sup-port from manned aircraft if needed (fig. 3.1).[86] The technical problems posed by distance had been resolved and the United States alone now possessed the ability to precisely target and kill individuals from

FIGURE 3.1 MQ-1 Predator unmanned aircraft, armed with AGM-114 Hellfire missiles, over southern Afghanistan, 2008.

cockpits located in trailers thousands of miles away. The only question was whether this capability would be taken off the "hot" battlefields of Afghanistan and Iraq and expanded to places where al Qaeda existed but the United States was not officially at war.

The Machinery of Killing

The first US targeted killing—of Sinan al-Harithi, a prominent al Qaeda leader in Yemen—occurred in November 2002. Although he had been placed on the CIA "kill or capture" list in 2001, al-Harithi was tracked down in the deserts of Yemen through a combination of human intelligence, aerial surveillance by Predator drones, and efforts by the National Security Agency (NSA) to track his cell phone use.[87] The Predator drone strike—launched after consultation with the CIA and lawyers affiliated with US Central Command (CENTCOM)— killed al-Harithi, Darwish, and several companions.[88] The Yemeni government consented to the killing, but only on the condition that it remain a secret. Both Yemen and the United States initially claimed that the deaths were due to a car bomb, but this fiction was shattered when Deputy Secretary of Defense Paul Wolfowitz admitted that the United States had killed al-Harithi.[89] The Yemeni government was furious that the promise of secrecy was broken.[90] Despite the Pentagon taking credit for the strike, the United States was reluctant to concede that this targeted killing created a precedent or represented a change in US policy. The State Department continued to argue that its opposition to Israeli targeted killings during the intifada remained intact, despite Wolfowitz's admission of US responsibility for targeted killings in Yemen.[91]

Inside the US government, the success of the al-Harithi killing was widely noted. The armed Predator program had already been expanded and developed for use in Iraq, although the United States would not resume targeted killings until the Nek Mohammed strike in 2004. But the most important shift was not in technology, but rather geography. With the al-Harithi and Mohammed killings, the United States was asserting a right to use targeted killings against al Qaeda and affiliated targets in countries where it was not at war. The use of Predators for targeted killing in declared wars like Afghanistan was uncontroversial: the United States had a clear legal right to strike targets in that country. It made no legal difference whether those strikes were conducted with manned or unmanned aircraft. But off the "hot battlefields" the law was different. Such drone operations are technically conducted under conditions of peacetime, and there is no presumptive right to use violence in such territories except under exceptional circumstances, like pursuing a terrorist across a border.[92] A different set of laws and human rights standards would apply to these circumstances, leaving the United States with no right to strike on the grounds of self-defense unless an imminent threat could be established.[93] The CIA would also not be considered a "lawful combatant" under international law.[94] Aware of the distance between accepted international law and their expansive interpretation of self-defense against the al Qaeda, both the Bush and the Obama administrations kept the program cloaked in secrecy and denied the jurisdiction of US and international courts.[95]

The al-Harithi targeted killing also illustrated the persistent bureaucratic divisions between the CIA and the Pentagon that would come to dominate the drones program. The ad hoc nature by which Predators were armed and later deployed for targeted killings left different agencies inside the US government in charge of different parts of the program. Today, the United States does not have a single targeted killing program, but rather two: one run by the CIA and another run by JSOC. These programs operate in parallel. CIA-controlled drone strikes in Pakistan, Yemen, and Somalia are conducted under the intelligence agency's legal authority (Title 50), which enables it to conduct armed, covert action to influence political and economic conditions worldwide.[96] In recognized, officially declared war zones, such as Afghanistan and Iraq, drone operations are officially conducted by the US military under a different authority (Title 10). JSOC operates under the same legal authority as other military forces, but its reach is worldwide and its targeted killings with drones can occur outside of recognized battlefields. This means that strikes outside declared war zones—for example, in Yemen—could be conducted by either the CIA or JSOC, or some combination thereof.

The CIA's drones program evolved organically from attempts to kill bin Laden and some of his top deputies. Although early CIA operations such as the al-Harithi attack often required the use of US Air Force drone pilots, the CIA has been fiercely opposed to surrendering control over its targeted killing program to the military.[97] Over time, the CIA's use of drones produced a change in the organization's orientation, away from intelligence gathering toward so-called "kinetic" or direct military action. Initially, the CIA's paramilitary wing, the Special Activities Division, wanted control over the CIA's drones. Yet the CIA leadership ultimately decided to give control over its drone fleet to the Counterterrorism Center (CTC).[98] The CTC was led by "Roger," a longtime CIA official with an unusual reputation in Washington as a chain-smoking, irascible convert to Islam.[99] A controversial figure because of his responsibility for the CIA's secret prisons and use of water-boarding against some senior al Qaeda officials, he nevertheless was able to build up a substantial reservoir of support for the CIA's drone programs on Capitol Hill, which allowed the program to grow substantially.[100] One estimate suggested that the CTC has expanded from only 300 employees in 2001 to over 2,000 within a decade of the September 11 attacks. [101] The CTC also commands substantial support from the CIA's analytic division as "targeters" for the drone strikes.[102] Under pressure to adapt its personnel structure, the CIA even created a career track for those specializing in targeting individuals for drone strikes. Although the CIA still uses drones for surveillance and non-lethal operations, the shift in its orientation toward paramilitary action has gradually turned the organization into "one hell of a killing machine."[103]

Although the CTC sets targets and commands the operations of drone fleets, it does not directly fly them. At Creech Air Force base in Nevada, there were a number of US Air Force units specifically designated for CIA use either for reconnaissance or targeted killings. One unit, the 17th Reconnaissance Squadron, flew between forty-five and eighty aircraft designated for CIA use and were responsible for more "kills" through targeted killings than any other unit in the US Air Force.[104] This unit was also responsible for flying many of the missions outside of declared war zones, such as those in Pakistan, Yemen, and Somalia. The fact that US Air Force pilots are flying CIA-commanded missions and the air force remained the owner of the drones themselves was not widely known, in part because the CIA's imprimatur acted as a shield to block Congressional and public scrutiny of the program.[105] Inside the government, the CIA is referred to as a "customer" of the US Air Force in that it sets objectives and mission parameters to be completed by trained drone pilots.

By contrast, the JSOC drones program is located in a military chain of command and reports to the US Special Forces Operations Command (SOCOM). JSOC's drone fleets are assigned to different regional combatant commands and work in different theaters in support of US military objectives. In theory, this should mean that JSOC is governed by traditional military rules of engagement and the law of armed conflict. In practice, JSOC is even more secretive than the CIA, and its compliance with traditional military rules of engagement and the law of armed conflict is uncertain.[106] JSOC's permissive rules of engagement were based on the Bush administration's executive order authorizing worldwide operations against al Qaeda. JSOC has been accused of torture and abuse of detainees in secret prisons in Iraq and elsewhere.[107] The scale of the abuses at JSOC-run prisons in Iraq was so severe that other government agencies, including the CIA, barred their personnel from participating in interrogations in those prisons.[108] JSOC operations are so highly compartmentalized and rapid that it is unclear whether all of them are subject to appropriate legal oversight. It also has limited Congressional oversight over its budget or activities.[109]

Under President Obama, JSOC expanded its kill or capture operations against terrorist targets around the world.[110] Although its operations are shrouded in secrecy, JSOC is widely acknowledged to have expanded dramatically over the last decade.[111] By 2015, it had a budget of nearly $1 billion and a staff of approximately 4,000 soldiers and civilians, supplemented by its own drone fleet and intelligence capacity.[112] Private contractors also assist with JSOC operations. Contractors from the private security firm Blackwater, which had played a controversial role in Iraq, were responsible for assisting drone operations, even to the point of loading missiles onto Predator aircraft, although they were not permitted to be in the so-called "kill chain," which authorized drone strikes.[113] With this elaborate infrastructure, JSOC has been alternatively described as a "self-sustaining secret army" and "an almost industrial scale counter-terrorism killing machine."[114]

The JSOC drone program operates in multiple theaters and across recognized conflicts, such as Iraq and Afghanistan, and undeclared battlefields in Yemen and Somalia. In the latter cases, JSOC conducted targeted killing operations with drones, sometimes in cooperation with the CIA. The relationship between the CIA and JSOC is complex. Although their drone programs operate simultaneously in these theaters, JSOC and the CIA sometimes clash on proposed targets and on information sharing. They also have different surveillance equipment and communications systems that are not fully compatible for sharing information about a target.[115] Reportedly, they have different standards for acceptable civilian casualties when identifying targets. Yet despite these differences, JSOC and the CIA have cooperated in implementing targeted killings. For example, JSOC ran secret drone overflights in Pakistan to allow the CIA to identify targets for targeted killings.[116] JSOC has "borrowed" CIA drones for targeted killings and sometimes closely coordinated their strikes in real time with CIA station chiefs.[117] This arrangement—where JSOC forces operate under the cover of the CIA—was employed by President Obama for the killing of Osama bin Laden in May 2011. It remains unclear how many drone strikes fell under the same blended legal authority, but some experts have argued that the distinction between JSOC and CIA is even more blurry than it appears at first glance.[118] The most aggressive counterterrorism operations outside declared warzones—such as manhunts or "snatch and grab" operations in places like Yemen and Somalia and targeted killings in Pakistan—have involved CIA, JSOC, other US military personnel from different branches, and private contractors working together, although their degree of their cooperation was kept secret.

These joint operations are especially controversial because JSOC forces would occasionally operate under the CIA's legal authority, effectively as seconded units to the CIA itself. This allowed JSOC to operate with looser rules of engagement and less legal scrutiny than would traditional military operations.[119] As a result, JSOC-run strikes or joint JSOC-CIA operations may operate with a higher tolerance for civilian casualties. For example, it remains unclear in joint JSOC-CIA operations whose lawyers—and whose standards for acceptable civilian casualties—apply. These operations may be effectively shielded from effective Congressional scrutiny, especially if it is unclear who is actually responsible for firing the missile.[120] An absence of public transparency about drone-based targeted killing compounds the problems surrounding responsibility and accountability: it remains as unclear today as it was a decade ago who is really pulling the trigger for some drone strikes.

One additional problem arising from the bureaucratic complexity of the targeted killing program is that different government agencies have developed different kill lists: at one point, the National Security Council, CIA, and JSOC each had their own.[121] In some cases, there was substantial overlap between these kill lists, but in others there were sharp differences of opinion about the degree to which an individual posed a threat or should be targeted. To resolve these differences, the US government has built a bureaucratic infrastructure to decide on targets and to administer its targeted killing program. Each of these actors could nominate a potential target for consideration at the so-called "Terror Tuesday" meetings. Biographical and threat-related information on the targets would be collected and condensed onto a short document dubbed the "baseball card" (BBC), which would be distributed to different stakeholders for discussion. In consultation with the National Counterterrorism Center (NCTC) and his senior deputies, President Obama had the final authority over who would be placed on the kill list and would personally approve the targets. The consolidated kill list even earned a different, more anodyne name: the Joint Prioritized Effects list. By 2013, the machinery of killing had become so complex that the US government constructed a master database—called, with Orwellian flourish, a "disposition matrix"—which included names, biographical details, suspected locations, and operational details.[122]

As some of these details slipped out to the public, the outcry over drones began to grow, and the Obama administration was forced to defend the legal basis of the program publicly.[123] In April 2012, Brennan offered a defense of targeted killings, arguing that they were justified on the grounds of self-defense against al Qaeda and they were conducted in ways consistent with the laws of armed conflict.[124] In a speech at the National Defense University in May 2013, President Obama described the US drone campaign in Pakistan, Yemen, and elsewhere as part of "a just war—a war waged proportionally, in last resort, and in self-defense."[125] Although he expressed some ambivalence about the expansive terms of the AUMF, President Obama still argued that targeted killings were authorized on the grounds of self-defense against al Qaeda and were legally and morally superior alternatives to using special forces or ground troops to capture or kill terrorist suspects in places where the United States was not formally at war. Yet the controversy continued to grow when legal rationales for targeting US civilians such as Anwar Awlaki became public. These secret memoranda produced by the Department of Justice suggested that killing US citizens was permitted if capture was infeasible and a threat was imminent. These memoranda allowed the US government to kill a citizen without presenting evidence or conducting a fair trial—clear requirements under the Bill of Rights—provided that the citizen joined the ranks of a declared enemy of the United States. These documents also left the standards for feasibility undefined and noted that imminence "did not require the United States to have clear evidence that a specific attack on US persons and interests will take place in the immediate future."[126] The Obama administration's response had brought some sunlight to the drones program, but had also illustrated how far the United States had drifted from its historical position of limiting the use of force to fighting foreign combatants on active battlefields.

In 2013, President Obama formalized these standards by signing off on new presidential policy guidance (PPG) for drone strikes with requirements for authorizing a strike much tighter than those his administration had used so far. This guidance required that: (1) the United States conclude that it could not feasibly capture a suspect; (2) that there be near certainty that a target is a lawful one and in the location where the strike is contemplated; and (3) that there is a near certainty that non-combatants will not be harmed by the strike.[127] It also required: that the US government certify that the target was a "continuing, imminent threat to US persons";[128] that the US government conclude, with legal review, that the government where the terrorist suspect was located could not or would not address the threat; and that the United States itself concluded that capture was not feasible. Once all of these standards had been met, a drone strike could—but would not necessarily be—authorized. In light of these new, stricter requirements, the pace of targeted killings in Obama's second term slowed considerably, with fewer strikes taken off the "hot" battlefields but more devoted to destroying ISIS forces in the declared war in Syria and Afghanistan.

As part of his May 2013 speech announcing the changes in the PPG, President Obama proposed that the CIA's drone operations be transferred to JSOC in order to subject it to traditional military oversight. Some prominent CIA officials, including President Obama's chief counterterrorism advisor and later CIA director John Brennan, were supportive of the plan on the grounds that drone strikes had diverted the CIA from its traditional mission of collecting intelligence.[129] The rationale also appeared to subject the CIA drones program to the military chain of command and its associated Congressional reporting requirements. Almost immediately the proposal ran into Congressional opposition from both parties, and Senator Diane Feinstein (D-CA) inserted a clause in a 2014 military spending bill that blocked the proposed transfer of authority. Part of the reason for this opposition was that many in Congress, from both parties, were convinced that the CIA was more careful and discriminate in launching targeted killing than JSOC. They also believed that the CIA, operated with higher evidentiary standards for selecting potential targeted killing and for avoiding collateral damage than JSOC. [130] In July 2016, the Obama administration released official data on the drones program claiming that it had killed only between 64 and 116 civilians between 2009 and 2015.[131] Almost immediately, outside groups called these estimates wildly implausible and demanded to know the methodology behind counting the deaths. By the end of the Obama administration, the CIA and JSOC's programs remained parallel operations and the true number of civilian casualties—as well as who bore legal and moral responsibility for them—remained unclear.

When President Donald Trump came into office in 2017, few people could anticipate what he would do with the targeted killing program that he inherited. On the campaign trail, he said little about it. His campaign rhetoric tended to be extreme—for example, he boasted that he would kill terrorists and their families—but it was unclear whether he would follow through, and if so, whether he would do so with drones.[132] In practice, Trump accepted the basic outlines of President Obama's polices, though he delegated more of the decision-making to the Pentagon and to lower levels. Although his administration has not released a revised PPG, it quietly revoked some of the new standards for avoiding civilian casualties that Obama had put in place in May 2013.[133] Trump made substantial changes to targeting standards, throwing out the standard that a target must pose a continuing or imminent threat to US persons.[134] He also lowered the standard for targeting individuals from "near certainty" to "reasonable certainty," effectively lowering the burden of proof for authorizing a drone strike and enabling drone strikes where the risks of civilian casualties were higher.[135]

In practice, this meant that the pace of drone strikes under President Trump ramped up considerably, although the geographic focus shifted. While the Trump administration continued to strike at targets in Pakistan, it has shifted more toward doing so in support of the counterinsurgency mission in Afghanistan.[136] In a tacit recognition of how these battlefields are intertwined, the Trump administration targeted high-level Pakistani militants in Afghanistan, including Mullah Fazullah, the leader of the Pakistani Taliban and the successor of Nek Muhammad.[137] Elsewhere, it accelerated the pace of drone strikes against ISIS in Iraq and Syria, although hard data on the number of drone strikes there is hard to come by. Outside the "hot" battlefields, the increase in targeted killings was more striking. In 2017, the Trump administration launched 131 drone strikes in Yemen, more than three times the number of strikes in the previous year.[138] These were allegedly against AQAP targets, though in practice it was hard to distinguish the actual purpose of these specific strikes amid the tumult of Yemen's brutal civil war. In that war, the United States found itself allied with, but not necessarily always attacking alongside, Saudi Arabia and United Arab Emirates in their campaign against Houthi rebels seeking to overthrow the central government in Yemen. The shadow war also spread deeper into Africa: in Somalia, the Trump administration conducted an aggressive drone strike campaign against al-Shabaab militants with thirty-five strikes launched in 2017.[139] By 2018, this had increased to forty-five so-called "precision strikes," although actual details on who was launching the strikes, and what targeting standards were used to allow them, remained shrouded in secrecy.[140]

President Trump has also followed Obama's lead in geographically expanding the shadow wars in order to allow for more drone strikes. Although hardly noticed at the time, President Obama ramped up the drone war in Libya, using nearly 300 drone strikes among hundreds of other attacks to force the Islamic State to evacuate Sirte in eastern Libya.[141] In December 2016, President Obama quietly designated eastern Libya as a zone of "active hostilities," thus exempting it from the tighter rules of his 2013 PPG guidance. This decision—effectively designating part, but not all, of a country a war zone—allowed more permissive standards to apply for drone strikes. This was revoked before President Trump took office, but he took his cue from Obama's practice.[142] He re-designated eastern Libya as an active combat zone, but also declared parts of Yemen and Somalia areas of "active hostilities" in order to enable more drone strikes.[143] These decisions went alongside the rapid creation and expansion of new drone bases in Niger and lower Somalia to allow the expanded air war to take place.[144]

The final step that Trump took to amend the drones program pushed it further into the shadows. In 2017, the Trump administration simply ignored the legal requirement imposed by the Obama administration in May 2013 that it should produce a report detailing the civilian casualties from drone strikes.[145] By 2018, his administration had taken steps to formally revoke this requirement, arguing that it led to unnecessary burdens on the government.[146] As a result, there is no official government data for the growing number of shadow wars that drones enable in Pakistan, Yemen, Somalia, Libya, and elsewhere. This makes it harder for anyone outside the closely guarded government bureaucracy running drone strikes—including those in Congress—to be informed enough to make judgments about its use.

As Trump's first term in office approached its end, it was clear that the finely calibrated killing machine built by the Obama administration had been handed over to a president who had placed it into overdrive and shielded even more of its operations from public view. In retrospect, the evolution of drone strikes is extraordinary. In less than twenty years, the US government had gone from scrambling to work out how to attach a missile to a Predator drone to developing a legal foundation and bureaucratic infrastructure for the practice of targeted killing in seven or more countries. This unwieldly bureaucratic infrastructure was remarkably different from the clear lines of authority that had traditionally governed military operations. It was internally divided between the customers (CIA and JSOC) and between the customers and those who actually fly the drones (the US Air Force). The entire process as to how someone in a foreign country is selected for killing is shrouded in secrecy and subject to relatively weak Congressional scrutiny and reporting requirements. For the most part, it remains entirely within the executive branch, effectively putting the US president in charge of deciding on the deaths of citizens in foreign countries.[147] Most of the key elements of the program—including the standards for selecting targets—remain unclear. Drone technology might have brought the United States the capability of continuously monitoring and striking targets remotely, but it also led the United States to lose sight of its goals and drift into a growing number of conflicts worldwide. The policy governing its use has also been allowed to travel some distance from what was once considered acceptable state practice.

Find, Fix, and Finish

The process for targeted killing is suffused with technology, procedures, and language that places some distance between those flying the missions and their intended targets. If the president approves a person for a targeted killing, a sixty-day window is given to "operators" for a potential strike.[148] Inside the government, the process for targeted killings became known in shorthand as "find, fix, and finish."[149] The organization that proposed the strike, typically CIA or JSOC, would generally be assigned principal responsibility for implementing that strike within that window. The first step—finding the target—involved a number of government agencies, including the CIA, NSA, and others, combing through multiple "streams" of information to locate the target. Among these streams were tips from sources on the ground, media reports, intelligence from foreign governments, and signals intelligence such as NSA intercepts of cell phones, emails, and other electronic records. These would be compiled into a coherent picture of the target and its environment. As chapter 4 will discuss, this process reflects the drive toward total battlefield awareness seen in the wars in Afghanistan and Iraq, where drones are helping to provide a level of detail about enemies and their movements that has never been achieved before in warfare.

The most difficult element of finding the target concerns human intelligence. The CIA had long traditionally struggled to develop human intelligence—dubbed HUMINT—from local sources in places like Pakistan and Yemen where tribal loyalty runs deep, and it is considered a violation of local custom to betray "guests" protected by one's family or tribe. Against these limitations, the CIA has offered cash payments to induce locals to provide information about the location of potential targets and, in some cases, to plant transmitters in their cars, homes, or even on their person so that they could be traced and later killed.[150] In other cases, the United States has engaged in social network analysis, tracking those who know or speak to the target in order to identify potential targets for surveillance. Information of the key personnel of a social network would then be passed on to ground sources or given to drone pilots to guide their missions. Many Predator or Reaper drones involved in targeted killings never fired a missile, but rather watched targets and their social networks continuously to understand their regular movements and pinpoint their locations.

Once the target was found, relevant operational details about a potential strike are collected into a "strike package," which would move up the military chain of command and ultimately arrive at the White House. This would "fix" the target at a location. Among these details were estimates about the feasibility of a potential capture operation and the degree of civilian casualties associated with a potential strike, with corresponding estimates of the degree of confidence in these casualty estimates. The location of the drone strike was conceptualized in three-dimensional space as a "killbox."[151] Inside this space, the target is centrally located, but the planners can also construct probabilistic estimates of how many others will be killed. The strike package would be approved by lawyers working for different government agencies and intelligence analysts who would estimate the degree of certainty of the intelligence behind the strike. One internal JSOC document suggested that targeted killings would not be approved without two pieces of confirming intelligence pinpointing a target and no contrary intelligence that the target was elsewhere.[152]

The degree of policy and legal scrutiny of both CIA and JSOC operations has varied across the different administrations. Under President Obama, the legal scrutiny was conducted through multiple rounds of review, beginning with the National Security Staff (NSS) and ending with the Deputies and Principals of the National Security Council. In some high profile cases, especially those involving US citizens, President Obama would personally sign off on the "disposition" of the target. The Obama administration required that lawyers be involved in the process of approving targets, but the Trump administration scaled back this requirement because it was too onerous. Trump instead opted for a more delegated approach, sacrificing much of the high-level vetting and allowing the decision to be taken at a lower level with considerably less legal scrutiny.[153] Only the decision to engage in direct action (either covert operations or drones) in a new country was taken to the highest level for approval.[154] This reflected President Trump's conviction that Obama had erred on the side of too much legal wrangling and that the process needed to be streamlined in order for the military to "take their gloves off" against enemies like the Islamic State.

Once the decision has been made to the target the individual and the legal review concluded, the final step is to finish a target. Although the precise details remain secret, it is clear that authorization for a targeted killing would come from the president, especially for risky strikes, or from senior deputies for more routine strikes. This authorization would be conveyed by a "customer," such as JSOC or the CIA, to the drone pilots. Once the authorization to launch a time-sensitive missile strike had been given, a team of government and intelligence officials from different agencies would view the drone's video feed to make assessments about the feasibility of a strike and the risks of civilian casualties. Communication between the participants in a strike would be done by chat messages, and the entire operation would be supervised by military lawyers who approved strikes.[155] Analysts would be able to watch the drone's video feed in real time, although senior Pentagon officials tried to limit this access to only those strictly needed for the strike. Once the missile strikes were approved and launched, the video feed from the Predator or Reaper drone did not stop, so the pilot and all others viewing that feed would be able to provide a damage assessment and conclude whether they had killed the target. Only at this point would the killbox be closed.

Distance and Intimacy

Drone technology is designed to allow states to escape what might be called "the tyranny of distance"—the fact that geography imposes real constraints on military operations and costs militaries valuable time that could be deployed to offensive activities.[156] It also does so without posing immediate physical risk to the pilots, who are located in cockpits on a military base thousands of miles from the strike. The process of targeted killings by drones is conducted with antiseptic language designed to create moral distance between the pilots and those killed. But it is clear that drones do not completely eliminate distance just as they do not banish risk. Drone warfare is a surprisingly intimate form of violence, as it allows the perpetrator of violence to observe the prospective target over time, thus breeding more familiarity than is often found with artillery, nuclear weapons, or other forms of violence conducted at a remove.

The issue of distance is crucial to the debate over the ethics of targeted killing. Critics have argued that drone warfare cheapens violence by making it feel and look more like a video game, rather than real combat. The UN Special Rapporteur on Extrajudicial Killings Philip Alston has remarked that drones produce a "PlayStation" mentality, through which the combatants are desensitized to the deaths that drones produce and treat it as a game.[157] To some extent, the technology itself reinforces this: the US military designed some of its controllers for drones to reflect those used by modern video game consoles, in part because so many young men were already adept with them. There is also some evidence from leaked documents that some drone pilots have described their victims in dehumanizing ways—for example, "bug splats"—and even called those running from their drones "squirters." Others described their job in routine terms as "mowing the lawn" by removing insurgents from the battlefield.[158]

Some pilots have described their operations as "like a video game" and seemed to enjoy the experience and power of delivering death from above.[159] Michael Haas, a former drone pilot involved in targeted killing, described how drone technology can desensitize pilots to the reality of targeted killings. "Ever step on ants and never give it another thought? That's what you are made to think of the targets—as just black blobs on a screen. You start to do these psychological gymnastics to make it easier to do what you have to do—they deserved it, they chose their side. You had to kill part of your conscience to keep doing your job every day—and ignore those voices telling you this wasn't right."[160] Although the violence was conducted without immediate risk to the pilots, many pilots expressed a willingness to shoot and to be seen as real warriors like the rest of the air force.[161] The warrior ethos that underlies much of modern militaries affects drone pilots, many of whom are derided by their peers as not fighting a "real war," and paradoxically produces coarse language and aggressive behavior designed to insulate them against this charge.

If war becomes "push button"—that is, like a video game and conducted by de-sensitized pilots—there is a danger that it may become too frequent. As chapter 1 suggested, drone warfare, especially targeted killings, is so different from traditional combat that there is a risk that governments will engage in it more often.[162] Drones are a seductive type of technology in that they appear bloodless from the vantage point of the person pulling the trigger. For the United States, targeted killings offer the possibility of controlling and limiting risk because its pilots cannot be shot down or killed. The safety of pilots thousands of miles away in an air-conditioned bunker might lead policymakers to conclude that they should use drones more often, perhaps engaging in riskier activities than they would if manned aircraft were employed.

Another danger of remote warfare is that the criteria for using drones for targeting killing will be "steadily relaxed" over time, so that it occurs more often and may not always be directed to those who are actual combatants.[163] The Obama administration came under criticism for this: for expanding drone strikes to targets in militant groups only indirectly related to al Qaeda and to lower-ranking operatives, rather than leaders. The Obama administration was accused of counting all "military aged males"—or MAMs, in official parlance—as terrorist operatives, even when there is no evidence for this designation. Some critics have argued that the absence of domestic political risk in targeted killing—specifically, that there is no prospect of downed or captured pilots—has made the Obama and Trump administrations careless in their selection of targets.[164] Some have gone even further, arguing that the practice of targeted killing itself shatters the assumption of mutual risk that underlies most combat, rendering it closer to a technologically sophisticated form of hunting, or even execution, rather than warfare.[165]

At the same time, many drone pilots report that the process of targeted killing is not as capricious or lawless as it is often depicted in the media. T. Mark McCurley, a former drone pilot, has written that his experience of flying drone missions did not show that drone pilots were careless; on the contrary, their fear of making a mistake made them exceptionally careful.[166] The fact that many people across the US government can pull up the video feed of a Predator drone ensures mistakes will be caught and properly punished. Drone pilots do not pull the trigger following a snap decision, nor do they make discretionary choices about who is targeted. Reckless killing via drones is possible, defenders argue, but difficult due to the number of people who have some supervisory or input role in the process. Others have noted that drone pilots are embedded in a decision-making process populated by intelligence analysts and lawyers watching their every move, so pilots themselves cannot be careless or indifferent to civilian casualties without consequences. They also argue that the process of targeted killing is more deliberative than the decision-making processes within war because participants have the time to reflect upon their actions before killing.[167]

Moreover, drone pilots often argue that they are hardly removed from the experience of killing in combat or targeted killings, but rather are intimately connected to their victims because they have been monitoring them for days. In contrast to combat, where violence emerges suddenly between strangers, drone pilots often are able to build up substantial amounts of time monitoring their target and developing some empathy with them.[168] Drone pilots have reported seeing their targets drinking tea, attending weddings, and spending time with family members. For this reason, many drone pilots reject the argument that drone warfare is desensitizing. They report having a vivid understanding of the death and suffering resulting from a drone strike.[169] With targeted killings, although the pilot was physically removed from the carnage that flows from a strike, he or she would see the human consequences of a drone strike, including the grieving families and sometimes even the burials of the victims. In the words of one unnamed pilot, "we're not disconnected from what's happening. We're not playing video games. With RPAs, you grasp your enormous level of responsibility. You witness it all."[170]

FIGURE 3.2 US Air Force ground control station at Balad Air Base, Iraq, 2007.

Some drone pilots report that the surprising intimacy of targeted killings has given them post-traumatic stress disorder (PTSD) because they witness the visceral effects of death and destruction up close.[171] Former US Air Force drone pilot Brandon Bryant reports that the experience of flying Predators over places like Afghanistan and Iraq was both alternately boring and horrifying, especially when he saw targets and those surrounding them in agony after a Hellfire missile strike.[172] Some pilots are also haunted by their mistakes, especially when their missiles hit women, children, and even animals. A 2011 air force study found that one third of all Predator and Reaper pilots suffered from burnout, while 17% suffered from "clinical distress."[173] Similar results were found for Global Hawk pilots. A later Department of Defense study found that drone pilots experienced the same levels of anxiety and depression as pilots of manned aircraft who were deployed to Afghanistan or Iraq.[174] These findings remain controversial, and at least one skeptic argued that the trauma experienced by drone pilots is due to boredom, rather than the voyeurism of virtual combat, and that claims of PTSD are designed to deflect pressure on the United States for engaging in war at a remote level.[175]

At a minimum, it is clear that the experience of drone warfare is leading some former pilots to become outspoken critics of the targeted killings program. In 2015, four former drone pilots involved in the program wrote a letter to President Obama calling it "one of the most devastating driving forces of terrorism and destabilization around the world."[176] Calling the program "morally outrageous," they argued that targeted killings are producing a deep wellspring of hatred of the United States around the world and that excessive civilian casualties are fueling recruitment to terrorist groups like ISIS.[177] While these dissenters remain a minority of the total number of drone pilots who flew Predator and Reaper drones, their criticisms cut to the core of the issue: does the targeted killing program work?

Effectiveness

The question of the effectiveness of targeted killings using drones is perhaps the most controversial question facing US policymakers.[178] There is considerable anecdotal evidence that drone strikes have decimated organizations like al Qaeda by killing their most dangerous operatives.

There is no doubt that drones have taken very senior figures in groups like al Qaeda, ISIS, and others off the battlefield. By one estimate in 2013, more than fifty senior al Qaeda and Taliban leaders have been killed by drone strikes.[179] In 2011, one CIA official remarked that "we are killing these sons of bitches faster than they can grow them now."[180] By 2015–2016, US and allied drones were also killing senior Islamic State members in Iraq and Syria, forcing the group to adapt to preserve its leadership structure. A secondary benefit of the drone program is that it places so much pressure on terrorist organizations that it becomes hard for them to operate, recruit, or launch strikes.[181] Amid the constant fear that their operatives will be taken out in a blinding strike, terrorist organizations can do little more than think of their own survival. For this reason, some government officials argued in 2011 that years of drone strikes had pushed al Qaeda to the point of "strategic collapse."[182]

In Washington, there is a consensus among both political parties that drones are effective in disrupting the operations of terrorist organizations. Former Bush counterterrorism advisor Juan Zarate argued that drone strikes had knocked al Qaeda "on its heels" as a result of the death of so many operatives.[183] In 2012, former Obama administration official Jeh Johnson publicly raised the question about what happens to the United States' war on terror once al Qaeda has been decisively defeated, in part due to the use of drones.[184] Even as criticism of the civilian casualties from drone strikes increased, the Obama administration insisted that their effectiveness was undoubtable. As President Obama remarked in his speech at the National Defense University in May 2013, "today, the core of al Qaeda in Afghanistan and Pakistan is on the path to defeat. Their remaining operatives spend more time thinking about their own safety than plotting against us. They did not direct the attacks in Benghazi or Boston. They've not carried out a successful attack on our homeland since 9/11."[185] In 2016, former CIA Director Michael V. Hayden described the drones program as "the most precise and effective application of firepower in the history of armed conflict."[186] President Donald Trump has said little about targeted killings, but clearly believes that they are an essential part of getting tough with terrorists.

There is anecdotal evidence that terrorist organizations are terrified of drones and that their presence imposes real costs on their ability to operate freely. The journalist David Rohde was held in captivity by the Taliban and reported that his guards were terrified of drones and took steps not to draw the attention of the Predators flying overhead.[187] In its publications, al Qaeda lamented the effects of drones, noting that its commanders had been snatched away by planes that are "unheard, unseen and unknown."[188] Elsewhere al Qaeda has discussed the carnage and destruction—as well as the pervasive fear—produced by the drones above. Osama bin Laden was very concerned about drone strikes and advised his followers not to gather in large numbers for fear that they would attract drones.[189] He even advised some of his followers to flee Waziristan to avoid their gaze.[190] He also recommended a range of operational security measures, such as traveling by road infrequently, carefully monitoring movements to not attract attention, and moving on overcast days to avoid being spotted. He recognized that drones had hollowed out the top leadership of al Qaeda and left the ranks of the leadership populated with younger, less experienced, operatives.[191] At the end of his life, he was also obsessed with operational security, particularly with the danger that the CIA would place a tracking device on a courier and reveal his location.[192] Perhaps the best evidence of the fear of drones has been the extent to which al Qaeda, the Taliban, ISIS, and other targeted groups have changed tactics and ruthlessly killed those that they suspect turned over information about their whereabouts.[193]

Because of the difficulties of getting reliable data on violence in places like Pakistan and Yemen and the secrecy surrounding the drones program, there have been relatively few empirical studies of the effectiveness of drones. One such study found a modest reduction in violence and reprisal attacks on tribal elders following US drone strikes in Pakistan.[194] Another found that drone strikes weakened the operational effectiveness of terrorist groups by causing such pressure that they experienced desertions and internal political struggles.[195] Yet a general problem with analyzing the effectiveness of drones is that there is no consensus on how the effectiveness of drones should be measured.[196] Reductions in violence following strikes are an intuitive way to do so, but in these environments the causes of fluctuations in violence are complex. Moreover, some policy advocates of targeted killing have conflated tactical success—that is, removing potentially dangerous actors from the battlefield—with strategic success. In fact, these are quite different: the United States could be eliminating dangerous terrorists in places like Pakistan and Yemen but simultaneously be weakening the government and putting longer-term goals, like regional stability, farther out of reach. Other arguments have confused efficiency—that is, the cost of drone strikes relative to manned aircraft strikes—with effectiveness, arguing that drones are a better, less expensive way to fight a war than a more expensive manned air plat-form. The problem is that the financial cost of a particular weapon is no guide to its effectiveness. A cheaper weapon is not, by definition, a more effective one. Many of the efficiency arguments are also problematic be-cause they assume that the goals of a targeted killing program are fixed and clear. But for years the Obama and Trump administrations have left the targeted killing program so cloaked in secrecy that it is hard to know what ultimate goals it serves. Does the United States have an end state to a campaign of targeted killing? Or is it simply one of maintenance of an unstable situation by removing "bad guys" from the battlefield? One danger is that drone technology may lead to goal displacement in which the objectives of a targeted killing program become more ambitious be-cause the technology enables new operations once seen as too difficult or risky.

At a minimum, if the goal of the targeted killings programs in Pakistan and Yemen is to defeat al Qaeda and affiliated forces, it cannot be considered a clear success. Instead of pushing al Qaeda to the edge of strategic collapse, drones have accelerated its fragmentation into a series of local affiliates, most of which are weaker than al Qaeda's central or-ganization was at its height before September 11. Here the effectiveness argument is murkier than it looks: drones have degraded the central organization of al Qaeda, but they have also turned one enemy into a series of loosely connected smaller foes. It is not yet clear whether this fragmentation will render al Qaeda strategically weaker than it was be-forehand, or whether in fact this dynamic will give al Qaeda a second life as a franchise outside the original theaters of operation. The prolif-eration of al Qaeda franchises—and plots flowing from these branches in Yemen and elsewhere—suggest that the claim of effectiveness is not yet proven. Similarly, the rise of the Islamic State in Syria and Iraq out of the ashes of an al Qaeda affiliate decimated by drone strikes suggests that the strikes themselves are not the kind of mortal blow that destroys the organization itself.

If the goal of the targeted killing program is to improve the sta-bility of governments in places like Pakistan and Yemen. While the cause and effect between drone strikes and stability is difficult to deter-mine because so many factors are at play, the countries most targeted outside of conventional wars—Pakistan, Yemen, and Somalia—are no more stable after years of drone strikes.[197] Despite the pressure of drone strikes, Pakistan remains a base for dangerous militant groups attacking US forces in Afghanistan, foreign fighters have continued to flow into the Afghanistan-Pakistan region,[198] and Yemen has collapsed into a vicious civil war. Perhaps the biggest difficulties involved in assessing these outcomes are that they involve a counterfactual: if the United States had not engaged in a targeted killing program, would these coun-tries have been more stable? In cases like Yemen, where the causes of militancy are complex and not reducible to drone strikes, this is a par-ticularly difficult question to answer. It is impossible to know whether the Houthi rebellion and ensuing civil war would have happened if the United States had not conducted drone strikes in the preceding years. The decision to launch a targeted killing is ultimately based on an ex ante judgment that the country was already unstable enough to justify a targeted killing. It is possible that the United States only engages in targeted killings in countries already sliding into chaos, thus making the adverse consequences of drone strikes hard to measure and to trace back to US "kinetic action." The relationship between drone strikes and stability remains extraordinarily difficult to disentangle, especially when it remains unclear whether drone strikes have disrupted potential insurgent attacks against the government or merely redirected them elsewhere.

The redirection problem with drones is a serious one because most insurgent and terrorist groups work across national borders. A sustained campaign of targeted killing may scatter an organization rather than cripple it. This contagion effect is recognized by senior government officials. Former CIA Director Leon Panetta argued that one measure of the success of the drones program is the fact that al Qaeda forces in Pakistan had sought refuge elsewhere to avoid the pressure from drone strikes.[199] But this may reflect a redirection of the violence rather than a reduction. The CIA found that the Tehrik-i-Taliban Pakistan (TTP) has fled out of the FATA and relocated to cities like Karachi, where they targeted Pakistani civilians with terrorist attacks.[200] In other cases, militants flee abroad but bring with them the skills, experience, and weapons needed to turn these local wars into even more fierce and long-lasting conflicts. As the aftermath of the civil war in Afghanistan (1979–1988) showed, the diffusion of trained foreign fighters to Algeria, Bosnia, and elsewhere worsened the civil wars in those countries. A similar dynamic is present in 2019 as foreign fighters that once joined the Islamic State in Iraq and Syria are fleeing to join conflicts around the Middle East and North Africa. If targeted killings push hardened terrorists out of Pakistan but relocate them to Afghanistan, or out of Iraq but into Libya, it is harder to consider targeted killings a net gain.

Most assessments of the strategic effectiveness of drone strikes consider the tactical benefits of drones—removing individual "bad guys" from the battlefield—but fail to consider the other side of the ledger: whether drones are also fueling terrorist organizations by stimulating recruitment of militant networks. Here the evidence is also murky. In Pakistan, there are accounts that drone strikes have convinced some locals to join the ranks of the TTP or other militant groups to fight the United States, or the Pakistani government for its complicity in their deaths.[201] One scholar has argued that US drone strikes in Pakistan motivate revenge attacks by Islamist groups against civilians, which can perversely lead to a hostile, uncertain environment which makes recruitment by those same groups easier.[202] Yet a comprehensive analysis of court records and of recruitment motives found no evidence that drones or targeted killings were fueling militant Islamism.[203]

In Yemen, local activists have claimed that drone strikes against AQAP have fostered anti-US sentiment in the tribal regions of the country and encouraged friends and family of killed civilians to join AQAP or other militant networks.[204] The drone strikes have allegedly bred "psychological acceptance" of AQAP among Yemenis, in part because they appear to confirm its narrative of a bloodthirsty United States dropping bombs from afar with no concern for who is killed.[205] As one local human rights leader put it, "the drones are killing al Qaeda leaders, but they are also turning them into heroes."[206] Because of inadequate data, it is impossible to know for certain whether drone strikes are creating more terrorists than they are eliminating.[207] At a minimum the steady drumbeat of popular criticism about drones from Pakistan and Yemen provides a narrative about the brutality of the United States and the fecklessness of their governments that is consistent with the recruitment pitches of dangerous militant groups.

Those recruitment pitches typically focus on one of the most controversial issues surrounding drones: civilian casualties. Advocates of drones compare the relative precision of drones favorably to other methods of warfare and note that drones cause fewer civilian or accidental casualties than attacks such as air strikes or ground assaults. Although the Trump administration has been largely silent on civilian casualties, the Obama administration offered a number of defenses when facing criticism about the human costs of its targeted killings campaign. It argued that the stringent guidelines for selecting drone targets ensure that relatively few, if any, civilians are killed in drone strikes. Former CIA Director John Brennan said that civilian casualties from drone strikes have "typically been in the single digits."[208] In January 2012, President Obama publicly argued that the drone program is "a targeted, focused effort at people who are on the list of active terrorists."[209] The Obama administration also insisted that drone strikes have a "surgical" character and are conducted with a "laser-like focus" on the targets alone.[210] In his May 2013 speech, President Obama acknowledged the reality of civilian casualties from drone strikes but insisted that "before any strike is taken, there must be near-certainty that no civilians will be killed or injured—the highest standard we can set."[211]

The problem is that casualties from drone strikes are notoriously difficult to count and verify. Most of the strikes are conducted in distant, sometimes ungoverned, territories of Pakistan and Yemen, where few have the ability to interview survivors or even count the dead. Moreover, it is well known that the Taliban, al Qaeda, and other local Islamist groups inflate the casualties for propaganda purposes. Many of the underlying newspaper accounts of drone strikes toss around words like "militants" and "civilians" casually, often without evidence. The underlying difficulties of reporting strikes in these countries has been compounded by the US decision to adopt a classification scheme which counts any male between the ages of eighteen and seventy killed in a drone strike as a "militant" unless posthumous evidence is presented to clear their names.[212] With the facts hard to come by, the perception of civilian casualties becomes almost as important as the strikes themselves. As Peter Bergen noted, the perception may be more important than reality, as more than half of respondents to a 2010 survey in North Waziristan believed that the strikes killed mostly civilians.[213]

A number of critics from inside the US national security establishment have come forward to voice concern that this perception of indiscriminate killing will generate backlash and anti-US sentiment. Admiral Dennis Blair, former Director of National Intelligence, has noted that while drone attacks did reduce the leadership of al Qaeda, it also "increased hatred of America" and harmed "our ability to work with Pakistan [in] eliminating Taliban sanctuaries, encouraging Indian-Pakistani dialogue, and making Pakistan's nuclear arsenal more secure."[214] Similarly, General James Cartwright, former Vice Chair of the Joint Chiefs of staff, said that with the drones policy in Pakistan, "We're seeing that blowback. If you're trying to kill your way to a solution, no matter how precise you are, you're going to upset people even if they're not targeted."[215] General Stanley McChrystal, who expanded the use of drones in Afghanistan, remarked that "what scares me about drone strikes is how they are perceived around the world. The resentment created by American use of unmanned strikes . . . is much greater than the average American appreciates. They are hated on a visceral level, even by people who've never seen one or seen the effects of one."[216] He also remarked that drones reinforce "perception of American arrogance that says, 'Well we can fly where we want, we can shoot where we want, because we can.'" While that blowback has not yet reached US shores, it is possible that it may in the near future. For example, the Times Square bomber Faisal Shahzad was allegedly trained and deployed by one faction of the TTP to attack New York in response to US drone strikes.[217]

The risk of backlash against the United States from drone strikes could be worse if drones produce a general climate of fear in the general population rather than just in militant groups. A controversial Stanford/NYU study of the impact of drones in rural Pakistan found that targeted killings had an invidious effect on the social fabric of the societies where they occur.[218] As Brian Glyn Williams has noted, drones are often described by local Pakistani villagers as *machays* (wasps) for their stings or *bangana* (thunder) for their ability to strike without warning.[219] While drones terrify their intended targets, innocent villagers are equally terrified of being in the wrong place at the wrong time when an attack occurs. Drones produce a "wave of terror" among the civilian population which has been described by some mental health professionals as "anticipatory anxiety."[220] In the words of David Rohde, the US journalist who was captured by the Taliban, drones strike fear because they are a "potent, unnerving symbol of unchecked American power."[221] This fear has led some ordinary civilians to refrain from helping those wounded in drone strikes for fear that they will be targeted in a second strike or to attend for funerals for fear that they will wind up in what are known as "double tap" strikes, which target mourners or those gathering at the scene of a strike.[222] Critics have also argued that drones have inhibited normal economic and social activity and even made parents reluctant to send their children to schools that might be accidentally targeted.[223] The problem, as critics have pointed out, is that these reports of terror from drones are based on select interviews with people in dangerous environments, and it is hard to know how representative their experience has been.[224]

At a minimum, there is more evidence that targeted killings have turned neighbors on neighbors and fueled communal mistrust in a society where overlapping family, tribal, and social ties are crucial. The targets of drone strikes are often pinpointed by paid informants who allegedly place small electronic targeting devices in the homes or vehicles of suspected terrorists.[225] Yet there is no way to tell whether these chips are left with real terrorist operatives or with those with whom the informant has a personal grudge. Rumors of these chips have produced mistrust in the community as "neighbors suspect neighbors of spying for the US, Pakistani or Taliban intelligence or using drone strikes to settle feuds."[226] The journalist Steven Coll reported that this fueled vicious internal power struggles within the societies under their watch as people turn on each other for being spies for the CIA.[227] While the drones circling overhead spread fear throughout the population and disrupt normal life, the suspicion produced by these chips and other means of nominating targets further erodes the social trust that underlies much of religious and political life in rural areas. At the same time, there are always some in every country who see drones as a net positive. Some Pakistanis and Yemenis support drones because they eliminate dangerous militants that kill more of their fellow citizens with terrorist attacks. The division between those who support drones and those who do not may fall along a number of lines, including region, class, and the degree of information that people have about drone strikes. For example, some scholars found that highly informed urban elites, and those deeply opposed to the agendas of Islamist networks, are more likely than less-informed and rural people to support US drones strikes even if they violate the sovereignty of the country.[228]

The depth of the degree of political protests over drones may still matter because it influences the stability of their governments and may make it harder to say "yes" to Washington. In countries like Pakistan and Yemen, drone strikes can corrode the governments' stability and legitimacy and deepen anti-US sentiment. Even when employed against local enemies of these governments, US drone strikes are a powerful signal to citizens of the helplessness or complicity of their governments and undermine the notion that these governments are a credible competitor for the loyalties of the population.[229] From a counterinsurgency vantage point, US drone strikes pose a critical problem, as governments facing ungoverned, unruly regions are usually engaged in a delicate dance with local tribes or clans for the loyalties of the population.[230] In September 2012, Pakistani Foreign Minister Hina Rabbani Khar captured this dilemma well, saying that "this has to be our war. We are the ones who have to fight against them. As a drone flies over the territory of Pakistan, it becomes an American war again. And this whole logic of this being our fight, in our own interest is immediately put aside and again it is war which is imposed on us."[231] The extent to which the United States has assumed the role of direct combatant and sidelined the Pakistani government through drone strikes undermines the authority and legitimacy of the government. This makes the establishment of a stable set of partnerships for counterterrorism cooperation difficult, if not impossible, to achieve over the longer term because such plans depend upon strong and legitimate governments as US partners. These political costs need to be weighed against the advantages accrued by removing dangerous militants in an analysis of the effectiveness of drone-based targeted killing.

Conclusion

The practice of targeted killings was not initiated by the rise of drones, but drones did begin to change the political and moral calculus of such practices. Following Israel's example, the United States has expanded the practice of targeted killing to a global scale, asserting for itself the right to strike al Qaeda and its affiliated forces anywhere in the world. Some of the unique features of drones—their ability to monitor targets over time, their precision, and the absence of physical risk to pilots—make them an attractive option. These technological features are not neutral, but rather affect strategic choice: if it becomes easier to engage in a targeted killing, there is a risk that states will resort to it more often. The danger is that the frequent use of targeted killings will corrode the traditional legal and ethical constraints on the use of force and create precedents that will be exploited by others.[232] The Obama administration was aware of this danger and its efforts to discipline the use of drones and improve transparency reflected an awareness that, as Brennan put, it the United States is "establishing a precedent that other nations may follow."[233]

There is some evidence that this has already begun to happen. Until recently, states with drones that could launch targeted killings have generally refrained from doing so, although China came close to using drones to kill a drug lord in Myanmar in 2013.[234] But now more states are joining the targeted killing club. In 2015, the United Kingdom, which had long supplied intelligence to and cooperated with the United States, authorized a targeted killing of a British national in Syria on the grounds that he was planning imminent terrorist attacks on his homeland.[235] In September 2015, Pakistan attached a missile to one of its Burraq drones, possibly with the help of Chinese technology or designs, and struck targets in the Shawal Valley in North Waziristan.[236] Less than a decade after secretly greenlighting a CIA drone to kill Nek Mohammed, Pakistan conducted its own targeted killings in the same tumultuous region from which he had emerged. Soon afterward other countries joined the club. Iraq used a Chinese-made drone in an assault against an Islamic State stronghold in Ramadi in December 2015.[237] In February 2016, the Nigerian military followed suit by conducting a targeted killing with a Chinese-made Rainbow drone against Boko

Haram, a radical Islamist group affiliated with Islamic State.[238] More recently, both sides in Yemen's civil war have used drones for attacks and have even begun to target each other's leadership as the United States has long done. In April 2018, the United Arab Emirates, part of a coalition led by Saudi Arabia to support Yemen's government in the civil war, used a Chinese-made drone to kill a senior Houthi leader who had expressed interest in a negotiating an end to the fighting.[239]

From its origins as a secret practice, hidden from public view in the remote regions of Pakistan and Yemen, targeted killing has now gone global, facilitated by the diffusion of drone technology and the corresponding change in what states can and are willing to do. Yet it remains unclear whether the spread of targeted killing—and the distance that it places between what drones can do today and the traditional respect for the sovereignty of other states—will produce a more peaceful world. The ability to inflict death from above in an almost surgical way reflects a desire among governments to target individuals, rather than groups, and to conduct war in a way that preserves the lives of civilians. It also reflects a desire to control risks to their own pilots, even if the risks to others on the ground increase. But it has also led to goal displacement: the United States may not have intended to become militarily involved in a growing number of countries around the world, but its pursuit of al Qaeda and now the Islamic State through targeted killing has led it do so.[240] In the end, the rise of targeted killings by drones may represent a significant technological accomplishment but it is hard to conclude that it is also a moral victory.

4

Eyes in the Sky

AS WORLD WAR I crept closer in August 1914, most European statesmen believed that the coming conflict would be short and glorious. In early August, German monarch Kaiser Wilhelm told his troops, "you will be home before the leaves have fallen from the trees."[1] The German army general staff believed that the war with France would be a rout that would last no more than four weeks.[2] Despite their different motivations and strategies, French officials largely concurred with this assessment, believing that a short, decisive war would forever resolve many of the tensions between the great powers of Europe. Some senior French officials almost seemed anxious for war. In 1913, one senior general declared, "give me 700,000 men and I will conquer Europe!"[3] Others were more circumspect about the prospect for gains from fighting but thought that the growing tensions in Europe could not be overcome in any other way. Across the continent, a sense of nervous anticipation prevailed, though only a few saw the disaster that would be World War I coming. The British foreign secretary Sir Edward Grey was among them. Looking across St. James Park at dusk on August 3, 1914, Grey quietly remarked to his companion that "the lamps are going out all over Europe, we shall not see them lit again in our lifetime."[4]

When the war began, events moved quickly. The German war plan was premised on the view that German forces could sweep through Luxembourg and Belgium and knock out France in less than five weeks before turning their attention to Russia. But like most war plans, this one did not survive the first brush with reality. German forces faced stiff resistance from Belgian forces and the unexpected entry of Britain into the war on the side of France and Belgium. This panicked some of the German leadership—the Kaiser, always prone to nervous fits, wailed in anguish that he had never anticipated British entry into the war—but the German forces marched relentlessly forward. By early September, they had advanced deep into northeastern France and pushed back exhausted French forces to within 30 miles of Paris. As Paris prepared for a siege, the French government fled to Bordeaux, while hundreds of thousands of French civilians left the capital. French General Joseph Joffre began to make urgent plans for a counterattack to break the German lines. Among the most important problems he faced was information about the movements of the enemy. To solve this problem, he turned to the fledgling aerial reconnaissance teams formed by the Grand Quartier Général (GQG), the command staff of the French army.

The decision to develop an aerial reconnaissance capacity only years after the Wright brothers discovered flight was a controversial one. In Britain and France, many dedicated army officers were convinced that aerial operations would never supplant ground forces in importance; in fact, General Sir Douglas Haig told one of his fellow officers in 1911 that pursuing aerial operations was a waste of time.[5] Most aerial reconnaissance units were placed under the control of the army even though there was a strong cultural and organizational bias against these operations among ground forces. Flying was described by senior army officials as a foolish indulgence and not something that real soldiers did. It did not help that aerial operations were also very dangerous. Early aerial reconnaissance involved pilots going aloft in rickety planes, landing on makeshift runways near their ground forces, and recounting orally what they had seen from the air. These reports were often "imprecise, because in the excitement of the first taste of combat, the observers' inadequate prior training frequently led them to misidentify troop nationalities and activities."[6] Army forces sometimes mistook their own planes for enemy aircraft and shot at them, an occurrence common enough to prompt militaries to paint flags and other colored markers on wings to allow them to be distinguished by ground forces. Bad weather frequently demanded the cancellation of flights, and turbulence was particularly severe against the thin wings that held the plane aloft. Some early planes were slow (roughly 26 miles an hour) and prone to either fail in mid-air or crash with little warning.[7] It is not surprising that many military officials initially saw aerial operations as a distraction that they could not afford as war approached. Marshall Ferdinand Jean Marie Foch, later supreme allied commander of Western forces, concluded in 1911 that "airplanes are interesting as toys but of no military value."[8]

The perception of aerial reconnaissance slightly improved when its advocates were able to take camera images of enemy positions from the air. These photos were naturally more precise and detailed than a verbal report and did more to convince skeptics that there was some value in the aerial enterprise. The Germans were the most sophisticated in terms of aerial photography at the onset of the war and had a wide range of cameras available to take pictures of enemy positions.[9] The French were the first among the Allies to take aerial photography seriously and by August 1914 had developed mobile labs for processing black and white images taken from the air.[10] By contrast, the British army treated aerial photography as a bothersome novelty and even made pilots buy their own cameras for this purpose. The earliest demonstrations of aerial photography were fraught with danger because they involved the pilot simultaneously flying the plane and using the camera. One British commentator remarked that the operations were so dangerous that observers of aerial reconnaissance exercises "came to scoff" but "remained to pray."[11]

The urgent needs of the coming war overcame the organizational barriers against the development of rudimentary air forces, starting with France in 1910.[12] At the beginning of the war, Britain had approximately 150 aircraft, France 160, Germany 246, and Russia about 150.[13] Not all of these aircraft were used for aerial reconnaissance; across all countries there were advocates of the strategic bombing of cities to break the morale of citizens, much in the way that the noted Italian general and theorist of air power Giulio Douhet predicted.[14] Other saw value in aircraft, but believed that their best use was to spot artillery fire or to direct ground forces. Germany even experimented with using planes for psychological warfare by dropping leaflets over Paris in late August 1914 warning that, "The German army stands before the gates of Paris. You have no choice but to surrender."[15] Yet for the most part the technology behind the aircraft operating in 1914 was not up to these tasks: the earliest attempts at strategic bombing were embarrassing failures, as the bombers were shot down or missed their targets by a wide margin. In 1914, the actual war in the air revolved around aerial reconnaissance. Although the first evidence of aerial reconnaissance occurred during colonial engagements in Libya and Iraq, it proved its worth during what came to be known as the "miracle" of the Marne.

As French forces considered options to defeat the encroaching German forces in late August 1914, the chief problem that they faced was detecting the location of the Germany army commanded by General Alexander von Kluck. At the Battle of Mons, information provided by two British pilots allowed the French to redeploy forces and hold off the German advance, thus allowing a fighting retreat against the advancing Germans.[16] Eager to cut off Paris from the main French forces, Von Kluck took his forces southeast and exposed the flanks of the German First and Second Armies.[17] His goal was to destroy the French forces entirely before moving against Paris.[18] British Royal Flying Corps (RFC) pilots were the first to discover this eastward shift of German forces and reported it back to the French ground forces. Although the reports were clear about the general direction of German forces, French military officials were reluctant to accept their findings because they did not correspond with other intelligence about the expected German war plan.[19] The chief problem was not intelligence but organizational inertia: GQG command were committed to a certain view of Germany's intentions, backed up by human intelligence, and the information coming from aerial reconnaissance seemed to contradict it. But additional French operations—including those by Corporal Louis Breguet, a pioneer in the design of aircraft who flew his own prototype for this mission—eventually confirmed that Von Kluck's forces were indeed shifting eastward.[20] French officials began to watch more closely, and over the next few days, British and French aerial reconnaissance provided irrefutable evidence of the movement of German forces. Crucially this information allowed the French forces to adapt and lay a trap for the Germans.

Under the command of General Joffre, the French military repositioned their forces to create a pocket where German forces would be caught and annihilated by the French 5th and 6th Armies and the

British Expeditionary Forces (BEF). To do this, the French needed to move the 7th Division northward from Paris to the front, but they faced a shortage of available rail lines to move the troops. The French quickly moved most troops and equipment by truck and the remaining train lines, but to supplement these efforts they also famously commissioned taxis from Paris to move nearly 4,000 infantry a distance of 30 miles to strengthen the French lines.[21] By the time the 750,000 strong German army walked into the trap, they were facing one million French and British forces.[22] The German forces, exhausted from previous battles and days of marching across the French countryside, found themselves enveloped by Allied armies. The battle began on September 6 and lasted for four days of intense fighting that saw the use of machine guns, small arms fire, and grenades for maximum carnage. The casualties were astonishingly high: one recent estimate suggests that as many as 300,000 soldiers perished during the First Battle of the Marne, mostly from French and German forces.[23]

The consequences of the First Battle of the Marne were strategic, psychological, and symbolic. The immediate strategic consequence was that the German army was forced to retreat and withdrew 40 miles to the River Aisne.[24] They settled in for another round of fighting by digging the trenches that would come to characterize World War I. The success of British and French forces at the Marne effectively set the stage for the brutal trench warfare of the Western front and guaranteed that the war would last for years. Its psychological consequence was the shattering of German optimism that they would indeed be home for Christmas. After the Marne it was clear that by assuming they could knock out the Western Allies before facing the Russians, the German command had made a grievous miscalculation. The grim fact was that the war would be conducted on both fronts, and the Central Powers would face encirclement by their enemies amid battlefields that resembled a charnel house. For Germany, this would contribute to the psychology of victimhood that coursed through its political history during the twentieth century. Many Germans saw the Marne as one of many examples when the nation was on the cusp of victory only to have it snatched away from them at the last possible minute by unexpected developments, even betrayals.[25]

The symbolic significance of the Marne went even further. The miracle of the Marne was the first time in which aerial reconnaissance played a decisive role in changing fortunes on the battlefields. It would not be the last. After the Battle of the Marne, no one could dispute the value of "eyes in the sky" for understanding events on the ground. The debate increasingly considered how best to see this way, not whether to do it at all. In a study of the effect of aviation on the war published in 1922, Walter Raleigh concluded that "reconnaissance, or observation, can never be superseded; knowledge comes before power; and the air is first of all a place to see from."[26] The information-gathering aspect of aviation was paramount and arguably more important than bombing or other functions of aircraft. In an interview published in the *New York Times* as the United States entered the war in 1917, Orville Wright acknowledged this development:

> I have never considered bomb-dropping as the most important function of the airplane and I have no reason to change this opinion now that we have entered into the war. The situation shows that, as a result of the flying machines' activities, every opposing General knows precisely the strength of his enemy and precisely what he is going to do. Thus surprise attacks, which have for thousands of years have determined the event of war, are no longer possible. When the United States sends enough airplanes abroad to bring down every German airplane which attempts to ascertain the disposition of the armies of the Allies—literally sweeps from the heavens every German flying machine—the war will be won because it will mean that the eyes of the German gunners have been put out.[27]

Wright saw that taking to the skies would transform warfare from a contest of brute force into one of information. As Raleigh noted, information precedes power and changes how power is enacted. In Wright's view, power is enacted not by destroying the enemy entirely but rather by rendering them incapable of acting—in his words, putting their eyes out—through achieving what would today be known as air superiority and information dominance. Although this line of thinking receded during the emphasis on strategic bombing during the Cold War, it has resurfaced with modern aircraft and drones over the last decade or more. With drones and other modern technology, the United States now establishes air superiority as its first priority and imposes "effects-based" operations to nullify the freedom of action of its enemies on the ground.

The emergence of aerial reconnaissance had a more subtle effect in changing what was seen by combatants. Having planes in the sky provided an unparalleled view of the dimensions of the battlefield and provided an enhanced ability to track battlefield developments over time. The rest of World War I would see aerial reconnaissance used to track enemy movements, the positions of vehicles, and trench lines, and also to create tiled photo mosaics of entire battlefields.[28] Perhaps more crucially it changed the way that we see the battlefield itself, shifting from a soldier's view of highly personalized combat like that described in the *Iliad* to a vertically oriented one where soldiers and tanks were seen even more as pieces moving on a chessboard. By its nature, aerial imagery offers a different vantage point on the features of the ground by mapping their relationship to each other spatially and portraying the earth anew in scale and proportion. Aerial imagery can also change how people respond to those elements of the terrain. As Paul Virilio noted, the battle between the GQG and those who saw the coming German advance from the air was not just a bureaucratic one, but one in which the advocates of aerial reconnaissance imposed a "point of view" on others.[29]

Today's aerial imagery from drones and manned aircraft does the same. It imposes a point of view, providing unprecedented levels of detail about the targets themselves, but it also changes how the features on the ground are perceived by different parts of the military. For drone pilots, the images on the ground are perceived with humanizing realism—for example, it is not unusual to watch targets perform daily tasks that become increasingly mundane to the viewer, or personal activities, for days on end—but with the remove and distance that comes from seeing from the air. For military commanders, drones hold out the hope of achieving a stereoscopic view of the battlefield that produces "information dominance" and allows them to fight while protecting civilians and their own personnel. It is these intertwined hopes—to learn enough about the battlefield that it overwhelms and paralyzes the opponent and to use those advantages in information to fight so precisely that the carnage of World War I becomes impossible—that led the United States and other countries to invest so much in drone technology since the Persian Gulf War. But this investment carries with it the risk of goal displacement: to learn more and more about the world until the US military is overwhelmed with images and data. The burden of this investment has gradually changed the organization and culture of the US military itself, while also pointing to even more changes in the way we fight in the years ahead.

Information Dominance

The Persian Gulf War of January 1991 is widely acknowledged to be the onset of what became known as the "Revolution in Military Affairs (RMA)."[30] While the concept is suffused with jargon, the underlying idea is simple: there has been a change in the nature of warfare due to a revolution in the speed of communication and information processing. As a result, the United States—the country most poised to take advantage of these changes—can now fight more precisely and humanely than ever before. Gone would be the days of mass-scale strategic bombing imagined during the height of the Cold War. In its place would be a lean US military that could fight from the air with a level of sophistication that made blunt force unnecessary. One early articulation of the RMA highlighted four essential elements of this transformation: (1) extremely precise, stand-off strikes, largely from the air; (2) dramatically improved command, control, and intelligence; (3) information warfare; and (4) non-lethality.[31] The promise of an RMA-inflected war was that it could be fought with such superior battlefield intelligence that accidents and "friendly fire" incidents could be dramatically reduced. In the words of one US general, the goal of the RMA was to "abolish Clausewitz."[32] The United States would be able to use data collected from satellites, aerial imagery, and signals intelligence to construct a richly textured picture of the battlefield and to estimate, even predict, the strategic behavior of the enemy.

Over time, the US military's impulse to learn as much as it could about the battlefield and to operate nimbly as a consequence has been enshrined as military doctrine. Although drones played only a small role in the Persian Gulf War, battlefield success against Saddam Hussein's forces was cited as a proof of concept for the RMA and expanded the ambitions of the Pentagon's elite planners to develop a superior knowledge of the battlefield to triumph over future enemies. This emphasis on the crucial role of superior battlefield information is reflected in different iterations of the doctrine and operational concepts of the Pentagon over the last two decades. In the mid-1990s, the Pentagon sought "dominant battlespace awareness" through its superiority in the information age; some of the Air Force planners went even further to anticipate "predictive battlespace awareness" that would allow the United States to guess what the enemy was planning to do before they did it.[33] By early 2000, the Pentagon had published a strategic plan for confronting conflicts in 2020 that called on the United States to use its intelligence-gathering capacity to maximize all of the elements of its power and to achieve "full-spectrum dominance."[34] Some optimistic analysts argued that the information technology revolution would now allow the Pentagon to act more like private companies that use their dominant knowledge of the marketplace to lock in permanent advantages for themselves.[35] What unites all of these formulations was the assumption that the United States would possess unique advantages over its enemies now that an unprecedented level of detail about enemy actions was in reach.[36] Alongside this assertion of opportunity for the United States was a conceptual shift that recast war as a competition for information. To win battles, the United States would know more than its opponents; to win the war of ideas and perception on a global level, it would deploy that knowledge to fight precisely and to spare civilians unnecessary harm.

Early efforts at capitalizing on the RMA with drones were not wholly successful. In the Persian Gulf War, Pioneer drones were used to locate targets for navy 16-inch guns, but otherwise they had relatively limited battlefield utility.[37] Throughout the 1990s, drones were deployed for a variety of reconnaissance missions in Iraq, Bosnia, and Kosovo, but they were seen as unreliable and could not always operate in poor weather. The technology lagged behind what was necessary for an efficient, precise use of airpower. In particular, the time lag between the collection of video of targets on the ground and the delivery of those images to military commanders was too long for the video to be used for effective targeting. When they were delivered, the images were also not sufficiently detailed to be useful. In some cases, these images were so narrowly cast that pilots described it as seeing through a "soda straw." Initially, the drive for information dominance from the Persian Gulf War onward was carried on the back of other technologies, such as manned aircraft, satellites, GPS, internet-based communication technologies, and advances in computing power. By the early 2000s, drones were able to play an enhanced role in the pursuit of Intelligence, Surveillance, and Reconnaissance (ISR) tasks because some of the technological problems, including most notably the time lag for video transmissions, had been overcome.

The growing importance of drones was seen at both the strategic and tactical level. At the strategic level, the United States began to experiment with substituting the U-2, a workhorse aircraft that had been conducting high-altitude surveillance for decades, with equivalent drone models, such as the Dark Star and the Global Hawk. The difference between the two was their operating environment. In development since the mid-1990s, the Dark Star was made of non-metal composites and was designed for stealth operations in a high-threat environment.[38] The Global Hawk, by contrast, was much closer to a conventional U-2 aircraft and was designed for surveillance operations in uncontested airspace (fig. 4.1). By the end of the 1990s, the Dark Star was abandoned due to problems with reliability and cost, effectively conceding that the United States did not have drone surveillance capabilities in places where manned aircraft or missiles might want to knock them from the sky. But the Global Hawk persisted and found a new life as the battlefield shifted to countries like Afghanistan and Iraq. Since the United States had air superiority in these environments and faced no risk of being shot down, the less-stealthy Global Hawk could thrive. Described as "the theater commander's around the clock, low-hanging (surveillance) satellite," the Global Hawk can fly as high as 65,000 feet, loiter for twenty-four hours or more, and monitor an area the size of the state of Illinois.[39] It takes both still images and video and can fly nearly autonomously between way points around the globe, all controlled from a ground station in the continental United States. The chief advantage of the Global Hawk is endurance: it can fly for more than twenty-four hours without needing to land because it has no

FIGURE 4.1 Global Hawk drone photographed in a hangar in South Asia in 2006.

pilot or crew who need to rest. It generally flies on autopilot between set waypoints, though it can be diverted to meet an urgent operational need. The Global Hawk is equipped with sophisticated radar and infrared cameras; it can track moving targets and collect signals from mobile phones and other electronic devices.

Although the Global Hawk surveillance aircraft were designed for high-altitude surveillance, to take pictures of fixed locations as the U-2 did, they were increasingly used in tactical operations. Between 2001 and 2003 in Afghanistan, the Global Hawk flew more than fifty missions and logged 1,000 combat hours.[40] As the journalist William Arkin noted, it gave "commanders something they never had before: a persistent, wide-angle view of the battlefield."[41] The soda-straw view was no longer. The infrared cameras and synthetic aperture radar were also used in Iraq to identify targets such as vehicles and groups of people through sandstorms.[42] One Global Hawk drone, nicknamed Grumpy, was used intensely through twenty-one straight days of fighting in Iraq in 2003 and was responsible for providing information that led to the destruction of 300 tanks.[43] The Global Hawk went through a number of technological developments and gradually absorbed more bandwidth. One estimate in 2003 suggested that Global Hawk drones would use 1.1 gigabytes per second of bandwidth per drone, ten times the total bandwidth used by the entire military in the Persian Gulf War.[44] The Global Hawk can take a picture every few seconds, but its chief limitation was bandwidth for conveying the image.[45] It was recording more information than could be recorded ever before and, as a result, an insatiable demand for overflights was generated across all levels of the military.

Medium-altitude drones, such as the Predator and Reaper, were soon deployed in the Iraq war for a variety of reconnaissance and targeting purposes. Although there has been more attention paid to the targeted killing functions of Predator and Reaper drones, these medium-range drones were just as often devoted to reconnaissance of the battlefield as they were to direct military action. The Pentagon developed two designations for Predators—RQ for reconnaissance and MQ for multi-mission—to reflect the different emphasis of various operations. Operated by different services, including the CIA, the US Air Force, and some National Guard units, Predator and later Reaper drones were flown over active combat zones (such as Afghanistan and Iraq) and over unofficial battlefields collecting still images and video. One Reaper pilot described a typical mission, working for a "customer" in the military or intelligence community:

> The typical mission for a conventional RPA is that you are in an [Air Tasking Order]. You are assigned to a ground entity. Somebody looks at your feed. You will get a brief and a chat with your customer. Once that happens, you are at the mercy of your customer. Some days [the mission] is to stare at a building, some days it's to scan a region or to escort a UN convoy. There is never a time when your commander in a squadron tells you what to do. It's always the customer.[46]

Many Predator and Reaper missions involved support of ground forces or surveillance of potential targets. The geographic scope of Predator and Reaper deployments expanded over time to active conflict zones like Iraq and Afghanistan, but also throughout the Middle East, North Africa, and East Asia. Over time, the sensor equipment on medium-sized drones grew increasingly sophisticated and was able to see in as many as thirty directions at one time, with plans to develop sensors that can see in as many sixty-five directions.[47]

The role that reconnaissance Predators and Reapers played in the hunt for IEDs that were killing and wounding US troops amid the insurgency in Iraq shows how they could have direct tactical value. In 2006, the Pentagon spent millions of dollars setting up Task Force Odin, a US Army specialized unit designed to detect and ultimately foil IED attacks in Iraq. To do this, they equipped Predators and other medium-range drones such as MQ-1C Gray Eagles with highly classified "black boxes" designed to track and detect evidence of IEDs being planted so as to harm US troops.[48] The goal was to achieve persistent surveillance of the area so those placing the bombs could be conclusively identified. In the US military, this was called getting "left of the boom" so they could find evidence of the emplacement of potential IED attacks and disrupt them before they affected troops on the ground. Visual imagery from these black boxes—for example, unusual movements by individuals and vehicles, or even signs of implantation of bombs like displaced brush by the side of the road—were combined with other sources of intelligence—such as spies on the ground and electronic intercepts—to prevent IED attacks. In some cases, Gray Eagle drones were affixed with a laser target that would locate or as pilots say "sparkle-" a target with infrared on the ground and allow other, manned aircraft to fire a Hellfire missile at it. Gradually, the Pentagon authorized a shift to "attacking the network" that produced the IEDs in Iraq. This meant that the drones deployed by the United States would convey imagery and intelligence to ground forces to allow them to identify and target those responsible for making and planting the bombs, but it also meant that armed drones such as Predators would launch missiles against these individuals if needed.[49] The result was a dramatic reduction in the number of IED attacks against US soldiers in Iraq, even if the tactic itself spread to other theaters.

Without much fanfare, the Gray Eagle became an essential tool for hunting and destroying insurgents in Afghanistan and Iraq. As journalist Sean Naylor revealed in 2019, the US Army's 160th Special Operations Aviation Regiment, officially called Echo Company and based out of Kentucky, developed a reputation for being the most lethal unit in the army due to their extensive use of Gray Eagle drones.

In less than twelve months between 2014 and 2015, they were responsible for killing 340 enemies in Afghanistan and the Iraq-Syria regions. They used Gray Eagle drones, each equipped with signals equipment and four Hellfire missiles to hunt ISIS operatives across the battlefield. Crucial to this effort was the fact that Gray Eagle drones could remain in the air for up to twenty-four hours, providing near-continuous coverage at a high altitude and allowing for close combat support. In the words of one former officer, "If you're looking to whack somebody who needs whacking, then send the Gray Eagle."[50]

One of the distinctive features of this effort was that it was conducted with a corresponding effort to destroy the insurgent networks, rather than just high value targets (HVTs) themselves. This was done with imagery coming from drones, but also with intercepts of cell phones. As the wars in Afghanistan and Iraq continued, the United States capitalized on the frequent use of cell phones by insurgents to track their locations through a variety of intercept methods, with a key role played by drones equipped with sensors. These drones captured imagery and signals intercepts from cell phones and other communications and combined all of this to track the movements of targets over time. An internal NSA document from 2005 described these sensors as the equivalent for the war on terror of the "Little Boy" nuclear bomb that destroyed Hiroshima in World War II.[51] Such a revolution required substantial changes in bureaucracy and the information infrastructure of the US military and intelligence establishment. To deal with the flow of data coming from drones, the United States built a database holding information on SIM cards for known insurgents, collecting metadata on who they called and deploying that data to build a picture of the network responsible for an attack. This data was known to be imperfect because many people in these theaters of war would trade off SIM cards to each other, making it difficult to know who exactly carried a particular number. Yet the effort provided a better picture of the networks themselves and could be combined with other data sources from signals intelligence, spies on the ground, and other sources to identify targets. The result was that the United States began to build a "counterterrorism machine" premised on the ability to sift through and merge this data for target identification. One way that this was done was through a novel NSA program called the Real Time

Regional Gateway, which scooped up data from drones and other surveillance resources and cross-referenced them to paint a picture of the
enemy and their intentions. This allowed the Pentagon to "collect the
whole haystack" rather than look for the needle.[52] Deployed in Iraq in
2007 and in Afghanistan in 2010, the Real Time Regional Gateway also
showed the degree to which the concept of information dominance became central to US thinking about war. The functionality of drones was
merely being folded into a wider US military effort to know as much
about the battlefield as possible.

Aside from these ISR functions, ground support—that is, using
an aircraft to assist with the progress of a mission on land—became
an important part of the medium-sized drone's repertoire and vastly
increased the demand for them. Only a few years after they had been
introduced in Afghanistan and Iraq, ground commanders were calling
for Predators and Reapers for battlefield reconnaissance so that they
could avoid sending their soldiers into dangerous situations. In 2008,
the US Air Force reported that it had passed 400,000 hours of Predator
drone use due in good measure to demands for ground support.[53] By
2019, the Predator and Reaper together had reached more than four
million flight hours.[54] One official remarked that "the warfighter can't
get enough Predator time."[55] To meet demand, the US Air Force began
to organize drone overflights as distinct combat air patrols (CAP).
According to a Pentagon estimate, the Predator alone surged its number
of CAPs by more than 520% between 2001 and 2009.[56] The problem
that the Pentagon faced was its supply of drones, which was limited by
its budget and by the natural time delay it takes to get sophisticated
military technology delivered to the appropriate battlefield. By the end
of the decade, Defense Secretary Robert Gates found himself under
pressure from Congress to increase the number of drones on the battlefield. This dynamic—with military commanders, Pentagon leadership,
and Congress responding to each other's demands for more drones—in
large measure explains the vast expansion of the US drone fleet over the
last decade.

This desire for more battlefield intelligence was also seen in the
extensive use of small drones for tactical reconnaissance in Iraq. Of
particular importance in this respect is the Raven drone. Made by
AeroVironment in California, the Raven drone is by numbers alone the

most important drone in the Pentagon's fleet, comprising over 7,000 of the 10,000 drones estimated to be owned by the US military in 2014.[57] The Raven, which derived from an earlier drone called the Pointer, is a lightweight small drone designed for "over the hill" reconnaissance.[58] Because it is so simple to assemble, it can be carried in a backpack by individual soldiers and flown using a simple screen and joystick interface that resembles a Nintendo Wii U controller. When later equipped with a more powerful camera and a lithium ion battery, Raven drones were among the most popular with US troops because they provided a quick and efficient way to detect whether insurgents were hiding in out-of-sight locations or planning an ambush. Crucially, they were adaptable to new technologies, and cheap enough that if they were destroyed they could easily be replaced. By the end of the decade, up to sixteen Raven drones could fly in tandem to relay information back to ground stations, producing an even more detailed account of the battlefield for ground commanders than was ever before available.[59]

The success of these drones led to a vast increase in demand for tactical drones in the US Army. The idea was that every soldier should ideally have a small, highly functional nano-drone which he or she could deploy at will. While the capabilities and cost of nano-drones lagged for a number of years, eventually the army got its wish. In 2019, the US Army awarded a contract for $39.6 million to FLIR Systems, an Oregon-based company, to make Black Hornet Personal Reconnaissance Drones available to soldiers on the ground in combat zones.[60] These drones are small, measuring only 6.6 inches across, and light (33 grams). Resembling a small helicopter, they can fly up to 1.24 miles away from their operator and remain aloft for 25 minutes (fig. 4.2).[61] Each Black Hornet has two daytime cameras and thermal imaging capability, and can fly at night and in heavy wind.[62] The drone is quiet, so those on the ground cannot hear it approach, and in theory it can operate indoors and even in caves. One of its most crucial advantages is cost: the Black Hornet only runs between $15,000 and $20,000 per unit, significantly less than a Raven or a comparable alternative.[63] As they become more popular and widespread, the Black Hornet and nano-drones like it will provide ground forces enormous tactical advantages in finding and striking enemies, but they will also begin to change the soldier's "point of view" on the battlefield.

FIGURE 4.2 Royal Marine carrying a Black Hornet 2 Remotely Piloted Aircraft System (RPAS), 2017.

This demand for Ravens and Black Hornet drones shows the extent to which ground commanders are demanding "eyes in the sky" for reconnaissance before approaching risky targets. Some pilots have defended this as natural given the cultural changes in the way the United States fights and the value that it places on the life of its personnel. In the words of one former Reaper pilot:

> The demand for information on the battlefield is crucial—and why not? If we knew where the bunkers were on Normandy, and we had the option not to drop so many bombs only to get 15% of the tunnels, why wouldn't we do that? We could have done it without losing that many lives. Now, we don't have the need to drop bombs like that anymore. If an RPA pilot can switch on an infrared camera, or do a pattern of life, to save the life of the guy kicking down the door of the bad guys, I'd want it. I don't blame ground commanders for asking for this.[64]

Yet this growing demand began to put a strain on pilots of medium-altitude drones who believed that it was now impossible to meet all of the battlefield reconnaissance demands they faced. The US military is increasingly attached to an ISR capability that is premised on extraordinary knowledge of the battlespace; this has shifted from a desired advantage to a requirement. In some respects, as key elements of the US military became accustomed to knowing more before they acted, the tempo of operations may have slowed. At a conference in September 2016, the chairman of the Joint Chiefs of Staff General Joseph Dunford noted that the military had increased the number of ISR aircraft by 600% since 2007 but was still only meeting 30% of the requests from combatant commands for overflights.[65] Such a path, he argued, was fundamentally unsustainable. As one senior Air Force commander and drone pilot put it, "what we really need to attack is how we make decisions, not how much information we have to make those decisions . . . So let's get off that glide slope and get at what is our risk calculus, how do we determine that, and make those decisions off the information that we have, not what we want to have."[66]

Despite these warnings, the demand for information continued to grow. Many drone pilots maintained that a slow and deliberative targeting process was best suited to avoiding mistakes and civilian casualties, as well as the institutional and legal penalties that flowed from them. This demand also began to affect the tempo of operations in contradictory ways. At a minimum, the speed of the demand for and supply of information on the battlefield had increased, although not equally. In the words of Lt. General Deptula:

> This is a result of modern technology, it has accelerated, to use a word on top of a word, an accelerated set of capabilities which has also resulted in a greater demand for information. So there's some good and bad associated with that. The good is the arrival of information in enough time for a commander or operator to use it. That's very beneficial. The bad is the overwhelming mass of data that can result in slowing down operations while people wade through the vast amount of data to determine its utility.[67]

The danger is that the tempo of operations could slow because ground commanders and pilots could always demand another overflight to know more about the battlefield and the risks they faced. Yet one senior drone commander noted that a slower tempo and increased situational awareness allow the United States to make war less of an "ugly thing" and should be defended morally on those grounds.[68] A slow war, he argued, is a precise war and one that is likely to be more humane.

Through the Global Hawk, Predator, and Raven, the United States was deploying drones to create an ISR capacity that was far beyond what anyone could have imagined a decade earlier. As late as the Persian Gulf War, pilots had to return to base to develop their film from reconnaissance missions before sending it for analysis and presenting it to decision-makers. The result was a time lag between when information was received and a decision on what to do was made. This time lag was important because it could allow enemies to move or to escape and lengthen what became known as the "kill chain," or the steps and time needed to move from identifying a target to moving against them. The emergence of drones allowed that kill chain to be shortened dramatically, but it also placed additional pressure on intelligence analysts to urgently confirm the identities and activities of potential targets. While the widespread availability of drones made war slower and more deliberative, it also made it more information intensive as analysts had to assemble an almost criminal case against those they wish to kill. This reinforced a model of combat within parts of the US military that was closer to hunting and even executing criminals than to warfighting in the conventional sense.[69]

To fight this way, the Pentagon sought to add more bandwidth to its existing networks and to retrofit drones for mobile communication. One reason for the growing demand for bandwidth was that communication devices and small drones like Ravens given to soldiers were becoming so complex that they were absorbing more bandwidth for data. For example, during the 1990s, the US military used 90 megahertz of bandwidth for approximately 12,000 troops. By contrast, it needed 305 megahertz for 3,500 troops in 2014.[70] In 2013, the United States leased bandwidth from a Chinese satellite to sustain drone operations and other communication tasks in Afghanistan and Iraq, even though that exposed them to considerable risks of espionage.[71] As early as 2005, the Pentagon noted that "airborne data link rates and processor speeds are in a race to enable future UA capabilities."[72] The problems were clear: poor connectivity, costly satellites, and stove-piped infrastructures that prevented the flow of information, which in turn led to poor information sharing.[73] Today, the Pentagon is working with private companies to construct a networked communications infrastructure that combines high capacity and robust data links (to prevent hacking) with powerful processing systems to collect, transmit, and even analyze visual images and data. An even longer-term goal is to build smarter drones that would transmit "the *results* of their data to the ground for decision-making."[74] In other words, the next generation of smart drones, perhaps enabled by artificial intelligence, will do the first cut of intelligence analysis before even sending its images onward.

By the end of the decade, the wartime experiences in Afghanistan and Iraq had created a vast infrastructure of data collection and imagery from drones and other sources. In 2009 alone, US drones collected the equivalent of twenty-four years' worth of video footage, more than could ever be processed by individuals from the intelligence community.[75] This was only a small portion of the total data gathered by the NSA, which collected every six hours as much data as is held by the Library of Congress.[76] But the amounts of data continued to increase. By 2012, the US Air Force was recording fifty days' worth of video every twenty-four hours and flying 1,500 hours of airborne ISR missions every day.[77] Another estimate in 2015 suggested that the air force's Distributed Common Ground System (DCGS) collected 100 terabytes of data every eighty hours.[78] The Air Force itself estimated that it had spent $67 billion on ISR operations since September 11 and that the number of ISR platforms increased 238% between 2008 and 2012. ISR platforms now constituted one-third of all US military aircraft.[79] If this was not managed properly, as Lt. General Deptula noted, the US military could soon be "swimming in sensors and drowning in data."[80]

One solution was to expand the ranks of the intelligence analysts employed to analyze the video feed from drones as well as other data. In 2010, the air force reportedly hired another 2,500 intelligence analysts to cope with the volume of data that it was generating.[81] By 2012, the air force had 20,000 airmen operating spy planes and analyzing the resulting intelligence.[82] To a much greater extent than their critics realized, drones are manpower-intensive, requiring fifty-nine people in the field doing launch and recovery, forty-five on mission control, and eighty-two analyzing the data gathered.[83] Many of these analysts are from the air force and the intelligence agencies (CIA, NSA, Defense Intelligence Agency, and others) but a growing number of these analysts are private contractors paid to wade through the thousands of hours of footage collected. This is one of the reasons why the operations of drones has been so often likened to a machine, because it is a vast, labor-intensive, complex system that collects information and finds what is useful for targeting.[84] One estimate suggested that the air force alone could need 117,000 people devoted to exploiting motion imagery by 2015.[85]

All of these analysts face a significant challenge. The chief problem with collecting such a haul of data is separating what is useful from the substantial level of noise in the data itself. According to Deptula:

> This produces another problem which is how you separate the wheat from the chaff. If you look at airborne surveillance capabilities nowadays, it can collect with an MQ-1 or MQ-9 with an electro-optical imagery ball that can look through a "soda straw," an area of the earth around 800–1,000 feet in diameter. With wide-area surveillance pods you can now look at an area about 8km in diameter. You have multiple orders of magnitude greater data involved in that look, but you also find that 99 percent of what is collected is not actionable data— it is "chaff." This has some impact on the issue of how you separate the elements you are interested in from the rest of the stuff. That is a growing problem.[86]

The shift toward information dominance had gone beyond collecting useful information about the battlefield and become an effort to collect everything that might be relevant, even if that produced real difficulties in curating and analyzing that information. The voracious appetite for information seen across the military and intelligence establishment resulted in an almost unchecked collection of data outside US territory, a fact later emphasized by Edward Snowden's NSA revelations. Such an approach does not always yield quality intelligence. As one former drone pilot put it: "we've been taking the 'Snowden approach'—collect it all and sort it out later. The problem is that different people—civilians, pilots, intel analysts—may all see different things with that data. We have tested this case with some imagery and we found that different groups saw entirely different things. Even the intelligence analysts from the military and civilian agency saw different things."[87] Noticed less in that controversy over NSA surveillance was the role played by drones in that effort and the unprecedented bureaucratic and logistic problems that emerged when converting that data haul into intelligence. For the Air Force, this effort was generating an incomplete cultural and organizational shift from its primary emphasis on destroying targets to becoming a flying intelligence service devoted to finding and fixing those targets.

The natural culmination of the drive for information dominance was an effort by the US government to track *all* signs of life in a small city or town. As chapter 1 noted, the United States has been developing the Gorgon Stare, a video capture technology that offers persistent, wide-area surveillance of small towns, since 2009.[88] Attached to the Reaper drone, the sensors of Gorgon Stare can provide a wide-angle view of the battlefield and record and store video for up to thirty days. It also has the highest resolution camera in the world, at 1.8 billion pixels.[89] With the Gorgon Stare, intelligence analysts can rewind or forward the video to learn where bombs appeared from or where trucks went; it offers a panoptic view of the battlefield over time unlike any produced by other sensors.[90] The expansion of the Gorgon Stare and the surveillance capacities of other drones has placed even more pressure on the Pentagon: in 2012, it was estimated that the program could collect the equivalent of 53,000 full-length movies per day of video. To deal with this unprecedented flow of data, the Pentagon installed new computer systems and turned to ESPN for their techniques for processing, analyzing, and preserving video images from sports.[91] The Pentagon also turned to Google, which agreed to help deploy artificial intelligence to interpret and decipher the video images in the hopes of improving intelligence analysis and targeting decisions. This project, called Project Maven, attracted considerable controversy and was canceled in May 2018 after a number of Google employees publicly objected to the idea of helping the military with drone strikes.[92] With both projects, the idea was to reduce the burden on the intelligence infrastructure by culling the millions of hours of footage but also to allow for instant recall to allow analysts to investigate events and people over time. It was an admission that the unrelenting quest for "information dominance" through drones and other means had stretched the capabilities of the military and subtly transformed its roles and responsibilities. It was also an indication of how this quest for information dominance had begun to change how the United States fights its wars.

Precision and Humanity

In contemporary US doctrine and strategic thought, the quest for information dominance has been accompanied by an emphasis on acquiring and using precision weapons. The dream of precision in warfare is an old one. As early as World War II, US strategists were hailing new aircraft and bombs as capable of hitting targets with unprecedented levels of precision.[93] But what counted as "precise" at the time was very different from now. Many World War II era precision bombs fell within hundreds, if not thousands, of feet from their intended target.[94] Even by this standard, precision weapons made up only a tiny portion of the overall aerial attacks during World War II. In the Vietnam War, interest in precision weapons grew because US planners were eager to avoid killing civilians and adding fuel to the anti-war protests in the United States. While US planners often touted their precision bombing campaigns to a skeptical public, precision bombs accounted for less than 1% of the total bombs dropped in the Vietnam War.[95] The technology and the policy for protecting civilians in aerial bombardment lagged behind the glowing rhetoric around precision weapons during the Vietnam War and many of the Cold War proxy wars that followed.

Despite this fact, the United States and other leading military powers never gave up on precision weapons because they offered the promise of a "surgical" form of warfare that could destroy enemies while sparing everyone else. The Pentagon invested heavily in precision weapons for decades following the end of the Vietnam War, all in the hopes of achieving a dramatic change in what was possible in targeting enemy forces and avoiding "collateral damage" in the civilian population. The search for these weapons was hardly altruistic; precision weapons could also shield the US military and public from some of the costs of war. Especially if employed with information dominance, precision weapons could spare the lives of US personnel by avoiding mistakes and friendly fire incidents while minimizing the backlash that naturally comes when bombs go astray. Yet perhaps the greatest attraction of precision weapons is that they could conform to the principle of humanity as defined under international law, which requires governments to fight as carefully as possible and to spare civilian lives whenever possible.[96] Seen in this light, the seductiveness of precision weapons becomes apparent: it reconciles what you have to do in war (i.e., to fight) with what you are morally obliged to do in most normal circumstances (i.e., to save lives). Advocates of precision warfare pointed to the overlapping strategic, legal, and ethical rationales and emphasized the twin benefits of a humane approach to fighting: that it was not only strategically wise but also a moral good worth pursuing on its own terms. These two imperatives—to make war precise and in doing so to make it humane— have been intertwined in almost all US debates over airpower since the mid-1970s.[97]

Following the Vietnam War, US military officials incorporated this principle into military doctrine and practice, subtly transforming targeting standards and ruling out the aerial strikes considered acceptable in prior decades.[98] The Persian Gulf War was the first time that the technology caught up with the doctrine of precision warfare. By capitalizing on 1980s advances in satellite technology, the United States was able to employ its superior knowledge of the battlefield to its advantage to generate overwhelming battlefield success against Iraqi forces. It did this though a targeted campaign to hit leadership and key infrastructure targets while avoiding the high levels of civilian casualties that accompanied Vietnam-style aerial bombardment. The Persian Gulf War also showed the extent to which efforts inside the Pentagon had internalized the concern for humanity in warfare by developing mechanisms of legal oversight and accountability that governed target selection. US lawyers carefully reviewed and authorized prospective strikes to ensure that they minimized harm to the civilian population while achieving maximum "effect." The United States touted the degree of precision in its war strategy, dazzling journalists with videos of its "smart bombs" and engaging in a sustained public relations campaign about how precise and humane their new way of fighting was. A number of prominent commentators saw the Persian Gulf War as a harbinger of a new type of warfare that was both precise and humane, in contrast to the bloody battles of attrition that ended the largest conflicts of the twentieth century.

The reality in the Persian Gulf War was more complicated than these laudatory accounts suggested. Without a doubt, it was a dress rehearsal for the new age of precision warfare, but the flaws in the production were evident. Many "smart bombs" turned out to be more propaganda than fact—some missed their targets or failed after launch. While the accuracy of precision bombs had vastly improved compared to previous decades, they were a small percentage of the overall bombs dropped— only 9% of bombs dropped in the Persian Gulf War were classified as precision munitions.[99] Some experts concluded that the greatest impact of precision weapons was psychological, in demoralizing the Iraqi army and convincing them to avoid direct military engagements.[100] Even with that, the war was hardly bloodless. Approximately 20,000– 26,000 military personnel and 3,500 civilians died during the Persian Gulf War.[101] In reality, what made the war look "clean" was not that few Iraqis died—General Norman Schwarzkopf remarked in a televised interview that actual number of Iraqi casualty figures did not matter— but that few US military personnel did.[102] The true promise of an RMA-inflected war was that it was clean for those who wage it. Only as a secondary issue did it matter whether the war spared the enemy or civilians caught in its midst. Perhaps the greatest impact of the Persian Gulf War was in language. While the war itself did not usher in a new model of conflict, it did show how concepts of precision and humanity were beginning to change how the United States talked about warfare. The United States now talked about the importance of saving civilian lives but also employed antiseptic euphemisms like "collateral damage" designed to obscure when this did not happen.

Subsequent conflicts—the wars in Bosnia (1990–1995), Kosovo (1999), Afghanistan (2001), and Iraq (2003)—showed a maturation of operational emphasis on achieving precision and preserving humanity in combat. The Pentagon got significantly better at fighting with precision weapons and portrayed itself as fighting wars that were restrained in purpose and scope. These concepts also influenced how these wars were presented to the public. The interventions in Bosnia and Kosovo were cast as "humanitarian" in nature because they were designed to save civilians from repression and civil war. The United States defended the principle of humanitarian war—waged from the air, largely blood-less for US personnel—against accusations that such a concept was a grotesque contradiction in terms. The emphasis on precision and humanity was gradually incorporated into the official doctrine, though not always in those words. One widely cited US Air Force formulation was to call for targeting the nodes of networks—so-called "net-centric warfare"—to force them to collapse while avoiding population centers entirely. Within the US Air Force, another influential conceptualization of this way of fighting, by Lt. General Deptula, called for "effects-based operations," which were highly targeted operations against nodes of the enemy's defenses to disable them to achieve "effects" without sequentially destroying all elements of their military power.[103] Deptula argued that the changes underway in developing battlefield intelligence and translating it into targeted attacks meant a change in the character of warfare "analogous to the difference in the world views between Ptolemy and Copernicus."[104] While this conclusion is debatable, many senior military officials were persuaded that the deployment of precision weapons represented some kind of step-change in the way that fighting was done.[105] By the Iraq war in 2003, 66% of US munitions were precision-guided munitions.[106] This embrace of precision warfare appeared to pay dividends. The rapid collapse of the Taliban in Afghanistan in 2001 and Saddam Hussein's forces in Iraq in 2003 lent support for the hypothesis that the United States had developed a new "way of war" which left grinding attrition behind and relied on the precise use of force against combatants through air power.[107] After seeing the success of drones against Iraqi armed vehicles in 2003, the US Air Force declared that it was now the age of "mass precision."[108]

The insurgencies that emerged after the overthrow of governments in Kabul and Baghdad dampened some of the enthusiasm for the revolutionary potential of precision aerial warfare but they did not eliminate it. Instead, air power advocates recast their arguments and maintained that counterinsurgency campaigns could use precision air power.[109] In these conflicts, air power could be used to assist ground forces, to enable strikes on individuals, or to intimidate, even unnerve, insurgents.[110] If precision air power could defend civilians from attacks against insurgents, it could win hearts and minds in counterinsurgency campaigns. By combining information dominance with precision, air power offered the possibility of pursuing "bad guys" while saving civilian lives even in messy counterinsurgency campaigns where the enemy was embedded in the civilian population. This newfound application of precision airpower to counterinsurgency began to enhance its moral hue. One advocate of precision air power described the new ethic of fighting as substantially different from traditional warfighting:

> Instead, the US public standard for military action now seems to resemble the ethic that prevailed on old TV Westerns: The good guy— the one in the white hat—never killed the bad guy. He shot the gun out of his hand and arrested him. Modern air power may not solve every military problem, but thanks to the innovations of the last decade, it is the weapon in the US arsenal that comes closest to fulfilling that goal.[111]

Instead of attempting to destroy the enemy or force them to surrender en masse, the United States should aim to locate the "bad guy" among the general population and kill that person without affecting the rest of the population. This is a radical departure from the traditional idea that warfare involves punishing the enemy as a collective until they surrender. Rather than warfighting, this approach is closer to the ethic of the hunter, according to French philosopher Grégoire Chamayou.[112] The advantage of a manhunting approach, according to one JSOC analyst, was that the United States could target key individuals of a terrorist network "without resorting to the expense and turbulence associated with deployment of major military formations."[113] While the United States attacks the target from the sky, it is invulnerable to a direct counterattack, a position which has left some calling this approach either assassination or "safe-killing."[114] This approach was a natural outgrowth of the US superiority in precision-guided weapons and aerial platforms. Only the United States was capable of employing its information dominance to find that "bad guy" and employ precision munitions in order to eliminate him. Only the United States, it follows, was capable of removing itself and much of the civilian population from harm's way and rendering the war humane.

As the technology for information dominance developed, the doctrinal and operational commitment to precision and humanity in warfare led the United States to apply this hunting model of fighting an enemy in an increasing number of theaters worldwide. Although the United States continued to perform traditional ground support and combat operations in theaters like Afghanistan and Iraq, it hunted "bad guys" during these wars and expanded the policy of targeted killing outside of declared war zones. The emphasis on precision and humanity gradually changed the tenor of the combat missions, as commanders demanded more drone overflights and imagery to be certain that their attacks were directed at the appropriate targets, rather than civilian bystanders. They also regularly requested overflights to ensure that they were not sending US soldiers into dangerous encounters with the enemy. Ground commanders sought to use the information dominance that the United States had acquired to reduce some of the uncertainties in confronting enemy forces and ensure fewer US personnel were killed in accidents and from so-called friendly fire.

Drones proved ideal for this approach to fighting. Without much fanfare, this hunting model, based on information dominance and dedicated to killing individuals and small groups rather than massed enemy formations, became a central operating ethos for the US Air Force drone fleet. Medium-sized drones, such as Predator and Reaper, were equipped with sophisticated gimble cameras that were useful for long-endurance missions watching and accurately killing targets. Over the last decade, thousands of US pilots have been trained to fly unmanned vehicles over declared and undeclared battlefields and to find militants from an array of organizations (including al Qaeda, ISIS, the Taliban, and others) located in over a dozen countries worldwide. With many of these militants hiding in rural regions or blended with the civilian population in cities, the hunt for these targets with drones was painstaking and careful. Yet many pilots with experience in both manned and unmanned aircraft came away with the conclusion that drones had notable advantages in avoiding harm to people and property in theaters of war. As one experienced drone pilot put it:

In general the precision itself is no different than it is for a manned weapons system. My ability to hit a target is no different. Our ability to reduce casualties and property damage is. Why? For one, it is time and endurance. We can strike at our own choosing. But we also have

intel analysts to help us decide the attack angle or vector to avoid civilian casualties. We have models to show how the blast damage from
a Hellfire missile on surrounding buildings happens. I can choose
how I come in towards the target. Or we can wait until the target
moves, to not take out other people.[115]

Some pilots have argued that drones allow for military operations
without the "urgency of time," which allows them to be more deliberative in nature.[116] As one pilot remarked, "[With RPAs] you can realize
that things are not time sensitive as they appear. A lot of the time when
civilian casualties happen it's because you are rushed and don't have the
time to make the decision. Sometimes it may not be the best decision
you can ever take, but that's not different with manned aircraft."[117] The
ability of intelligence analysts and commanders to watch the video feed
of a drone remotely means that their missions naturally come equipped
with more legal oversight. In the words of one former Reaper pilot:

> If we are comparing manned and unmanned aircraft, conservatively
> I'd say it's the same level of operational precision in manned and un-
> manned aircraft. More aggressively, I think that the RPAs are better
> because we have 3–4 people looking at an operation to make sure it's
> right. Why people think that an RPA could not distinguish a target
> when a manned aircraft that is flying thousands of feet above can
> mystifies me. At my disposal, I have an intel coordinator with me
> that is capable of talking to me. I can telephone other crews. I have
> the ability to talk to anyone I want to. I can make a decision without
> the pressure of having to stay alive in the line of fire.[118]

Others highlighted the accuracy of drones relative to other options. Lt.
General David Deptula noted that:

> It is not widely known that the accuracy of a 150mm howitzer is
> around 1,000 feet. The accuracy of the most accurate mortar is
> around 250 feet. What's the accuracy of a projectile launched by an
> 18 year old scared shitless coming under fire? The difference is with
> remotely piloted aircraft we record every single second of not just
> engagement time, but also flight time. Whereas the flight path and

results of artillery rounds or mortar rounds or rifle rounds are not recorded. What if you put a camera on the shoulder of every soldier and Marine on the ground carrying a rifle in combat and said we are going to record everything you do. Or before you can fire we are going to have to review and approve your engagement parameters before you can fire. That's what's going on with RPA operations. There's a microscope on RPA engagement that is enhancing attention paid to these operations.[119]

Due to these tight rules of engagement and the level of oversight, one US Air Force drone commander concluded that, "we are fighting the most precise air campaign in the history of warfare."[120]

The language of precision and humanity became an essential element of the Obama administration's political defense of the use of drones worldwide. Many senior Obama administration officials touted how precise drones could be relative to other options in conventional wars. In one of the first public comments on drones made by the Obama administration, State Department legal advisor Harold Koh argued that drones "have helped to make our targeting even more precise."[121] In 2012, CIA director Leon Panetta called drones "one of the most precise weapons we have in our arsenal."[122] President Obama emphasized how drone operations produced a rate of civilian casualties lower than any other conventional option and insisted that he was choosing "the course of action least likely to result in the loss of innocent life."[123]

The evidence for these claims is less clear than these accounts suggest. Some of the Obama administration's arguments for the precision and humanity of drones relied on false comparisons—for example, comparing drones to World War II style bombardment or indiscriminate artillery shelling—rather than to realistic options.[124] Concerns have been expressed over the reliability of the government's data on deaths from drones outside war zones, with suggestions that their assertions that casualties were in the "single digits" cannot be taken at face value.[125] One independent study by Micah Zenko and Amelia Mae Wolf found that drones were thirty-five times more likely to cause civilian casualties than comparable operations by manned aircraft.[126] There is also evidence that President Donald Trump's "gloves off" approach to fighting terrorism has produced more civilian casualties despite the use of precision weapons and drones.[127] At a minimum, drones are not bloodless; accidents and the "fog of war" are inevitable with all forms of warfighting. It is clear that the intertwined concerns for precision and humanity have produced doctrinal changes in the US Air Force and inadvertently spurred the increase in the use of drones on the battlefield. They have led the United States to slow its tempo of operations for some ground operations and to hunt for its enemies on an ever increasing number of battlefields. But whether they have actually rendered US-led air campaigns as clean as these accounts suggest is harder to know given the secrecy surrounding their use and the lack of accurate data. At this point, a precise and humane war remains a promise of the drone age, but it may not be the reality.

Consequences

The development of information dominance and the emphasis on precision and humanity in combat have produced a series of organizational and cultural pressures on the US Air Force itself. The shift toward a hunting model inside and outside conflicts is placing stress on the air force and changing it in subtle ways. At a minimum, the US Air Force has been under considerable pressure to meet the increased demand for drone overflights. At the beginning of the decade, the United States had only a few drones and conducted almost no daily or weekly missions with them. By 2015, the US Air Force was flying approximately sixty CAPs daily, while the Army was flying sixteen and Special Operations Command an additional ten.[128] For the US Air Force's CAPs, at least 300 drones were required, sustained by thousands of personnel.[129] Yet even this was not enough, as the United States estimated that it would need to deploy ninety CAPs daily by 2019 to meet the demands by military commanders across the world.[130] US officials admitted that the drone pilots were at a "breaking point" due to the increased demands for drone overflights.[131] The geographic range of the CAPs was wide, with overseas bases located in over a dozen far-flung locations, including Seychelles, Chad, Nigeria, Italy, Japan, Turkey, Guam, Ethiopia, and the United Arab Emirates.[132] The result was that US military personnel were responsible for staffing CAPs with drones taking off all hours of the day, necessitating shift work on a twenty-four-hour-, seven-day-a-week schedule.

One organizational consequence of this growing demand for drone surveillance is that the US Air Force has struggled to train enough pilots. Under pressure to deliver drone pilots quickly, the Air Force created a streamlined training process to get trainee pilots through a shortened one-year training course, as opposed to a thirty-month course for manned aircraft. The goal is to produce cadres of specialized drone pilots who can help the Air Force meet the insatiable demands for ISR, combat support, and manhunting missions. The Air Force has also struggled to retain drone pilots, because of the demands placed on them. In response, it is considering developing a second "wing" of drone pilots to provide opportunities for promotion within a favorable career trajectory and creating a series of bonuses for retention and performance, as well as performance medals. Yet the Air Force has been mired in an internal debate about whether to reward, and hence valorize, drone pilots for their performance as for pilots of manned aircraft. The culture of the Air Force traditionally portrays manned aircraft as more risky and authentic than unmanned aircraft. For this reason, the Pentagon has struggled with the decision to offer medals to drone pilots—and faced an internal revolt when it did so in 2013; they were dubbed "Nintendo medals" by some pilots and regarded as inferior to those earned by traditional pilots.[133]

Drone operations are also scrambling the command structures within which air force pilots typically operate, because they are remotely controlled. In theory, drones are subject to the same command structure as manned aircraft, but in practice this breaks down because the video feed is conveyed to so many different screens in Washington, DC, and elsewhere. Drone feeds can be monitored by as many as 200 people, including intelligence analysts, military officers, civilian officials from the Pentagon and intelligence agencies, and government lawyers. For many pilots, the extensive remote control over drone operations can be a surreal experience that is remarkably different from flying in manned aircraft. In a manned aircraft, the pilot's actions are unsupervised and only audio communication links the pilot back to those commanding the mission. In a Predator or Reaper drone, an unknown number of people may scrutinize the pilot's actions in real time and criticize them or penalize them later for actions seen as excessively dangerous. Far from the media image of the pilot as unrestrained and free to fire in "push button warfare," drone pilots often report feeling micromanaged by the number of people watching their remote feeds. The result is that they will become remarkably careful in following their "checklist" of approved steps for the mission in order to avoid professional penalties. As one drone trainer put it:

> For pilots, this means that you know your actions will be monitored and you will be called to account for your actions. The result is that you follow your checklist. Even a small deviation from the checklist which would allow you to achieve the mission is a risk. You know that you will be blamed for violating the checklist, even if the fact you did is unrelated to the mistake itself. When you fly RPAs, you know that a lot of people will be watching. Some of them are civilians. The problem is that high level military officials can more directly interfere in the targeting process. There are stories of 4-star admirals calling a pilot of an RPA saying "Attack this guy." The problem is that this person may be in your line of command, but you are supposed to listen to the Air Operations Command. This puts pressure on the pilots. The proper response is to say "No, sir, I am sorry but I have to follow my checklist."[134]

This picture of closely watched pilots stands in contrast to the media depiction of drone pilots with broad discretion to pull the trigger. While many pilots emphasize that the ultimate decision to launch a strike remains their own, the fact that commanders, lawyers, and intelligence analysts watches them may subtly affect their behavior and make them less prone to taking risks. It also reveals the double irony of remote-controlled warfare. While drones are controlled remotely by their pilots, the pilots themselves are monitored and sometimes controlled remotely by their commanders and other senior government officials with access to their video feeds on screens and laptops worldwide.

Among the dangers of senior military commanders and civilian officials becoming immersed in the video feeds of drones are micromanagement and a loss of strategic perspective. As one former Predator pilot put it, "if someone wants your feed, they can get it. From a pilot, you are helping with a piece of the puzzle that you can. But when you get the tactical side influencing and being shown to the strategic thinker, it produces a danger of micromanagement."[135] With drones, senior military and civilian officials are now capable of ordering adaptations in the flight patterns or attacks on targets in real time, but they may not have the full complement of information available to the pilot. They may also make such orders in contravention of the mission's objectives or unaware of the larger strategic issues at play. It is at this point that the disjuncture between less-constrained political decision-making and the constrained rules of engagement of the pilots matters. One former Predator pilot remarked that, "because I can pass the video out to everyone, losing sight of the larger strategic picture is a danger."[136] Drones allow the visual immediacy of the battlefield to be beamed into a number of different screens, but this in turn exerts a unique pressure on the Air Force not to lose sight of the purpose of the mission and to ensure that messages to the pilot are proper and conveyed through appropriate command structures. The diffusion of the battlespace—with the pilot viewing the targets on the ground and hundreds of eyes on the pilot around the world—is posing an organizational and cultural challenge for militaries more comfortable with clean and hierarchical command structures.

Conclusion

The United States is now confronting a situation in which drone operations are arguably on their way to becoming the most important component of its Air Force's mission. This dramatic increase in the use of drones is due to two intertwined hopes: that the United States could develop "information dominance" with a panoramic view of the battlefield and that doing so would allow it to fight precisely and humanely in a way never before achieved. The irony is that these two impulses gradually led the United States to adopt a manhunting approach to warfare that now permeates its operations both inside and outside declared battlefields. While drone technology only incompletely delivers on the promise of a precise, humane way of fighting, the aspirational language around these priorities combined with popular view that drone technology is clean and antiseptic serves to insulate this hunting approach from sustained public criticism. The lack of transparency concerning drone operations means that the average person has little ability to assess if drone technology is as precise and humane as its advocates suggest.

Drones are also exerting an effect on the militaries that are adopting them. In part due to the size and sophistication of its drone fleet, the US Air Force has been among the first to grapple with the organizational and cultural pressures involved in making drones a centerpiece of their operations. But as other countries begin to develop larger and more powerful fleets, their militaries will transform as they convert to a model of unmanned piloting and scramble their command structures in response to remote control. It is not self-evident that they will adapt or behave as the United States has. At a minimum, it is clear that this transformation will affect their behavior by shifting their calculations of risks and giving rise to what the military calls "mission creep." Just as the arrival of reconnaissance aircraft changed the doctrine and culture of militaries in World War I, the spread of drones is changing how military organizations see their missions and exercise control over their own pilots. Drone surveillance and combat imposes a "point of view" on the battlefield, as early theorists of air power acknowledged, but it also changes the field of vision for the pilots and the militaries that use them. By changing their approach to piloting, command structures, and targeting thresholds, drones may ultimately exert a more lasting impact on the military and its conduct of war than those concerned with "push button" warfare currently imagine.

5

Terrorist Drones

ON MARCH 11, 2011, a massive underwater earthquake struck approximately 43 miles from the northeastern coast of Honshu, the largest island in Japan. Registering 9.0 on the Richter scale, this earthquake was the fourth largest recorded since 1900.[1] It generated a ferocious tsunami with waves as high as 127 feet[2] that rapidly devastated the eastern coast of Japan, killing 15,890 people and leaving another 2,590 missing and presumed dead. While it caused over $220 billion in damage in Japan alone, its ripple effects were felt thousands of miles away in Indonesia, Papua New Guinea, Hawaii, and even California. What made this event particularly devastating was the Fukushima nuclear disaster that followed it. The Fukushima Daiichi nuclear power plant—a six-reactor nuclear plant first commissioned in 1971—lay directly in the path of the tsunami. Although the reactors at Fukushima shut down as soon as the earthquake struck, the emergency generators cooling the reactors were badly damaged. In the days that followed, three nuclear reactors melted down and radioactive material began to leak into the surrounding area. Approximately 300,000 people were rapidly evacuated from the disaster zone. Whole towns and villages were left as ghostly reminders of the life that existed there before the meltdown. The surrounding region was found to be so radioactive that the Japanese government declared it an exclusion zone, effectively rendering it off limits for decades.

Following the disaster, the Japanese government came under harsh scrutiny for inattentive oversight and regulation and poor communication during the crisis. An independent inquiry later revealed that

Japanese officials urged calm while secretly fearing a "demonic chain reaction" through which the Fukushima radiation would affect other nearby nuclear reactors, leading to cascading crises and the eventual abandonment of Tokyo.[3] Facing protests and calls to forever end Japan's reliance on nuclear power, the Japanese government had temporarily shut down all fifty of its nuclear reactors for maintenance and safety inspections by 2012. It also committed billions to decommissioning the reactors and to the decades-long environmental clean-up. Yet it would not bow to the protests and end Japan's reliance on nuclear power. In 2014, a new government, dominated by the Liberal Democratic Party (LDP) and led by Prime Minister Shinzo Abe, made it official policy to embrace nuclear power as part of its balanced energy plan. By 2015, reactors were coming back online despite a vocal protest movement that no longer trusted the Japanese government to conduct proper and safe oversight of these facilities.

Among these critics was Yasuo Yamamoto, a forty-year-old unemployed man who lived in the town of Obama in Fukui prefecture in western Japan. On April 9, 2015, Yamamoto flew a small DJI Phantom 2 drone, one of the most popular commercial models available, onto the rooftop of Prime Minister Abe's office. The drone was painted black and marked with a small red symbol for radioactive material. It carried a camera and a small plastic bottle containing radioactive sand from the Fukushima region. The amount of radioactive material, cesium, was ultimately found to be so small that it was unlikely to affect humans or the environment.[4] But this event marked the first time that anyone had gotten radioactive material that close to a world leader. It was also not quickly discovered: the drone sat on the rooftop of the prime minister's office for two weeks until it was discovered by an employee giving tours to visitors.[5] If the drone had carried more dangerous radioactive materials, the threat to lives of Prime Minister Abe and others in the area could have been much worse.

On April 25, Yamamoto turned himself into the police and admitted that the drone incursion was designed to protest the government's decision to restart Japan's nuclear reactors. The point was clear: if the Japanese government was going to restart nuclear power plants in his prefecture—the location of nearly one quarter of all Japan's reactors—then Yamamoto was going to bring those risks home to those making the decision.[6] After his arrest, a series of blog posts by Yamamoto were discovered in which he showed off the next drone he planned to launch and remarked that "in order to prevent the restarting of nuclear reactors, one cannot rule out terrorism."[7]

Although Yamamoto's protest did not cause any direct harm to Prime Minister Abe or anyone else, it provided an alarming illustration of how terrorists might use drones for assassinations and even terrorist attacks. The Japanese government reacted swiftly by banning drones over high-profile government sites and nuclear reactors, and later it enacted a wider ban of drone flights over most of Tokyo. It also began researching a number of different technologies—including radar and sonar detection, as well as police drones deploying nets to capture suspicious drones—to protect critical sites in Japan.[8] But none of these measures could wholly eliminate the risk of a drone attack by a disgruntled person or group. Yamamoto's "drone protest" showed that a small commercial drone packed with explosives or radioactive material could access well-defended sites, and even possibly spread radioactive material, without notice or clear attribution.

The potentially grave risks associated with terrorist drones were highlighted by President Barack Obama at the Nuclear Security Summit in Washington, DC, in April 2016. At a meeting of fifty heads of state, President Obama sketched a scenario where radioactive material was purchased on the so-called dark web—the uncharted part of the internet, where illicit material is often sold and exchanged with little scrutiny—and attached to drones to launch attacks in Western cities.[9] His point was that if governments could not control drone diffusion, they should at least try to control access to radiological material. Although President Obama's "dirty drone" scenario was hypothetical, British Prime Minister David Cameron warned that the threat of a group like the Islamic State deploying dirty bomb drones was "only too real." British intelligence officials also revealed that the Islamic State had been seeking low-level crop duster drones which would be ideal for spreading radioactive material.[10]Given that radiological material exists in governments and research labs in 130 countries, and that drones are spreading like wildfire across the world, some officials worry that it is only a matter of time before a drone-based terrorist attack, possibly involving nuclear material, becomes a reality in a Western city.

In September 2017, FBI director Christopher Wray testified that drone terrorism is "coming here, imminently."[11]

While some of these dire predictions may be overblown, it is clear that the drone age is one in which non-state actors—such as terrorist groups, but also rebel groups fighting for different reasons—will embrace this technology for their own ends. To some extent, this was inevitable. Given how easy it is to acquire and fly a drone, and the difficulty in enacting surveillance over all potential targets of attack, it would be impossible to stop non-state actors from turning drones into weapons. There is also no doubt that they would be interested: many non-state actors, especially terrorists, are naturally risk-taking and would have no compunction about using drones to advance their cause. But the ultimate effect of the ability of non-state actors to access drone technology remains unclear in two respects. First, will drones expand the goals of these organizations? Will terrorists or rebel armies choose to do more or do things differently as a result of having access to drones? Second, will the introduction of drones make a difference? The wars between governments and non-state actors—for example, the United States versus the Islamic State—are typically marked by stark asymmetries in power, with the governments having a decisive advantage in most respects. While it is possible that drones will only be an occasional threat, even a nuisance, which does not materially change the outcome of the battle between them, it is also possible that drone technology may enable non-state actors to begin to level the playing field against their more powerful opponents in new and surprising ways.

Terrorist Drones

To answer these questions, it is first important to understand what non-state actors like terrorist organizations will be able to do with drones. The most widely known military drones—for example, the Reaper and the Global Hawk—are far beyond the reach of even powerful terrorist groups like Hezbollah or the Islamic State due to their high cost and restrictions on their sale.[12] But with hundreds of smaller drones available on the commercial market, there is no shortage of drones that could be used for terrorist attacks. Given this fact, governments around the world are beginning to prepare for these attacks. The US government has identified risks associated with both commercial and hobbyist drones being converted into vehicles for attack.[13] In 2013, the National Counterterrorism Center convened a sixty-five-member working group tasked with analyzing the threat and preparing a response.[14] In 2015, the Department of Homeland Security (DHS) issued a worldwide alert that commercially available drones might be used in attacks on crucial infrastructure, such as airports.[15] Similar concerns have also been expressed in Britain, which has established a cross-government working group to address this risk.[16] Some have described the current state of play as a "new global arms race" between terrorists seeking to turn drones into weapons and the governments that are struggling to develop effective countermeasures.[17]

One of the most common fears is that terrorist organizations will pack explosives onto a drone and drop the bombs on a populated area, causing mass casualties. A number of different types of explosives— mortars, hand grenades, and other IEDs, for example—could be dropped from modestly sized drones and injure people or damage buildings. Although it would pose a technical challenge, more than one explosive device could also be attached to a drone for maximum damage.[18] Multiple drones could also be used at the same time to scout targets, drop multiple payloads, or even confuse defenders. The degree of the damage caused by such an attack is likely to vary based on the payload and type of explosive employed. A small hobbyist drone could drop a small payload—estimated to be 5–10 kg of TNT, roughly the equivalent of a pipe bomb or suicide vest—to devastating effect if the target was hit precisely.[19]

In an alternative scenario, terrorists turn the drones themselves into bombs. An explosive-laden drone could also be flown directly into a crowd, causing casualties and generating mass panic, or into a fixed target at a symbolic or high-profile location. There is already some evidence that these attacks have been planned. In 2010, Rezwan Ferdaus was arrested in Massachusetts for planning to slam remote-controlled planes, each containing 5 pounds of plastic explosives, into the Pentagon and US Capitol building.[20] In 2014, a Moroccan national, El Mehdi Semlali Fathi, was arrested for plans to pack explosives onto radio-controlled airplanes and fly them into a school and a federal building.[21] Most of these attempts were amateurish, but there is a real danger that the use of drones on the battlefield in Iraq and Syria by the Islamic State and others will produce expertise and technical adaptations needed to make such attacks feasible elsewhere. Such attacks would not need to have pinpoint accuracy to be effective; the degree of the panic they would cause would be enough to render the attack a success.

For terrorists, there are also clear benefits to deploying a large number of explosive-laden drones at the same time. The basic principles of deploying drones in coordinated arrays has been illustrated by the drone light shows at major sporting events and Disneyland. But the tactical benefits of such an approach are also compelling: multiple drones moving in tandem could overwhelm defenses, confuse defenders, and ensure that a target is struck. Similarly, swarms of drones—that is, coordinated drones which can move semi-autonomously and adaptively to hit a particular target—would be ideal for terrorist use. For example, terrorists could launch a swarm of small drones, packed with explosives, into a major sporting event like the Super Bowl.[22] A swarm of drones would represent the equivalent of "multiple suicide bombers launched at a single target at the same time."[23] Swarms are particularly hard to neutralize as a threat with conventional weapons like firearms given their adaptability and persistence once one has been knocked down.[24] Swarms are out of the reach for terrorist groups without a deep bench of technical expertise at present, but as the technology to create swarms becomes more common in the commercial sphere the risk of swarming drone attacks will naturally increase.

Another risk is of drone assassinations. In 2013, retired Admiral Dennis Blair, who previously served as Director of National Intelligence for President Obama, warned that explosive-laden drones could be used to kill high-level political and military figures.[25] Al Qaeda has long had an interest in drone assassinations, having plotted to kill President George W. Bush with a remote-controlled airplane at the G-8 summit in Genoa, Italy, in 2001.[26] The fears that such an attack might be possible were heightened in September 2013, when a protester managed to fly a small quadcopter within a few meters of Chancellor Angela Merkel and other German officials at a public event.[27] Other events have demonstrated the vulnerability of world leaders to such attacks. In 2018, Saudi Arabian security forces shot down a small drone approaching a royal palace in Riyadh.[28] While King Salman was not present at the time, the shooting and subsequent confusion raised the specter of a coup attempt.[29] While neither of these attempts ultimately came close to harming a world leader, they showed that a drone assassination attempt was in fact possible.

Nothing illustrated this fact to the wider public more than the attempted assassination of Venezuelan president Nicholas Maduro with a drone on August 4, 2018. On a Saturday afternoon at 5:30 p.m., Maduro was giving a speech in Caracas to celebrate the eighty-first anniversary of the formation of the National Guard. This was a reasonably well-protected speech: the embattled leader, blamed for the country's economic chaos since his election in 2013, was flanked by high-ranking officers and his security detail at a military parade. In the middle of his speech, Maduro and others on the dais noticed a drone coming toward them and a sudden bang as it exploded. Another drone was following along and crashed into a building two blocks away, exploding as it landed on the ground.[30] Maduro was protected by blast shields and rushed off the scene, but the first drone explosion injured seven members of the Venezuelan military and caused panic among those assembled. The drones were identified as DJI Matrice 600 commercial rotor drones, estimated to cost between $5,000 and $8,000 each, equipped with cameras and 2 pounds of C4 explosive.[31] These drones are capable of moving very fast and covering hundreds of yards quickly as they zoom toward their target.

It is unclear why the drones exploded where and when they did, although most experts speculated that they were equipped with remote detonators. The Venezuelan military subsequently claimed that they had disrupted the drones with "special techniques and radio inhibitors," disorienting them and making them crash before they could hit Maduro.[32] It is unclear whether this was in fact the case, and whether the second drone malfunctioned or was otherwise knocked from the sky. While one anti-Maduro group implied that it had knowledge of the attack, the ultimate perpetrator of the drone attack remains unknown. The Maduro government, however, wasted little time pointing fingers at the opposition and arrested a number of those it claimed were responsible.[33] The attack also inflamed international tensions, as the Venezuelan government blamed Colombia and the United States for funding and supporting "ultra-far right" figures behind the assassination attempt.[34]

This event also illustrated that the risk of drone assassinations is growing. The small commercial drones used in the Maduro attempt are easy to use, even by those with limited flying skills. As drones get smaller and more lethal, they will make assassination missions available to a wider range of groups. One option would be for terrorist groups to attach what are sometimes called "mini-munitions"—essentially small missiles—onto smaller drones. The US military has already found that attaching mini-munitions could make drones such as the RQ-7 Shadow capable of very precise killings of specific individuals on the battlefield.[35] Some military-grade drones are specifically designed for this purpose.[36] One such kamikaze drone is called the Switchblade; it was introduced by AeroVironment in 2012 and sold widely to the US military. Weighing only 6 pounds and fitting into a backpack, the Switchblade can be quickly assembled and thrown in the air for a short flight. It can also carry small munitions, enough for an explosion comparable to a grenade, and could be flown at an individual or a vehicle in an assassination attempt. While most mini-munitions and the Switchblade drone are not sold to individuals or groups other than the US military at this point, similar technology will inevitably spill out into the public domain and make assassinations easier for a larger array of non-state actors.

Another potential type of attack possible with drones involves attaching lightweight small arms, such as a pistol or a rifle, to a drone and flying it over a crowd of people to strafe them. There is some evidence that these types of drone attacks are possible, but they are harder to launch and operate than comparable attacks with IEDs. In 2015, Austin Haughwout, an eighteen-year-old college student in Connecticut, proved this was possible by developing a handgun-firing drone; the video showing its operation he uploaded to YouTube was viewed 3.7 million times within a year.[37] Another one of Haughwout's videos—called "Thanksgiving dinner"—showed a flamethrower attached to a drone roasting a turkey in his backyard. Both videos attracted interest from the Federal Aviation Administration (FAA) and local police; he and his family found themselves in court explaining why their drones did not violate FAA safety regulations.[38] Although it is possible to equip drones with firearms, and this has been demonstrated in controlled conditions, such drones take more skill to use than it might appear. The recoil of the weapons tends to destabilize the drone and throw it off course, rendering them less accurate than even a terrorist might hope.[39] They are also more likely to crash, and to be thwarted by countermeasures, than one which drops an explosive payload and quickly retreats.

One much-feared scenario involves terrorists loading crop-dusting drones with radioactive material or chemicals and spraying them over population centers or crowds of people. To some extent, this is a long-held dream of terrorists. Before their June 1994 attack on the Tokyo subway, the apocalyptic cult Aum Shinrikyo tested the use of remote-controlled mini-helicopters equipped with spray mechanisms to distribute sarin against their targets, but their tests failed.[40] By 2003, the Bush administration was concerned that UAVs equipped with crop-dusters could be used to spread chemical or biological weapons over Western cities.[41] Some analysts have argued that spraying drones would be more effective than dropping explosives and would affect a wider area.[42] Al Qaeda has long had an interest in these types of attacks.[43] Several of the ISIS-inspired operatives who attacked the Brussels airport in 2016 had first sought information on how to conduct a "dirty drone" attack.[44] Yet the "dirty drone" scenario relies on a number of things going exactly right to work. Aside from acquiring the right kind of dangerous or radioactive materials, terrorist organizations would need to pack the material safely onto a drone, develop and test an effective dispersal system, and conduct test runs to ensure that dispersal happened in the way that was predicted.[45] All of these steps would need to be conducted without attracting the attention of the security services. Even if the tests were successful, such an attack would more likely cause panic rather than many direct fatalities.[46]

Finally, hijacked drones are a potential threat. In a sky filled with thousands of new drones, it is possible for terrorists to find ways to seize control of drones flown by private companies or the police and redirect them to nefarious purposes. While military and some police drones have a secure connection between the drone and the operator, many commercial and hobbyist drones have an insecure communications link that could be interrupted or hacked by terrorists. If a drone hack were successful, terrorists could crash a drone or drones into a potential target or reverse engineer it for nefarious use later.[47] Some have suggested that widely available commercial drones, like the quadcopters soon to be employed by Amazon, could be retrofitted to deploy explosive devices if hijacked by a terrorist group.[48] While seizing control of a drone is difficult, it is possible if a hijacker is able to "spoof" the drone and send commands that the drone believes is coming from its legitimate operator. This has already happened: in 2011, Iran claims to have "spoofed" an RQ-170 Sentinel drone, sent it fake GPS signals, and convinced it to land in Iran itself, rather than Afghanistan.[49] Iran then proudly displayed the captured US drone on televisions and harvested its technology for their own use. Today, these hacks are not only the province of governments. One US-based security researcher has seized control of police quadcopters with insecure lines of communication using a laptop and a cheap radio chip connected by a USB port.[50] At a minimum, hijackers from terrorist groups will be able to break into drone signals and collect data and imagery from them.[51] In December 2009, Iraqi insurgents managed to hack into a US Predator feed, but they did not succeed in changing its flight path.[52] To do so, they used cheap, off-the-shelf software called "SkyGrabber" that is widely available online.[53] Others have followed suit. In 2016, it was reported that a twenty-three-year-old hacker from Palestinian Islamic Jihad had hacked and monitored Israel's drone feeds for at least two years.[54]

Drones and Aviation

Some potential terrorist scenarios also rely on turning the drone into a weapon, but in a more creative way, for example directing a drone toward an engine of a manned aircraft.[55] As early as 2004, German intelligence found that al Qaeda was planning on conducting these attacks, though they never got far beyond the concept stage.[56] Much of the fear around this potential attack has to do with the increasing level of near-misses between drones and civilian aircraft. In the United States and many other countries, there is a clear exclusion zone around major commercial airports, and in the United States drones are not allowed to be flown above 400 feet so that they do not endanger other air traffic. Yet the enforcement of these standards has been slow and inconsistent, and drones are regularly reported flying near or inside exclusion zones and close to manned aircraft. An analysis by Bard College of 921 unexpected drone encounters reported in US airspace between December 2013 and September 2015 revealed that 90% of them occurred at over 400 feet and that a majority occurred within 5 miles of a commercial airport.[57] Some of the commercial and consumer drone accidents are due to human error, but a portion are due to a malfunction with the communication signal which leads the drone to fly into congested or otherwise banned airspace. Another estimate of drone accidents between 2006 and 2016 found that 64% of incidents were caused by technical problems.[58] FAA data indicates that this is a constant threat: in 2015, there were 3.5 near misses a day between drones and manned aircraft, compared with less than one a day in 2014.[59] In 2016, the FAA counted 583 near misses between drones and airplanes, a threefold increase from the number of potential accidents in 2014.[60] In 2017, the FAA reported a reduction in such incidents, but there were still 385 near misses with aircrafts in the United States alone, with drones accounting for more than half of reported events.[61] Part of the problem is verifying whether a near miss was with a drone, a bird, or something else. While the FAA reported an estimated 6,000 sightings of drones near manned aircraft or airports between 2014 and 2018, it emphasized that it could not verify all or even most of these were actually drones.[62] A more detailed analysis of near misses between December 2013 and September 2015 found that in 158 cases a drone came within 200 feet or less of a manned aircraft, and in twenty-eight cases a pilot had to rapidly maneuver to avoid a collision with a drone. Although more than half of the near misses occur near airports, some occurred further afield and above the altitude of 400 feet, where drone flights are forbidden.[63]

Drones are particularly dangerous around civilian aircraft because they are hard to detect on radar and because their owners are difficult to trace once an incident has been reported. In 2013, the FAA launched investigations of twenty-three incidents of illegal drone use near civilian airports or in proximity to aircraft, but in most cases the owners of the drones were never found.[4] While most near misses were minor incidents, some drones came close to crashing into large passenger aircraft near major airports. In March 2013, a small private drone came within 200 feet of an Alitalia commercial jet over JFK airport, in New York.[64] One year later, an American Airlines jet had a near miss with a drone in Florida.[65] In 2015, a Jetblue A320 pilot reported a near miss with a drone at 6,000 feet near JFK airport, while in 2016 a Lufthansa A380 super-jumbo jet came within 200 feet of hitting a drone over LAX airport in Los Angeles.[66] A prominent airline pilots association has argued that the widespread introduction of drones into domestic airspace carries could "profoundly degrade the safety of both commercial and general aviation flight operations" unless they are integrated into the FAA systems in a comprehensive way.[67] Although the FAA is working on implementing regulations to integrate drones into commercial airspace in a safe way, it has not settled on a clear, uncontested set of regulations, and questions remain about tracing drones effectively in US airspace.

Britain has a similar problem: near misses with drones are increasingly common at their major airports. In 2015, the United Kingdom reported twenty-three incidents in which drones flew too close to manned aircraft in its airspace. In one instance, a drone flew within 25 meters of a Boeing 777 after it took off from Heathrow airport, and no one ever found the drone operator. An Embraer jet came within 60 feet of a drone over the Houses of Parliament.[68] In 2018, a Virgin Atlantic airliner flying from Delhi to Heathrow was nearly struck by a drone on its descent: the drone came within a few meters of the aircraft, making it the closest near miss yet recorded.[69]

Despite these incidents, the overall odds of a collision between a manned aircraft and a drone remain low.[70] Most drones fly well below the altitudes of commercial and private flights. There is also considerable debate over what would happen if a terrorist managed to make a drone collide with a commercial airliner.[71] Some analysts believe that the kinetic energy of the drones would be sufficient to take an aircraft down, while others argue that a plane could easily survive a "mechanical bird strike" if the drone did not come close to a plane's engine, or took out only one of them.[72] Although there have been lab tests of this scenario, none have obviously occurred in real life, so no one knows what would happen to a plane in midflight. It also matters whether the drone would strike the aircraft at a vulnerable point in its flight, such as during take-off or landing. The conventional wisdom is that interruptions of the aircraft's operation at those points are far more perilous than during cruising, but the size, speed, and angle of strike for the drone aircraft would still influence the severity of the damage caused.

Part of the reason why what might happen with a drone hitting an aircraft remains unclear is that near misses—or at least reports of them—are now common, but actual strikes of drones into manned aircraft are extremely rare. A widely noted report of a drone striking a British Airways jet near Heathrow airport in April 2016 turned out to be a false alarm.[73] There are only two verified accounts of a drone striking an aircraft and neither resulted in the catastrophic loss of the aircraft. In Canada, a small drone collided with a Skyjet plane making an approach to Quebec City in October 2017, but the plane landed safely after the pilots heard a loud bang.[74] In 2017, a drone was flown into a Black Hawk Army helicopter over New York Harbor, but caused little damage.[75] Although accidents are always possible, it is perhaps even more difficult for terrorists to deliberately strike an aircraft flying at 500 feet per minute with a drone unless the navigation, accuracy, and speed of small commercial and hobbyist models improve significantly. DJI Technology, the largest commercial drone manufacturer, has argued that these attacks are unlikely, because it would be "like trying to hit a bullet with a bullet" and would require "an unprecedented act of marksmanship."[76]

It would be far easier to interrupt commercial flight operations with drones, while not actually striking an aircraft. Over the last few years, drone sightings near airports have become commonplace and have shut down airports across the world. Some of the biggest airports in the world, including Frankfurt, Dubai, Singapore, London Heathrow, and Newark have been shut down temporarily due to drone overflights. In many cases, a pilot or other airport official will report seeing something like a drone near the airport, but verifying that it actually was a drone is difficult.[77] This was illustrated by the sighting of one or more drones near London's Gatwick Airport over Christmas 2018. The airport was shut for three days with multiple reports of drones over landing strips, and over one hundred flights were canceled over the busy holiday period, affecting approximately 140,000 passengers. But the event remains a mystery: despite multiple credible reports of sightings, no video has emerged to confirm that it was a drone, and no drone has been found despite an extensive search of the surrounding area. One Sussex police officer even remarked that it was "always a possibility that there may not have been any genuine drone activity in the first place."[878]

The mystery deepened when the subsequent investigation revealed no obvious culprit. Two people were arrested but were quickly released when British police admitted that they were no longer suspects.[79] The police reportedly suspect that this may have been an inside job by someone who once worked at Gatwick because the perpetrator knew the airport layout and the blind spots—for example, behind buildings—where counter-drone technology would not work.[80] Part of the problem the police faced was finding the drone: signal-jamming counter-drone technology is risky to use near an airport, and some more crude ways of getting the drone—for example, shooting it down—would be prohibited near airports. Although the military was called in to find the drone, they had no more luck than the police. In the end, the motive was equally obscure: police were uncertain whether the event was an act of terrorism, a protest of some kind, or criminal mischief.[81] Ultimately, it did not matter, because the Gatwick incident revealed that drone disruptions like this could be accomplished without the perpetrators getting caught provided that the attack was well planned. The attack was also cost-efficient: for the price of a single drone and related equipment, the attacker cost Gatwick airport £1.4 million and cost the airlines £64.5 million.[82] Although London Gatwick and other major airports have rushed to buy counter-drone technology that will prevent such attacks in the future, the success of the Gatwick airport disruption has surely been noticed by terrorist groups.[83]

How Likely?

How likely are these scenarios? Any calculation of risk from terrorist drones is partially determined by the number of drones available. As drones wind up in the hands of hundreds of thousands of new people each year, the risk that someone will find a way to misuse them and to harm others naturally increases, as it does with many other types of technology. One reason why government officials have been so concerned is that many had vastly underestimated just how many people would acquire drones for commercial and personal uses. The rapid proliferation of cheap, small drones—particularly the increasingly sophisticated commercial off-the-shelf drones, available for only a few hundred dollars from Amazon and other retailers—means that it is inevitable that the technology will wind up in the wrong hands and be put to criminal purposes. This can already be seen in the use of drones for delivering contraband into prisons. In the United States alone, there were twelve attempts to drop mobile phones, drugs, and pornography into prisons between 2012 and 2017.[84] It is likely that this is a substantial underestimate and that the number today is far higher. For example, Mexican drug gangs have been using drones to move drugs and contraband and to monitor smuggling routes.[85] They have also weaponized small commercial drones by equipping them with explosives to use against their rivals in the drug war.[86]

For terrorists, the barriers to entry are generally higher but they are not insurmountable. Many attempted drone attacks will fail or be thwarted, but if enough attempts are made, at some point the odds are that an attack of some kind will succeed. It is also likely that terrorist interest in drones—and hence attempted attacks—will increase as their availability and capabilities increase. There are at least five reasons why terrorist organizations might want to turn to drones over the next decade or more. First, for organizations like al Qaeda and ISIS, who struggle to get their operatives into the United States, it is inefficient to use suicide attacks and waste those assets; ideally, a reusable means of attack, like a drone, is better.[87] Terrorist organizations would prefer not to have any operatives near an attack, but would instead prefer to program the drone to attack on a predetermined flight path, as happened with the Maduro assassination attempt. This is now possible through freely available open source software such as ArduPilot.[88] Second, if small drones could reach their target quickly, it would dramatically shorten the time that law enforcement authorities have to react. One senior British government official noted that "once you start to think of a large hobbyist drone carrying a standard payload, most of the traditional defenses against terrorist penetration of high-value places become irrelevant."[89] Third, the lack of adequate regulation and control of drones also provides an opportunity for terrorists. An off-the-shelf hobbyist drone can be bought without attracting the attention of most regulatory officials or law enforcement. The sheer volume of drones available on the commercial market, and the inadequate records of who has purchased a drone, will mean that attribution for an attack will be difficult. The time lost in finding who really owns a drone might give terrorist operatives the opportunity to escape detection and capture. Fourth, drones also allow for forward planning. The increasingly sophisticated cameras on drones allow for terrorist operatives to scout out future attack locations and conduct dry runs to ensure that their attacks work as planned. Fifth, drones are adaptable: they can be amended with new cameras or payload, and new technology, like 3-D printing and Openware, provides an opportunity to tailor the drone's operations for an attack. As off-the-shelf drones become smaller, more portable, and more capable, they can be adapted more quickly to a wide range of purposes, including terrorist attacks. This is particularly the case given the wide availability of information on adapting drones available through online forums like DIY Drone. Similarly, the integration between small commercial and hobbyist drones and widely available consumer electronics—WiFi, GoPro cameras, and even iPads and mobile phones—will lower the technological barriers to effective drone use for terrorists and enhance adaptability.[90] The more that drones adopt "plug and play" technology, the easier terrorist drone attacks will become, especially for novices or lone wolves.

There is no doubt that evidence of terrorist interest in drones is growing. One unconfirmed report suggested that authorities in the United States, Germany, Spain, and Egypt foiled at least six potential terrorist attacks with drones between 2011 and 2015.[91] There is no estimate of the number of foiled attacks more recently, but given the spread of the technology it is likely to be higher. There is evidence that drones are in the hands of a number of terrorist organizations, including al Qaeda, the Taliban, Fuerzas Armadas Revolucionaria de Colombia (FARC), Hezbollah, Hamas, and the Islamic State, although mostly on foreign battlefields.[92] Evidence of actual attacks, or even scouting for attack sites, is fragmentary. There have been reports of unauthorized use of drones in sensitive sites in the United States, the United Kingdom, and France, but it is unclear how many of these were genuine attempts to scout locations for future attacks. The most famous drone accident in the United States looked for a brief moment like an attack. A drunken US government employee accidentally flew his drone onto the lawn of the White House, and then fell asleep.[93] The incident

was comical, but the implication that a terrorist drone could penetrate the protected airspace above the White House was widely noted.
In the United Kingdom, the Metropolitan Police has acknowledged
twenty suspicious drone incidents in and around London, though most
were violations of airspace or criminal activity.[94] In France, unidentified drones have been flown over a number of sensitive sites, including
the Eiffel Tower, Place de la Concorde, and Elysee Palace, and over
multiple nuclear power stations.[95] An unidentified drone even allegedly
struck the Sydney Opera House in October 2015.[96] Yet despite these
near misses, no terrorist organization has successfully used a drone for
an attack against a fixed location or major grouping of people, such as
a sporting event, to date.

This situation—that drones are widely available, interest is high
among terrorist organizations, but attacks have not yet occurred—is due
in part to the formidable but underappreciated technological obstacles
to launching these attacks. Put simply, imagining a "terrorist drone" attack scenario is easy, but conducting that attack is harder than it looks.
Even if drones are readily available and terrorist interest in them is
growing, there are practical obstacles that must still be overcome. First,
drones have limits in range and endurance that make it hard to plan
an attack at a great distance. Most commercial off-the-shelf drones are
able to sustain a datalink of 1–10 km from a ground station, with many
of the less regulated models on the lower end of that range.[97] What this
means is that with current drone technology terrorists would need to be
relatively close to the site of the attack for it to work. These communications and data links are often unsecured, which leaves the plot at risk
of being detected by the police or security services. Although higher
end commercial drones have an endurance limit of several hours, many
hobbyist drones have only an endurance of under an hour.[98] While
this can be modified, doing so will affect payload and limit what kinds
of explosives might be carried. Given these technical limitations, most
terrorists would need to get somewhat close to their proposed target
and assume some personal risks that they will be intercepted before the
attack proceeds. Especially in cities, where their activities may already
be monitored by the police and security services, this presents some real
dangers. This is obviously not insurmountable: the perpetrators of the
Maduro and Gatwick events got away with it, but clearly not everyone

else will. Finally, some contemporary drones are slow and could be seen on approach before an attack. Most hobbyist rotary wing drones are not optimized for speed and can travel at a maximum of 15–20 meters per second (49–65 feet per second). This leaves them susceptible to anti-drone technology and other cruder ways of stopping a drone in mid-flight if they are launched a decent distance from the target. Until drones get faster or move in swarms, current generation drone attacks are likely to be spotted or even heard, given their whirring in-flight noises, before they can reach their target.

Arming drones also presents its own challenges. Many of the alarming scenarios of terrorist drones tend to underplay the real difficulties involved in deploying dangerous materials—for example, radiological material, chemical weapons, and others—onto mobile flying machines. The chemicals and radioactive materials are under varying degrees of scrutiny and control; purchasing or otherwise obtaining them can draw the attention of the police or the security services. Many of these payloads are unstable and carry real risks for those installing and deploying them; some, such as radioactive materials, require careful storage and handling that is difficult to provide outside a lab. Even simple explosives are difficult to handle and can backfire, leading to the death of the planners themselves. Moreover, these heavier, more unstable explosive payloads can reduce the stability and the payload capacity of a drone. One estimate found that the popular DJI Phantom 3 model lost 14.4 minutes of flying time for each 1 kg of payload it assumed.[99] Large payloads of explosives can destabilize the drone and cause it to crash; destabilization and crashes are even more likely when the drone is rigged to firearms and must remain stable during recoil.

A terrorist contemplating using a drone for an attack would have to weigh the advantages conferred by the technology against the technological obstacles to their use. It is one thing to simulate a terrorist attack with a drone in a laboratory or in controlled conditions, but it is quite another to conduct it in real life. Much of the planning for a terrorist drone attack—for example, buying the necessary components or conducting dry runs against fixed targets—may attract the attention of the police and security services, especially in inhospitable environments like major cities in the developed world. In other words, "terrorist drone" attacks are possible, but they are harder than some dire warnings suggest. It would obviously be foolish to declare that terrorist drone attacks will never happen given the explosive growth of drone technology worldwide and the evident terrorist interest in the technology. But equally taking the nightmare scenarios at face value and hyping the threat flowing from them is wrongheaded. The risks posed with current drone technology are real but not insurmountable, especially given the growing attention and investment among governments, private companies, and others in counter-drone technology. Given these risks and barriers, successful terrorist drone attacks will generally require some basic technical knowledge and organizational capacity by the perpetrator. Among non-state actors, these characteristics are more likely to be found in disciplined, well-resourced rebel groups in today's theaters of war than among resource-constrained terrorist cells or angry "lone wolves" in the developed world.

Rebel Drones

Much of what has been written on terrorist drones conflates their use by rebel groups in theaters of war in Central Asia, the Middle East, and elsewhere with their potential use by terrorists in cities like New York, London, and Paris. Government officials and experts often point to evidence of the use of drones by rebel groups fighting wars abroad as showing why terrorist drone attacks on targets in the developed world are likely today. Yet the context surrounding these rebel drones is very different. The terrorist organizations that have been most successful in their use of drones—for example, Hezbollah in Lebanon and the Islamic State in Syria—have deployed them locally in their capacity as rebels engaged in a violent struggle against a government. In these wars, the targets are generally less well defended and the risks of being caught are lower than they are in well-monitored cities. War is a naturally permissive environment that enables experimentation with and testing of the means of violence; that testing is further possible because rebel groups remain in regular communication with military command and supply lines that put the technology in their hands. Many of the difficulties present in launching a terrorist attack in the developed world are simply not present in states experiencing an active armed conflict. For these reasons, it is far easier to fly a rebel drone than a terrorist drone, and it is far more likely that drone deployment will help level the playing field for rebels against their government opponents.

The emergence of "rebel drones" signals a significant change in warfare in three ways. First, for many years, militaries from the developed world have made achieving air superiority their first goal when undertaking any military operation. For example, the US Air Force insists on air superiority in counterinsurgency, peacekeeping, and other missions and destroys any enemy aircraft before they can even take to the skies. This gives the United States complete freedom to attack its enemies with air power and enables ground forces to undertake maneuvers that would otherwise be too risky. But with drones, rebel groups and insurgents can now take to the skies themselves and challenge the assumption of air superiority that underpins most of their opponent's military strategy. Most commercial drones are small enough and fly at such low altitudes that they evade detection by normal radar. Second, most asymmetrical conflicts do not distribute vulnerability evenly but instead generally leave rebel groups or insurgents at greater risk of death and injury than troops from the developed world. This is particularly the case with US forces that have high standards of force protection. While rebel drones do not level the playing field completely, they shift vulnerability toward soldiers on the ground by leaving them subject to attack from the air, sometimes for the first time in decades. Third, these rebel drones have a psychological impact on their opponents by illustrating their vulnerability to an attack which may come without warning. The psychological aspect of drones—that they can come anytime, unexpectedly, and cause damage—is as important as their immediate tactical value, if not more so.

The earliest example of a rebel use of a drone provides an illustration of how a permissive environment and organizational infrastructure is an essential foundation for rebel drone use. A Colombian army unit stumbled across nine remote-controlled planes belonging to the FARC rebel group in the remote jungles of Colombia in August 2002.[100] The purpose of the remote-controlled planes remains unknown, but army officials suggested that they were intended to launch IEDs against government targets.[101] This example is often cited as the beginning of the terrorist drone threat, but in fact the FARC was hardly a normal rebel group. At the time, it was at its most powerful, controlling vast swaths of territory inside the country and managing a significant portion of the drug trade. It was, in other words, not a ragtag group on the margins of the society, but rather a powerful army with substantial membership, funding, and infrastructure.[102] It is not surprising that such a well-equipped and capable armed group would be among the first to truly experiment with launching drone-based attacks.

The same could be said of Hamas, the terrorist organization that has effectively controlled the Gaza strip since the Israeli withdrawal in 2005. Its military wing, the al-Qassam Brigades, has a small number of drones and has attempted to penetrate Israeli airspace and to launch attacks on Israeli targets with them.[103] In November 2012, Israeli military officials discovered and destroyed a drone workshop run by Hamas in Khan Yunis in the southern Gaza strip.[104] In 2013, Hamas operatives attempted to pack a drone full of explosives and send it into Israel, but the plot was disrupted.[105] In July 2014, Israel shot down a Hamas drone with a Patriot Missile, although Hamas later claimed that two additional drones made it through undetected. Hamas has clearly exploited drones for their psychological impact, even flying a drone over a military parade in December 2014.[106] Hamas claims to have a wide range of surveillance and attack platforms and releases grainy images of its drones after successful flights. They have continued to send small drones from the Hamas-controlled Gaza Strip over Israeli territory. For Hamas, drones are remarkably cost-effective against the more powerful and well-resourced Israeli government. For only a few hundred dollars spent on a commercial drone, they can force Israel to scramble its jets and spend thousands or more to shoot them down. Unsurprisingly, Israel has retaliated by regularly striking their drone factories and allegedly by killing a Tunisian man who led a double life as a Hamas drone engineer.[107]

The armed group that has been most successful at using drones is Hezbollah, one of the most organized and disciplined non-state actors in the Middle East. Hezbollah is a complex organization, operating more like a powerful political and military player in Lebanon, but also employing terrorism against Israel and other targets in the Middle East. By some estimates, Hezbollah now has more than 200 platforms for reconnaissance and combat missions, with a significant portion of its drone fleet provided by Iran.[108] In September 2013, Hezbollah became the first non-state actor to launch its own successful drone strike, drop-ping either explosive warheads or air to ground rockets on the Jabhat al-Nusra forces in an attack that killed twenty-three al-Nusra fighters.[109] It is also the only non-state actor with its own dedicated drone airfield. In April 2015, it was reported to have a constructed an airfield and size-able drone fleet in the Beka'a Valley in Lebanon.[110] By August 2016, Hezbollah was dropping Chinese-made cluster bombs from drones on buildings and cars occupied by Syrian rebel forces.[111] In doing so, Hezbollah became the first rebel group to do what the United States has long done: use drones to eliminate its enemies through targeted killings. It has continued to launch drone strikes against its enemies in the Syrian civil war, but its experience with drones predates that by some time.

The first evidence of Hezbollah mastering a drone came shortly after the onset of the second intifada against Israel. In December 2003, a Hezbollah cell was caught by Israeli security officials working with the al Asqa Martyrs Brigade, an armed wing of Fatah, to plan an IED drone attack on Jewish settlements in Gaza.[112] After that attack was foiled, Hezbollah shifted toward deploying drones as an almost symbolic show of force over Israeli territory. In November 2004, it launched a Mirsad-1 drone for a twenty-minute reconnaissance mission over northern Lebanon, though the drone crashed shortly thereafter. Hezbollah's leader, Hassan Nasrallah, boasted that the group could strike "any-where, deep, deep" inside Israel and could deploy as much as 200 kg of explosives.[113] While this may have been an empty boast, Hezbollah suc-cessfully continued its drone program with a successful drone launch in April 2005, when another Mirsad-1 drone hovered over the Israeli city of Acre and returned to Lebanon before it could be destroyed.

During the Lebanon War in 2006, Hezbollah launched several drones against Israeli targets and even rammed an explosive-packed drone into an Israeli warship, causing a small fire.[114] By August 2006, Hezbollah's capabilities had improved and it was able to launch three small Ababil drones—packed with 40–50 kg of explosives in warheads—against targets in Israel.[115] While these were shot down, they illustrated that small drones could evade detection by Israel's sophisticated air defense systems.[116] These attacks were negligible from a military vantage point, as Hezbollah's drones were shot down by manned aircraft and paled in comparison to Israel's sophisticated drones. Yet they proved to be a powerful signal of capacity and source of propaganda for Hezbollah. Matthew Levitt of the Washington Institute for Near East Policy has remarked that, "they love being able to say 'Israel is infiltrating our airspace, so we'll infiltrate theirs, drone for drone.' "[117]

Although Hezbollah may love claiming credit for its drones, it is not clear whether it would have been capable of using drones in the absence of external sponsorship. The Ababil drones—named for a mythical race of birds which dropped stones on an army invading Mecca, according to the Quran—have been exported by Iran widely to both Hezbollah and Hamas, both determined enemies of Israel.[118] Iran has also sold the Mirsad drone—an updated model of a reconnaissance drone flown since the Iran-Iraq war in the 1980s—to Hezbollah.[119] There are unconfirmed reports that Hezbollah has Shahed-129 drones, which are like smaller, less-capable versions of the Predator.[120] Hezbollah has insisted that these drones were not provided by Iran and were developed by its own engineers, but the available photographic evidence shows a clear resemblance between Iranian models and their Hezbollah counterparts.

FIGURE 5.1 Pieces of a fixed wing Hezbollah UAV downed by the Israeli air force in 2006.

There are unconfirmed reports from Russian and Israeli sources that Iran sent eight Mirsad drones to Hezbollah and even trained thirty Hezbollah operatives near Isfahan to fly the aircraft.[121] Expertise from Iran is also flowing to Hezbollah's engineers, allowing them to build a stronger capability for attacks against Israel. This close relationship between Hezbollah and Iran led Milton Hoenig to conclude that "drone launches by Hezbollah into Israel are planned and carried out to meet the political agenda of Iran, while shielding Iran's involvement and allowing a measure of deniability."[122]

The relationship between client and sponsor may be more complicated than this depiction suggests. While Hezbollah may do the bidding of its Iranian sponsor in some instances, in others it is clear that Hezbollah is using its drone fleet for its own priorities, specifically deploying them in an almost theatrical way to show off its capacity to threaten Israel. None of the drones deployed by Hezbollah are comparable in capacity to those flown by Israel; in fact, many are somewhat low tech, flying at such a low speed and elevation that they are hard to detect on radar.[123] These relatively simple drones have been detected by Israel's sophisticated Iron Dome defense system and shot down. While their military utility is limited, they have some value in scaring Israeli civilians and reminding them they remain as vulnerable to Hezbollah's drones as Lebanese and Palestinian civilians are to Israel's drones. In other words, Hezbollah's drone incursions are more about psychological warfare than military strategy. They serve as a symbol of Hezbollah's persistence and a signal of its capabilities, even if those capabilities are vastly outmatched by its opponent.

This approach was clearly apparent in Hezbollah's second wave of drone flights in 2012. After a six-year moratorium, Hezbollah launched an Iranian drone near the Israeli nuclear reactors in Dimona, approximately 35 miles within Israel's territory across the Negev desert.[124] It was shot down by Israeli aircraft, but not before it was rumored to have taken pictures of sensitive nuclear sites.[125] Within days, Hezbollah was exploiting the operation for propaganda purposes. Hassan Nasrallah, the leader of Hezbollah, noted that Israel frequently sent drones into Lebanon's airspace and that "it is our right to send other drones whenever we want." He further warned, "It was not the first time and it will not be the last."[126] As if to prove his point, a further drone incursion was reported in April 2013.[127] Neither of these drones posed a military threat, but both had the effect of allowing Hezbollah to claim a propaganda victory and to deliver a shot across the bow of its enemy Israel. This psychological impact is only possible because actors like Hamas and Hezbollah are connected to the funding, infrastructure, and supply lines needed to make these drone operations possible.

On the Battlefield

It is often said that experience is a harsh teacher.[128] The experience of war has proven valuable, however, for rebel groups who have begun to experiment with drones. While the use of drones by non-state actors in the mid-2000s revolved around sporadic tactical attacks and signaling, rebel groups have more recently shifted to using drones as a tactic in a range of conflicts, including Afghanistan, Libya, Syria, and Ukraine. For some rebel groups, this meant stealing or borrowing a Western military grade drone for their own battlefield use. In May 2012, a NATO raid in Helmand province in Afghanistan discovered that the Taliban had captured a small drone, possibly modeled on NATO's own Desert Hawk.[129] In Libya, government and militia forces have deployed a wide array of commercial and hobbyist drones for battlefield reconnaissance, some of which came from the West. In 2011, Libyan anti-government forces even managed to acquire a $120,000 quadcopter from the Canadian firm Aeryon Labs for battlefield reconnaissance.[130] But drone use—by the United States, Turkey, United Arab Emirates, and others— escalated throughout Libya's civil war, and their practice was matched by non-state actors. By 2019, Libyan National Army (LNA) forces were being supported by armed Chinese Wing Loong drones operated either on their own or by their allies, the United Arab Emirates.[131] In June 2019, LNA forces also destroyed a Turkish drone that they claimed was targeting their positions.[132]

The rebel groups that manage most consistently to deploy their own drones for strategic effect have a strong organizational base, a steady flow of funding, and one or more external sponsors. These groups use battlefields to hone their skills, to experiment, and to learn from their failures in a complex, violent environment. But beyond just using battlefields as incubators for drone expertise and development, rebel groups fighting in Ukraine, Iraq, and Syria have also used the technology to level the playing field against their more powerful opponents, turning asymmetrical wars into wars where governments and rebels are in a race to deploy drone technology in more innovative ways.[133] In the case of the Islamic State, it also led them to expand their goals and made them willing to undertake attacks against US forces that would have been unthinkable before.

When Russian forces invaded Crimea and later eastern Ukraine in 2014, combatants on both sides turned to drones for a wide array of reconnaissance activities. Both sides were seeking battlefield awareness: in the fog of a proxy war, with pro- and anti-government rebel groups operating without uniforms in an environment with civilians present, getting information on the location of the enemy becomes paramount. Pro-Russian separatist forces turned to Moscow to provide them with drones. A number of Russian-made drones, including the Orlan-10, Eleron-3SV, Granat-1, and ZALA drones, have been shot down in Ukraine.[134] One estimate suggests Russia has deployed as many as sixteen drone prototypes there.[135] Most of these drones are broadly similar to the MQ-11 Raven drone and are highly effective reconnaissance platforms. They allow pro-Russian rebels to identify targets, such as pro-government rebel positions, and direct artillery fire or other types of bombardment.[136] A senior US military official observing the conflict found that this tactic of drone spotting and bombardment had been brutally effective, with 85% of Ukrainian casualties coming from rocket and cannon fire.[137] Pro-Russian separatist rebels had also purchased commercial off-the-shelf drones and found ways to strap explosives to them and even drop grenades with them.[138] According to one analyst, the successful exploitation of drones poses a real risk that the conflict would be determined by "technical overmatch."[139] Knowing that Moscow's supply of drones would reveal its hand behind the conflict, pro-Russian rebels have also deployed GPS spoofing technology and signal jammers to block Ukrainian and Organization for Security and Co-operation in Europe (OSCE)-funded drones from collecting evidence of Russia's direct involvement in the conflict.[140]

Ukrainian military forces have been at a substantial strategic disadvantage because the government had underinvested in drones and had relatively few ready for the conflict.[141] To compensate, the Ukrainian government turned to commercial, off-the-shelf drones as a way of evening the score.[142] It has mostly employed quadcopters with commercial cameras for reconnaissance of the battlefield and identification of potential targets. The Ukrainian military also created a special military unit, the Aerorozvidka, to train pilots in the effective use of commercial drones.[143] To more quickly make up the "drone gap," Ukrainian pro-government forces have turned to crowdfunding to get drones to the battlefield. For example, one organization, the People's Project, operated its own version of a Kickstarter fundraiser to buy drones.[144] One US organization, the Chicago Automaidan, has also been purchasing and retrofitting DJI Phantom drones for military use and sending them to pro-government forces.[145] Other more locally homemade drones have been also deployed by Ukrainian military forces. The United States also provided some RQ-11 Raven drones to Ukraine's military, but these have proven less effective than hoped.[146] Although they remain outmatched by pro-Russian forces in tactical terms, the Ukrainian rebels have skillfully used drones for propaganda purposes and even revealed the existence of a long-denied Russian encampment in eastern Ukraine.[147] They have also periodically succeeded in dropping grenades from their drones against enemy positions and fixed targets.[148] Although the introduction of drones has not leveled the playing field between pro-Russian separatist and Ukrainian government forces, it has reduced the gap in capabilities between them.

Similarly, the regional war in Syria has also proven to be a fertile testing ground for rebel groups to try their hand at flying drones. In many respects, this brutal conflict, stretched across the borders of Iraq and Syria, has been a laboratory of rebel drone innovation, with actors on all sides deploying drones for increasingly complex tasks. Within this battlespace, multiple armed actors—Kurdish Peshmerga, Jabhat al-Nusra, Islamic State, Hezbollah, and an array of militias allied to the Syrian government and to Iran—have employed drones on the battlefield. The frequency and sophistication of drone use has varied, but together these rebel organizations have shown that drones can redress some inequalities on the battlefield, broadcast their cause to the wider world, and in some cases expand the ambitions of their users.

One way that drones can redress inequalities on the battlefield is by enhancing reconnaissance of the enemy's positions. This is a crucial development because most rebel groups face chronic manpower shortages and are reluctant to lose personnel when they are overmatched by their opponent. Without drone technology, many rebel groups lacked the ability to look around a corner and see the enemy's position from the air, a skill that the United States and other developed countries have had for years. But this is now changing, giving them a point of view that they never had before. In Syria, pro-government militias were using DJI Phantom quadcopters for battlefield reconnaissance as early as 2013.[149] Iranian-backed Syrian militias, such as Saraya al-Khorasani, have also used simple hobbyist drones to provide reconnaissance on Islamic State positions.[150] The Kurdish Peshmerga have used small, fixed-wing drones to assist their operations and even worked with a US entrepreneur to acquire an LA-300 drone that boosts their capacity to see the battlefield.[151] They have supplemented these with drone imagery provided by the US military, although they complain that it often arrives too late to be useful.[152] In May 2015, the Kurdish Peshmerga shot down an Islamic State drone that was monitoring their position.[153]

All sides in the Syrian conflict have turned to drones—including commercial and hobbyist drones—as a way of advertising their successes on the battlefield and drawing support to their cause. The war in Syria is one of the first in which multiple combatants are directly filming and producing propaganda videos for upload to YouTube and other file-sharing sites. These videos are often striking in their visceral detail of destroyed buildings, combat operations, and even deaths. They are producing a new kind of spectator war, in which battlefield imagery is broken down into vignettes of video-game-like destruction for propaganda purposes. For this reason, drone imagery is now often slickly edited and produced to maximize its propaganda value. Russia Works, an organization closely linked with Russian state television, has used DJI Phantoms to produce videos of Russian soldiers and tanks in the middle of the fight in Syria to sell the war to the Russian public.[154] Some Russian-made videos emphasize the careful, humane use of force that Russian forces allegedly use while others show drones sweeping over the destruction of cities like Palmyra to highlight the barbaric behavior of the Islamic State.

Non-state actors have been equally adept at making propaganda videos with drones. In 2014, al-Qaeda affiliated Jabhat al-Nusra released a well-crafted propaganda video called "Breaking the Siege," which showed the rough outlines of their operations to rescue prisoners held by the Islamic State.[155] By 2015, al-Nusra had produced slick top-down footage of its operations in Aleppo in that year, and even produced a video showing the execution of a vehicle-borne suicide attack.[156] Both appeared to be taken with a standard hobbyist quadcopter and nothing more sophisticated than a GoPro camera. By 2016, they were producing widescreen shots of battles which allowed the viewer to track the movements of vehicles and the impact of cruise missiles.[157] The growing technological and artistic sophistication of their videos shows that al-Nusra was aware of the need to rise above the din of the other videos of the Syrian war available online and to match those made by others, including the Islamic State. The Islamic State began releasing propaganda videos with "The Clanging of the Swords, Part 4," a May 2014 video that played like a Hollywood trailer and showed an Islamic State drone over Fallujah in Iraq.[158] It gradually developed a sophisticated infrastructure to build drones and to show off their battlefield exploits. The Islamic State has released videos of its fighters controlling drones and directing ground forces and even suicide bombers, while carefully editing the videos for propaganda purposes.[159]

While drones were initially used more for propaganda than tactical advantage, this has now changed. Today, there are numerous reports of the Islamic State using small, commercial, off-the-shelf drones for battlefield reconnaissance purposes and support of artillery fire.[160] Of all of the rebel groups fighting in Syria, the Islamic State has deployed the greatest variety of rotary and fixed-wing drones and has produced a large number of videos interspersing drone footage with other battlefield imagery. Banned from receiving most military-style drones, the Islamic State has concentrated on acquiring Chinese-made commercial drones (such as DJI Phantom FC40) and also X-UAV and Skywalker X8FPV fixed-wing models[161] and have also worked with external suppliers in Europe to get their hands on commercial drones.[162] The United States has been destroying Islamic State drones more frequently as they approached the location of US troops and allies, often by jamming their signal or just shooting them down. The United States has also begun killing IS drone-makers and targeting trucks with IS drones inside.[163]

The Islamic State has also been experimenting with attaching improvised explosives to drones. Since they acquired surveillance drones in mid-2014, experts have worried that it was just a matter of time before they could "jury-rig surveillance drones into flying IEDs."[164] By December 2015, these fears were realized: the Islamic State packed a drone full of explosives for an attack, but it was shot down by Kurdish forces before it could be successful.[165] These bomb-laden drones were little more than cheap quadcopters, but they nevertheless represent a growing threat.[166] In August 2016, the US Army identified small quadcopters packed with explosives as "the greatest challenge for Army forces" when it comes to air defense.[167] In October 2016, these fears were vindicated. Two Kurdish soldiers were killed and two French special forces operators were wounded when a booby-trapped drone operated by ISIS exploded near Mosul in Iraq.[168] By early 2017, it was reported that the Islamic State had killed about a dozen soldiers and wounded more than fifty in approximately eighty missions in which bombs were dropped from quadcopters.[169]

Explosive-laden kamikaze drones have also been used by Houthi rebels in Yemen to even the playing field with Saudi Arabia. In Yemen's civil war, Saudi Arabia and the United Arab Emirates have been fighting alongside Yemen's military to restore its government after an overthrow attempt by Houthi rebels. The Houthi rebels, clearly outmatched by Saudi-backed forces and attacked from the air, resorted to Iranian-made drones to destroy missile defense systems provided by Saudi Arabia. [170] Over time, they expanded their ambitions with drones and brought the war home to their enemies. In July 2018, Houthi rebels claimed to have attacked an oil refinery in Saudi Arabia, and they continue to penetrate Saudi airspace to threaten its government.[171] They also claimed an attack on Abu Dhabi airport in the United Arab Emirates.[172] By January 2019, they had launched an Iranian-made Ababil drone equipped with a bomb to kill Yemeni military officials, including the head of the military intelligence division.[173]

The next step in the evolution of rebel drone attacks will involve swarming fixed locations with small, disposable drones packed with explosives. In January 2018, Russian forces in Syria reported that two of their bases there fell under simultaneous attack by DIY drones packed with explosives.[174] Thirteen drones, all fixed wing, made of plywood and with a crude engine, swarmed Russian air bases and had to be disabled or shot down. Russia has blamed Syrian rebel forces for the attack, but also pointed a finger at Turkey for backing the rebels responsible for the attack.[175] These swarming attacks will continue in the future. In February 2018, Houthi forces claimed to have been experimenting with drone swarm attacks to disable air defense systems, although it is unclear whether these attacks have been successful.[176] Notably, in both cases, it was the external sponsorship of a government—Turkey and Iran—which allowed the rebel groups to target their enemies in such a creative way.

Conclusion

We are now living in a world where terrorists at home and rebel groups fighting on distant battlefields have as much access to small drones as consumers in the developed world. There is no natural limit to their ingenuity: with these tools in their hands, they will find creative ways to learn about their enemies and target their weaknesses. With drones diffusing rapidly across the world, the greatest advantages will flow to those non-state actors with the organizational infrastructure and budget to acquire and deploy drones effectively. Just as is the case with states, there is a limited first-mover advantage: those actors, like Hezbollah, who master drone use earliest will be more capable of using them to strategic ends than those who come to them later. Similarly, those rebel organizations with a generous external sponsor (like Russia, Turkey, or Iran) will have better access to sophisticated drones, and thus better results, than those who can only purchase them from commercial outlets.

At this point, drones are not a total game-changer for most non-state actors. They will have the greatest impact in leveling the playing field in conflicts where the asymmetry of power is relatively slight between the major players. In these cases, the use of even small commercial or hobbyist drones for reconnaissance will cut into the advantages of more powerful fighters. This dynamic could be seen in Nigeria, where the Islamist group Boko Haram acquired drones that matched or even bettered those held by the government, which allowed them to conduct more sophisticated attacks against the Nigerian military and civilian targets and expand their geographic reach across the countryside.[177] They may also affect the duration of the conflict by allowing weaker rebel groups to protect their forces, harass their enemies, and survive longer even when they are obviously overmatched. Retrofitting commercial drones with explosives may provide a short-term tactical advantage to rebel groups in evenly matched wars, but it is unlikely to turn the tide in one direction. In other cases where that power asymmetry is stark—for example, Hamas attacks on Israel—drones will be of less importance. In these cases, drones may enhance the vulnerability of a powerful government by directly threatening its civilians with sporadic attacks and have an array of psychological and symbolic consequences. Against a well-armed, well-droned opponent like Israel, drones in the hands of a group like Hezbollah will not tilt the strategic balance, but may allow them to compete more effectively and contemplate doing things that they would not have done otherwise.

This suggests that the ability of non-state actors to use drones is highly context-dependent. Much of the discussion of the threat emanating from terrorist actors possessing drones conflates their use on open battlefields with their use against well-policed targets in the developed world. These are environments with different opportunities and constraints: a "rebel drone" is not necessarily a "terrorist drone." That the Islamic State can conduct battlefield reconnaissance in Syria does not imply that it would be equally capable of doing so in Paris or London. That Hezbollah can launch drone strikes in Syria does not imply that it could do so against Western targets or even well-defended areas of Israel. With current drone technology, the ability of most terrorist organizations to use drones in non-permissive environments is likely to be limited. Even organizations like the Islamic State which fight on a battlefield and conduct attacks in Europe and the United States will find barriers to transferring their expertise and technology between these theaters of activity. While it would be a mistake to rule a terrorist drone attack out, or to dismiss the threat entirely, it is equally a mistake to give in to some of the hype surrounding this threat and assume that a terrorist drone attack on a Western city is imminent. Instead, we should expect that swarming attacks with cheap drones will become regular events in asymmetrical conflicts around the world but that terrorist drone attacks will be rare events in the developed world.

In the long run, perhaps the most important factor influencing the likelihood of a successful terrorist drone attack is the development of counter-drone technology. As more terrorist organizations are developing drones, the US government and private industry have turned to new technologies to block, track, and disrupt future attacks. In July 2016, the US Department of Defense requested an additional $20 million from Congress to address the new drone threat from the Islamic State.[178] The US Treasury is now sanctioning foreign companies involved in transferring drone materials to terrorist organizations like Hezbollah.[179] The Department of Homeland Security (DHS) is actively investigating counter-drone technology and even reportedly cooperated with the Department of Defense and the New York Police Department on a secret test of microwave-based counter-drone technology on New Year's Eve in 2015.[180] The FAA's efforts to create a master drone registry, with traceable serial numbers, will go some way toward undermining one of the chief advantages of drones for terrorist groups: the lack of attribution to a person responsible for the conduct of that drone. If drones have serial numbers equivalent to Vehicle Identification Numbers (VINs) and are registered with the government, it will be harder to sell them on the black market or to buy them in an untraceable way for an attack.

For both the domestic and foreign market, a number of private companies are producing counter-drone technology designed to detect drones and, in some cases, knock them out of the sky. Some, like DroneShield, rely on sensors that can detect the audio signature of drones and warn police of a drone heading to an unauthorized target.[181] Others, like Dedrone, are working with radio frequency (RF) sensors to detect drones in the sky, provide video and documentary evidence of the incursion, and enable counter-measures. Others are designed to knock drones out of the sky through any means necessary. The Departments of Defense and Homeland Security have also purchased one hundred DroneDefenders, "non-kinetic" rifles that disable and knock drones out of the sky.[182] A British company, Selex, developed a technology called Falcon Shield which takes control of a rogue drone and lands it safely.[183] The US Army is also working with companies who propose to jam the signal of the drone or disable it with lasers.[184] The most widely used drone detection technique is still jamming, but doing so still presents obstacles for interdicting drones and bringing them down safely, especially in urban environments.[185] Some drone producers are even experimenting with a form of in-built deterrence. In 2015, DJI, the China-based company responsible for the most popular hobbyist drones on the market, installed a firmware update to its Phantom 2 and 3 models that prevents users from flying drones near sensitive sites such as airports, nuclear facilities, military installations, and across national borders.[186] This process—called geo-fencing—can be circumvented or hacked by those with advanced programming skills, but the average user will find it more difficult to fly drones in places forbidden by the authorities.

The race to find effective counter-drone technology is an important one, but it will not be easily won. An analysis by the Center for the Study of the Drone found 230 different counter-drone products available across a wide range of manufacturers and countries. But none of these are foolproof.[187] Not every drone works off the same signal or can be detected by a single counter-drone system; as drones proliferate and new varieties from more manufacturers emerge, more drones will be able to evade detection and get closer to their targets. Most electro-optical (EO) and RF counter-drone systems must have a direct line of sight to the drone to jam its signal. Even so, there is always a risk of false positives if counter-drone systems mistake something benign, such as a police drone, for a threat and knock it from the sky. Even if a drone can be detected, interdiction poses a particular problem because it is hard to bring a drone down without risking harm to people beneath it. Even if that were not the case, many police forces and other law enforcement organizations lack the legal authority to bring down threatening drones in US airspace.[188]

This race between terrorists acquiring and deploying drones and governments finding sufficient defenses will be crucial, and perhaps dispositive, for the number and frequency of terrorist drone attacks in the future. If counter-drone technology develops rapidly into a highly effective shield against drone attacks, terrorists will gradually become deterred from seeking drones for attack and may turn to other means. If this counter-drone technology lags, or is riddled with holes, the incentive for a terrorist organization to use drones for an attack will increase. There is a risk of a cascade effect: if one successful terrorist drone attack occurs and illustrates the weaknesses in anti-drone technology, others may be tempted to follow suit and strike before the technology can adapt. These dynamics mean that governments will remain under pressure to prevent attacks and, if possible, to anticipate when these attacks are likely. To do this, they will need to turn to surveillance technology of a scope and level of sophistication never before available. And they will likely turn to drones to make sure that their societies are properly surveilled and safe, even if the use of those drones poses a risk to their democratic character.

6

The All-Seeing Drone

IN JANUARY 2016, A small, white Cessna aircraft began to hover over the city of Baltimore.[1] Although the plane was indistinguishable from hundreds of ordinary private aircraft that are seen high above many US cities, this one had a very different purpose. While it was not acknowledged at the time, this Cessna flight was part of an effort toward placing a major US city under the kind of persistent, wide-area surveillance that had been pioneered by the US military in conflict zones around the world. Like many drones, this manned aircraft was equipped with sophisticated, real-time video and communications capabilities that would allow it to capture grainy, black and white images and convey them to analysts on the ground. These analysts, tucked behind a bank of computer monitors in an unremarkable office building in downtown Baltimore, could produce new evidence, and sometimes photos, of crimes reported to police. And the cameras were capturing a lot: when the Cessna was aloft for five to six hours a day, its wide-angle lenses could capture 32 square miles of ground per second.[2] At a typical height of 8,000 feet, it could capture a significant portion of the city each day provided it went on one or more flights. The analysts could see cars moving and people walking around the city in real time, but the images were not yet detailed enough to see features of people's faces or their clothing. For the Baltimore Police Department, which had sponsored these flights as test runs, this was enough. The images coming from the Cessna allowed them to track the movement of cars from hit-and-run incidents or to find which buildings people fled into after crimes had taken place. All of these were clues that allowed the police to hunt down suspects, even days later, who might otherwise have escaped capture. Wide-area surveillance seemed to offer a vision of the future for law enforcement and gave the Baltimore police department some hope of stemming the tide of violent crime that had been afflicting the city.

Like most developments associated with drones, the origins of these surveillance flights lie with the US military's investment in research and development of unmanned aerial technology. After the invasion of Iraq in 2003, the US military faced a vicious, multi-faceted insurgency that threatened to destabilize the new government. Perhaps the greatest threat for US troops was IEDs: the Iraqi insurgents proved resourceful in developing crude explosives to target US convoys and personnel to devastating effect. Roadside bombs were often buried along major routes taken by convoys or concealed under piles of trash and the carcasses of dead animals. Hiding in the civilian population, insurgents detonated the bombs by remote trigger as US forces moved through towns and villages. As the insurgency wore on, the IEDs became more sophisticated and deadly, in part because Iran was supplying technology to aid the bomb makers.[3] As the death toll grew, the United States began to uncover bomb factories across Iraq where insurgents tried to make these explosives on a near-industrial scale. After its success in Iraq, the tactic spread to other theaters, including Afghanistan, where even more US troops found themselves thwarted by crude improvised explosives. One Pentagon estimate in 2013 suggested that between half and two-thirds of all US casualties in Iraq and Afghanistan were due to IEDs.[4]

Facing such a serious threat, the Pentagon began to invest billions of dollars into counter-IED technologies that either prevented attacks or shielded US troops from their consequences. Among the many contractors that the Pentagon worked with was a private company called Persistent Surveillance Systems (PSS), which was owned and operated by Ross McNutt, an MIT-trained engineer who had worked for the Air Force. He developed a wide-area surveillance system that would allow the Pentagon to record real-time video over an entire city. One crucial aspect of the system that he developed, initially called Angelfire, was that it could rewind through hours of video and identify who placed an IED once it had been located. This meant that US intelligence analysts could rewind to find the insurgent who placed it, no matter how long ago it was placed. The movement of the bomb placer could then be traced over time back to safe houses or the IED factories that had sprung up all over the country. Analysts could then figure out who this person spoke to and what places he visited in order to draw a picture of the social network lying beneath the insurgent cell. This capability was ideally designed for rolling up the networks of insurgents who had made the emplacement of IEDs so skillful and deadly. The program, pitched by McNutt as "Google Earth with TiVo capability," was purchased by the Pentagon and transferred to Los Alamos National Laboratory in New Mexico, where it was combined with other prototypes and put into service.[5] Like many military surveillance aircraft, it was also equipped with night vision and EO sensors, as well as signals intelligence and geolocation sensors, to provide a full-spectrum stream of data on the insurgents placing bombs in the way of US troops.[6]

The surveillance program for IED detection reduced combat deaths for US personnel, but it did not last long.[7] Once US personnel began to leave Iraq at the end of formal combat operations in 2011, the program fell into disuse. Like many former military contractors, McNutt began to explore uses for surveillance inside the United States itself. He ultimately settled on offering Baltimore a pilot program, in part due to the severity of the violent crime problem that the city faced and in part because he had personal contacts there.[8] Baltimore's rising crime rate made it an attractive potential test site, and PSS was convinced that they could lower crime rates by 20%–30% if their system was implemented on a full-time basis.[9] Private philanthropists donated the funds to the Baltimore Community Foundation, a non-profit involved in civic work in the city, which in turn activated the contract with PSS and began feeding the data to the police.[10] Their contribution was important because it was an expensive operation, costing between $1,500 and $2,000 per hour of flight.[11]

The secret to making the pilot program work in Baltimore was to keep the program as quiet as possible. The contract with PSS was never publicly announced or discussed at government hearings. Baltimore's mayor, Stephanie Rawlings-Blake, admitted that she never heard of the program until the story was broken by *Bloomberg* news in August 2016.[12] The Baltimore police department retitled the program, in anodyne terms, the "Community Support Program" and its offices were located in a nondescript building in the center of the city. In some cases, police officers who received tips or even images from the Cessna did not know where the information was coming from. By some measures, the surveillance worked: police officers were able to track individuals and vehicles, provide briefings with leads to investigators, and increase the percentage of crimes "cleared" by the Baltimore police department. The independent Police Foundation reviewed the program and concluded that it was effective, but conceded that it was not well explained to the population at large.[13] The Baltimore Community Support Program even argued that the cameras were capable of deterring crime, although measuring whether crime has not happened for a particular reason like the presence of a camera is difficult.[14] The Police Foundation also concluded that its success could be replicated elsewhere.[15]

Despite its apparent effectiveness, the program remained highly controversial. The presence of a Cessna aircraft collecting video on a near-constant basis over a city violated what many saw was a reasonable expectation of privacy. The PSS program captured video of thousands of people who were not charged with a crime and who expected (incorrectly) that many of their actions at that point were entirely private. In particular, it violated an expectation of privacy in non-public places. While many people have accepted that they are now being filmed by closed-circuit TV cameras (CCTV) in shops and public places, they do not accept being filmed in their backyards or other private venues. As critics have noted, people engage in activities in presumably private venues that differ significantly from their behavior in public, thus rendering surveillance of private and semi-private locales more revealing and perhaps invasive.[16] Persistent surveillance also reveals people going to specific locales—such as mosques, doctor's offices, AA meetings, gun shows, abortion clinics, and so on—which can reveal their religious and political preferences. Some of this information could be personally damaging if turned over to an employer or an insurance company.[17] The program was supposed to hold video footage for only forty-five days, but in practice the Baltimore police broke this promise and kept the video for much longer, with no notice.[18]

Another potential danger is that persistent surveillance could reinforce the biases already existing among law enforcement organizations, rather than correct them. When PSS was revealed, it immediately raised questions about whether it had been used disproportionately over African American neighborhoods. Defenders of the system argued that the cameras did not target any one neighborhood in particular and that its coverage was balanced between the white and black areas of the city, as well as wealthy and poor parts.[19] McNutt argued that the cameras could not yield facial or skin details to make such racial profiling relevant; however, the PSS data is used in tandem with other CCTV cameras and could, at least in theory, be combined for racial profiling.[20] The danger that surveillance technology could be used disproportionately over "suspect" areas makes it particularly alarming from the perspective of fairness and equality.

The PSS has also raised new questions about freedom of protest and government accountability. With no public notice and operating entirely in secret, the PSS took as many as one million photos of daily life in the city of Baltimore.[21] Most of those photos were of routine life, but the American Civil Liberties Union (ACLU) noted that McNutt and his employees were alert to the movements of protesters, such as those coming from Black Lives Matters groups, and monitored the city for signs of protest after some controversial verdicts in criminal trials.[22] Watching for signs of protests goes far beyond delivering videos of existing crimes and poses a danger to the right of peaceful assembly guaranteed under the First Amendment. Much of this surveillance was also done without any evidence of democratic consent. There were no city council meetings to approve what was essentially a privately funded venture.[23] The Baltimore Police Department tried somewhat evasively to defend the PSS program as being part of CitiWatch, a network of stationary CCTV cameras already in use. But the scope and mobility of the PSS camera—in effect, the fact that it can see anything anywhere—suggests instead that this technology was different, and perhaps more dangerous, than a fixed camera system.[24] Entirely in secret, and without a single public hearing, the Baltimore Police Department had put the entire city under watch but left no one officially accountable for misuse of the imagery flowing from their cameras.

All of this was done using a Cessna aircraft operated by a pilot. In many respects, the surveillance was therefore limited: the plane could remain in the air for only six hours at a time, and the practicalities of human endurance meant that night flights and flights in bad weather were often ruled out. But when surveillance capabilities are combined with the endurance and adaptability of unmanned aircraft, many of these natural limits fall away. In the future, it is likely that cities will attempt this kind of "eye in the sky" level of surveillance with drones, which are cheaper and can operate for longer periods of time than their manned counterparts. There is already evidence that drone manufacturers like DJI and AeroVironment are specifically targeting police forces that cannot afford their own small planes and helicopters and offering them specialized law enforcement drones for surveillance and other tasks.[25] While much of the debate over the rise of drones in law enforcement has revolved around issues of privacy, there are two other equally important issues at stake. The first is goal displacement. Does the availability of surveillance cameras and videos combined with drones make users of all kinds—law enforcement, dissidents, journalists, and even governments—"throw the dice" and engage in activities than they otherwise might otherwise have been reluctant to undertake? The second issue concerns the risks of employing this technology for political life more generally. If governments are equipped with surveillance drones, will this state of affairs erode the right to privacy? And will it lead to greater repression of dissent, and a darker future ahead, for those living under authoritarian governments?

Surveillance in the United States

In US skies today, drones are used for a growing variety of purposes from surveying real estate to tracking wildfires and delivering pizzas. The long-heralded commercial boom of drones—in which Amazon will send drones for "last mile" package delivery and most commercial goods shipments from DHL, FedEx, and others are operated by drones—is not yet here but will probably get underway within the next decade. The current pilot programs for drone delivery are generally restricted over cities or other high population areas due to FAA regulations and public nervousness about what would happen if they came down unexpectedly.[26] Still, at present, the US government estimates that 110,000 commercial drones are in the skies and that up to 450,000 may be operating by 2022.[27] Although progress has generally been slower than expected so far, the FAA is under intense pressure from lawmakers to open US airspace to commercial drones and to enable retailers like Amazon to realize the cost savings that "last mile" drone deliveries will yield. In 2018, the FAA authorized a new series of pilot tests for drone deliveries and proposed new, as yet unapproved, regulations governing how and where commercial drones could move through US airspace.[28]

Yet while the skies will soon hum with more and more drones, that does not necessarily mean that surveillance of the US population will also increase in equal measure. Most domestic or civilian drone use has little, if anything, to do with surveillance, and rather remains commercial or scientific in nature. One estimate of non-violent drone usage from 2009–2015 found that surveillance constituted only 9% of the global usage of drones, far less than the commercial or scientific deployment of the technology.[29] Even a police or other law enforcement drone will not necessarily always be conducting surveillance, or even in most cases. The drone that is hovering over someone's backyard is far more likely to be a hobbyist drone or a drone mapping local real estate than a surveillance drone.

What has changed is the number of actors capable of launching a drone with surveillance capability. The spread of drone technology has effectively democratized the US airspace and opened up opportunities to fly independently to those who would never otherwise have it. Until recently, the ability to fly one's own aircraft was a privilege given only to those who could afford it. Only the rich with private planes and powerful companies could field their own aircraft; almost everyone else who wished to fly was forced to take commercial aircraft. This meant that only a select few—predominantly government agencies, but also some private entities like universities and private companies—could get "eyes in the sky" and spy on activities below. As a result, aerial surveillance was largely limited to the government, law enforcement, and select businesses such as high-end real estate. That is not the case today. Now virtually anyone can get "eyes in the sky" and conduct some form of crude aerial surveillance. Everyone from real estate companies to paparazzi to private investigators can now get a bird's-eye view of their target and use drones to take video and film from the air. So, while the drone hovering over the backyard is probably not a surveillance drone, it will be harder to tell who it belongs to if it actually is.

At the same time, the use of drones for surveillance in domestic airspace is increasing. Inside and outside the United States, law enforcement organizations are now seeking drones for a variety of uses, one of which is getting an aerial overview of their territory. In April 2017, a study by the Center for the Study of the Drone at Bard College estimated that 347 state and local police, fire, and emergency services have purchased drones.[30] Most drones are held by sheriff's offices, police agencies, and fire departments, though other first responder organizations have also bought drones.[31] At this point, the kind of sweeping surveillance conducted by military grade technology, as seen in Baltimore, is relatively rare. The vast majority of drones purchased by law enforcement organizations are relatively small quadcopters, which can remain in the air for only a short period of time and have a limited capability for capturing images and video. Unarmed, camera-equipped drone models such as the Skyseer, the Skyranger, T-Hawk, WASP III, Shadowhawk, and a number of DJI models have all been used by domestic law enforcement agencies.[32] Law enforcement agencies in more than a dozen US states have sought approval from the FAA to fly drones. A number of large cities, including New York and Los Angeles, have also purchased drones for their police forces. As one New York Police Department spokesperson remarked, drones "aren't that exotic anymore."[33] Private companies are also eager to capitalize on the demand for drones among law enforcement. For example, DJI has partnered with a company called Axon to sell drones directly to police departments and other public safety agencies.

Many of the missions envisaged for police drones have little to do with constant surveillance but rather involve their use for time-limited, specific tasks. For example, some small police forces have purchased drones to search rural areas for endangered or missing children, to survey wildfires and other natural disasters, to hover over active crime scenes (for example, hostage situations), and to provide intelligence to officers on the ground. Drones can reach remote crime scenes relatively quickly and alert emergency services perhaps before the police even arrive; in other cases, they might also be able to look into windows of buildings to find active shooters and other threats.[34] Drones can also be used to watch large gatherings of people—for example, music festivals and events like the Superbowl—in order to guard against terrorist attacks.[35] With some adaptation, they can also be deployed as mobile phone base stations and intercept calls before they go through.[36]

Some law enforcement uses go beyond the limits of a simple camera-equipped drone in the sky. For example, Customs and Border Protection (CBP), an agency which operates under the auspices of the DHS, has been deploying Predator drones and their maritime equivalent, Guardian drones, for border surveillance since 2005. The CBP drones are also sometimes flown for the benefit of other government agencies including the Drug Enforcement Agency (DEA).[37] Operating out of Texas, Arizona, and North Dakota, the CBP's fleet of ten drones has focused on routine patrols over the southern border, especially for surveillance and interdiction of drug shipments, but it has also been used for emergency and disaster response. In 2008, the CBP began to use drones to monitor the northern border with Canada.[38] By 2010, the United States had expanded drone patrols to watch the entire southern border with Mexico.[39] The CBP's Predator and Reaper drones are capable of filming video and detecting signals on the ground, and at least one Reaper was equipped with wide-area surveillance video capability that could capture an area approximately 3.7 miles wide.[40] Although the CBP planned to expand their drone fleet, they ultimately ran into criticism over the effectiveness and cost of the program. The CBP's drones were involved in only 0.5% of the total apprehensions and drug arrests conducted by the agency between 2013 and 2016, at a cost of more than $60 million a year.[41] The drones were also expensive per flight hour. In 2013, an internal investigation found that they actually cost $12,255 per hour—not $2,468 as had been estimated by DHS—and even this excluded the costs of pilots, equipment, and overhead.[42] For this reason, the inspector general of DHS concluded that expansion of the medium-altitude drone program was unwise. There have been also concerns over data protection and privacy violations from CBP's drone surveillance, especially as the drones can fly as far as 60 miles inland from the southern border and 100 miles from the northern border.[43]

Under pressure to cut costs, the CBP has also experimented with using smaller drones, such as Shadowhawk drones, to monitor the southern border. But this has turned policing the border into a game of drone versus drone, especially against the Mexican cartels. Since 2010, Mexican cartels have used commercial drones, like the DJI Phantoms, to bring small packages of drugs across the border. In 2012 alone, US law enforcement interdicted 150 different drones carrying cocaine, heroin, and marijuana.[44] Drones are also used as spotters to identify safe routes for traffickers to cross the border. The advantages of commercial drones for this practice are clear. Unlike drug runners, drones can tell no tales, and there is little risk to the organization if one is shot down. Drones are particularly cost-efficient now that Mexican cartels have invested in indigenous drone production. Building their own drones also saves the cartels the risks involved in buying commercial models made by other countries, which might have in-built GPS trackers and other revealing technology. The Mexican cartels have also been upping their game in terms of what drones are capable of doing. In 2017, a drug runner was caught with a commercial drone packed with a homemade bomb, the first time that law enforcement saw the use of weaponized drones along the border.[45] The United States has been deploying smaller drones as a surveillance substitute for the medium-altitude drones like the Predator to stop these threats along the border, but have been confounded by GPS disruption and spoofing by the more sophisticated Mexican cartels.[46]

Although this shows how drone use can expand over time, it is worth noting that this sweeping surveillance effort along the United States-Mexico border is beyond the reach of most normal police departments. DHS has given grants to local law enforcement to jumpstart their drone programs, but many grants are not enough to buy some of the sophisticated technology on offer. The result is stark variation, with some well-funded police departments purchasing military-grade drones and others getting by with much less. For example, the Arlington, Texas, police department purchased the Leptron Avenger, a large helicopter drone that can fly and take pictures at up to 12,000 feet.[47] Similarly, the Miami Dade police department has been using Honeywell T-Hawks, which can fly at up to 10,000 feet, record images in daytime or with infrared cameras, and zoom to follow suspects.[48] Small and rural police departments have been buying commercial drones equipped with thermal scanners and biometric tools not widely available to the public.[49] Some of these drones are also equipped with night vision cameras.[50] These purchases are driven in part by the need to cover large amounts of ground on a small budget.[51] For example, in 2011 the police department in relatively small and rural Mesa County, Colorado, was among the first to win FAA approval to fly Dragonfly drones all over their jurisdiction.

For police departments with the fewest resources, the best available option has been to buy off-the-shelf commercial models like the DJI Phantom and DJI Matrice drones. In 2017 the Center for the Study of the Drone found that most of the drones purchased by law enforcement and other public safety organizations fall into this category.[52] Even so, the spread of drones across law enforcement organizations has been wildly uneven. Some major cities have been foregoing drones because of tight budgets, while smaller police departments are investing in them when they have a surplus budget or just a high level of interest among their officers. The result is a patchwork of drone coverage: some counties will have law enforcement drones while their neighboring counties, equivalent in all other ways, do not. It is then hard to know whether a specific police organization will have drone assets that they can call in. Some law enforcement organizations who lack their own drones may be able to borrow drones from private citizens or other government departments to respond to emergencies. Some areas will have no drone coverage at all. It comes down to resources, local initiative, and the degree to which high-ranking law enforcement authorities want a particular city or region—like Baltimore—under close watch. Over time, the United States might become even more unequal regarding aerial surveillance, with citizens in some areas living in full privacy but others under surveillance, either by their own choice or because of the choices of those more powerful than themselves.

Outside the United States, law enforcement organizations have been using drones for an increasing variety of tasks, including monitoring crowds at high-profile events, pursuing criminal inquiries, and even for routine work like stopping speeding motorists. Some European governments have begun to experiment with using drones for more invasive law enforcement tasks, which the US public would not support, such as for issuing speeding tickets.[53] For example, the London Metropolitan police plan to use drones to watch for dangerous driving, such as road racing, which may endanger other motorists.[54] Police in Kent, England, have begun working with BAE Systems, a British defense manufacturer, to conduct routine monitoring of anti-social motorists, agricultural thieves, and even those who dump trash in protected public areas.[55] Similarly, in Ireland, the government is contributing funds to purchase drones to catch those dumping waste illegally, especially in rural areas.[56] In France, drones are used to take aerial images of reckless drivers for future prosecution.[57] In general, European law enforcement officials have been more willing to experiment with using drones to encourage good social behavior, such as obeying the speed limit and not littering, than simply stopping bad behavior as their US counterparts have.

Law and Privacy

In the European Union and the United States, the deployment of drones for law enforcement tasks has triggered a renewed debate about the balance between security and privacy. The EU position on this balance is complex and multi-layered: in general, EU regulations impose more legal limits on the use of data by private companies and permit citizens to have damaging stories removed from the internet in the interest of their privacy.[58] The European Convention on Human Rights explicitly recognizes the right for "private and family life."[59] At the same time, many EU governments have responded to fears of terrorism by increasing the surveillance power of the state and reducing the privacy protections of their citizens. Drones are one element of this surveillance power, but in terms of raw numbers they have been eclipsed by the growth of CCTV and other more intrusive forms of surveillance, including the bulk collection of communications data. For example, in Britain alone, one 2011 study estimated that there was one CCTV camera for every thirty-two people in the country.[60] Similar efforts to ramp up CCTV camera coverage of public transportation hubs and other major tourist spots have been made in France and Germany. In these countries, the CCTV cameras are not always owned by the government, as many private businesses have extensive CCTV networks whose footage can be pulled and used by law enforcement in criminal investigations. In countries where CCTV cameras are more dispersed, such as Italy, other means of surveillance are common. For example, Italy features more bulk collection of communications than many other EU countries, according to a 2008 estimate by the BBC.[61] In recent years, most major EU governments have passed new legislation boosting their capacity to spy on their citizens and to share data with EU authorities and sometimes the United States.[62] The uneven spread of police drones across the European Union, while important, pales in comparison to the surveillance collection capabilities that these other tools yield for EU governments.

In the United States, the adoption of drones by law enforcement has been challenged at the federal and state level on the grounds of privacy. Although the US Constitution does not expressly offer a right to privacy, the Supreme Court has ruled that various amendments—specifically the Fourth and Fifth but also the Ninth—together yield a right to privacy for US citizens. This right is also enshrined in Anglo-Saxon common law and incorporated within a number of federal and state laws.[63] The minimum standard for privacy protects the security of one's person and property from search and seizure without a warrant. This protection generally extends beyond one's immediate bodily person to include one's home, vehicle, or place of business. The interpretation of the right to privacy held by the Supreme Court has traditionally been a physical one, deemed violated only when police directly apprehend people or conduct searches of their homes without a warrant, although some states may interpret this right more broadly.

In applying this standard, the Supreme Court has generally been willing to permit the government to conduct surveillance provided that it does not lead to a direct or illegal search or seizure of a person or property. For example, in *Olmstead v. the United States* (1928), the Supreme Court upheld the government's decision to tap the phones of Roy Olmstead, a notorious bootlegger who has been importing and selling alcohol in violation of the government's laws during Prohibition.[64] The Supreme Court found that wiretapping surveillance did not violate Fourth Amendment rights provided that they did not directly search or seize the person or effects of the subject. In a dissent, Judge Louis Brandeis argued that this narrow interpretation of a violation of the rights to privacy did not account for technological change and the capabilities and powers that wiretapping and other new technologies offered the government.[65]

In fact, Brandeis had recognized the core of this problem—that technology can radically challenge one's expectations of, and rights to, privacy—in an influential article published three decades earlier. In a scholarly article co-authored with Samuel D. Warren in 1890, Brandeis had offered a reinterpretation of the right to privacy and recast it as the right to be "left alone" from eavesdropping and other intrusions into private spaces.[66] Their interpretation was a reaction to the invention of the camera, which enabled a celebrity-obsessed press to photograph and record movements in or near one's home or place of business. Their goal was ultimately to locate a standard of privacy that would "protect the privacy of the individual from invasion either by the too enterprising press, the photographer, or the possessor of any other modern device for recording or reproducing scenes or sounds."[67] Applied to today's technology, the "right to be left alone" standard would protect individuals from drones taking photos and videos of them in private spaces, even if the drone never landed on their property or violated the airspace above. But as *Olmstead v. the United States* and other cases demonstrated, the government and courts were slow to adopt this broader standard and generally deferred to law enforcement in cases concerning surveillance.

In 1967, the Supreme Court finally overturned the Olmstead ruling and offered a new legal test for applying privacy standards in the case of *Katz v. the United States*. The Court found that the FBI violated the Fourth Amendment rights of alleged bookie Charles Katz by attaching an eavesdropping device to a public telephone that he was using.[68] The court found that wiretapping constituted a search and hence was a violation of Fourth Amendment rights if conducted without a warrant. In a famous concurrence, Judge John Marshall Harlan II offered the "reasonable expectation of privacy" test, later known as the "Katz test." This test requires the subject of surveillance to believe that their activities were private at that moment in time, even if it fact this was not completely true. What this means in practice remains debatable— ultimately, it is hard to know when someone has a reasonable (but still subjective) sense of privacy—but it suggests that privacy can be violated even when no physical search or seizure is conducted. Yet the Supreme Court has acknowledged limits to this standard as well. For example, activities which are clearly in the public view—for example, gardening in one's backyard—would fail the reasonable expectation of privacy test. In general, the Supreme Court has found that one does not have a reasonable expectation of privacy for behavior which may be seen accidentally by bystanders or others in the immediate vicinity.[69] It has also ruled that there is no "reasonable expectation of privacy" in public areas, such as train stations, airports, and open roads. In these environments, one may be openly subject to CCTV or other forms of aerial surveillance, such as drones, without notice or a warrant.

The implications of the Supreme Court's "reasonable expectation of privacy" standards are important for understanding what role drones may play in extending surveillance. In public areas, the absence of any privacy protections means that law enforcement can fly drones overhead and take photos and videos of individuals without consent. There are few, if any, restrictions on what can be filmed in public areas. In private areas, one has a reasonable expectation of privacy except when criminal activity falls into plain view. In a number of prominent cases, aerial surveillance did produce evidence of criminal activity inside a person's home. For example, in *California v. Ciraolo* (1986), the Supreme Court ruled in a five to four decision that police officers acting on an anonymous tip did not need a warrant to fly a plane over Dante Ciraolo's backyard to search for marijuana plants.[70] This case implies that the taking of pictures and videos of the public areas of private residences and places of business during aerial surveillance, conducted either by manned or unmanned aircraft, would not necessarily be prohibited, even if carried out without a warrant. Similarly, in *Florida v. Riley* (1989), the Supreme Court found that no warrant was needed to fly a helicopter only 400 feet over the property of a marijuana grower's property. But this again exposed the degree to which new technology is throwing up challenges to existing standards of privacy. In a famous dissent to the *Florida v. Riley* case, Justice Brennan wrote:

Imagine a helicopter capable of hovering just above an enclosed courtyard or patio without generating any noise, wind or dust at all—and, for good measure, without posing any threat of injury. Suppose the police employed this miraculous tool to discover not only what crops people were growing in their greenhouses, but also what books they were reading and who their dinner guests were. Suppose, finally, that the FAA regulations remained unchanged, so that police were undeniably "where they had a right to be." Would today's plurality continue to assent that [t]he right of the people to be secure in their persons, houses, papers and effects, against unreasonable searches and seizures was not infringed by such surveillance?[71]

As Matthew Feeney from the Cato Institute has argued, this description of the hypothetical device is eerily close to today's drones.[72] It is not unthinkable that police drones could be found hovering in one's backyard, or just outside a window, collecting incriminating information about a person. Even the collection of that information without a warrant would not stop police from getting a warrant for a full search of the property under existing US law.

A natural reaction to this state of affairs is to insist that police and private drone operators—for example, real estate agencies, private investigators, and so on—would be limited by property rights, especially for private homeowners. A homeowner might assume, for example, that they would have the right to knock a drone out of the sky if it was hovering just over their backyard. But would someone be able to shoot down or disrupt a drone over their own house because it has trespassed on their property? The original interpretation of property rights grounded in common law would presumably permit such an action based on the traditional principle of what is known as *cuius est solum, eius est usque ad coelum*, which translated from Latin means "to whomever the soil belongs, he owns also the sky."[73] It is this principle that asserts ownership of not only the skies above one's house but also whatever lies underneath the ground of the property itself. But this principle was tested in the case of *United States v. Causby* (1946) and was found, in the words of Justice Douglas, to have "no place in the modern world."[74] In this case, Thomas Lee Causby owned a chicken farm outside Greensboro, North Carolina, that happened to be near a military airport. Regular military flights passing 83 feet above the ground flew too close to his home and startled the chickens, causing them to panic and sometimes injure or kill themselves. The Supreme Court found that Causby did not have full ownership of the entire airspace above his property, as the traditional principle would hold. In other words, a homeowner does not have unlimited control of the airspace above their house. However, the Supreme Court held that the overflights over his property did constitute a violation of Causby's property rights provided that it did substantially interfere with his use of the land and his enjoyment of the property.[75] The result of this verdict was to affirm commercial and private flight over private land but also to insist that it must be conducted at a sufficient height not to materially affect the owner of the land itself.

The central question that emerges here is at what height must an aircraft, manned or unmanned, be flown to violate the *Causby* standard? Some subsequent interpretations of the *Causby* standard held that there must be some designated navigable airspace—perhaps 400 feet and above—where commercial and private aircraft and drones may freely fly, though the courts have not always upheld this strictly.[76] At a minimum, the FAA insists that 400 feet is the minimum safe altitude that aircraft and drones must maintain over congested areas and exercises its own regulatory power on aircraft in that range and above. Below that altitude, there are exceptions made for model and toy aircraft, which can include some drones depending on their size. The FAA permits small aircraft like toy planes and some small drones to fly freely up to 400 feet with a more relaxed set of regulations (such as not requiring pilot's licenses for an owner). From a surveillance vantage point, this suggests that law enforcement and government drones flying in navigable airspace above 400 feet are in the clear, legally, and even some small drones may be permitted overflights over private property at a lower altitude. All of this suggests that ordinary citizens' privacy protections against drone incursions are relatively weak.[77]

In response to this weakness and uncertainty, a number of state and local governments began to pass laws concerning drone usage and violations of privacy. As of 2018, forty-one states had passed laws regulating the use of drones in their territories, and three more had passed resolutions.[78] Most have been motivated by the sense that FAA regulations and legal protections for privacy do not go far enough given the advance of drone technology. Yet state and local officials lack the appropriate jurisdiction to regulate the airspace, clearly given to the FAA under federal law.[79] Although there are relevant privacy laws at the state and local level, these do not supervene the FAA's jurisdiction surrounding the use of drones in the airspace. Some states have also passed legislation permitting homeowners to shoot down drones over their properties. At a minimum, permitting drones to be shot down is dangerous because individuals may miss the drones and send ammunition off into the distance to harm others.[80] But the FAA has plainly stated that it is a violation of federal law for an ordinary person to shoot down a drone, much in the same way that it is illegal to interfere with a manned aircraft.[81] From a legal vantage point, a drone is classified as an aircraft just like a private plane is. In fact, the DHS, among other federal government agencies, is seeking authorization by the FAA to shoot drones out of the sky if a "credible threat" from a drone emerged.[82]

In the end, this suggests that the opportunities for government or even private surveillance using drones are substantial. Although it is commonly held that individuals have a right to privacy, the Supreme Court has generally sided with the government when it conducts surveillance of a target without a warrant using new technology such as wiretapping devices, GPS trackers, and now drones. The courts have also created a generous interpretation of navigable airspace that suggests surveillance above 400 feet is permissible even over private property. While there are greater protections for privacy below that altitude, it remains unclear what protections, if any, exist for those who wish to stop routine drone incursions. The laws permitting drone overflights are likely to be clarified and possibly expanded once commercial drones from Amazon and other companies take off. At a minimum, the FAA's default position—that drones should be treated as equivalent to aircraft—restricts the ability of subjects of surveillance to interfere with drones that are hovering over their backyard. This fact, coupled with the absence of restrictions on drones in public spaces, suggests that the drone technology itself will gradually weaken the presumption of privacy currently assumed by many people.

Drone Activism

In some cases, this may be a good thing. A world full of drones, and one in which privacy is harder to come by, may be a world in which light falls on injustice and corruption. A number of drone activists have argued that with drones in the hands of more citizens government officials of all kinds, including police officers, will be less likely to exercise their power capriciously. In the United States, a number of press organizations have begun to use drones to cover news stories, though they have run into regulatory and legal barriers imposed by the FAA over the commercial use of the technology.[83] Once "drone journalism" flourishes, it is not hard to imagine the technology being used to monitor government or corporate misdeeds, or police abuses. Just as cell phones have been used to record evidence of corruption and abuse, drones equipped with cameras and video equipment may also be put in the service of keeping government honest. For example, in 2017 protesters against the Dakota oil pipelines deployed drones to monitor the progress of the pipeline but also to expose evidence of police using tear gas and water cannons to disperse them.[84] Their activities led to a number of proposed state laws declaring the use of drones in this way to be an attack on critical infrastructure.[85] Journalists have also used drones to fly over and photograph immigrant detention camps along the southern border of the United States to defy the Trump administration's attempts to limit scrutiny of the camps by Congress and members of the press.[86]

Outside the United States, there is also opportunity for drones to be put in the hands of human rights activists to record government repression, violence, and corruption. Deployed correctly, drone activism can shed light on human rights abuses and possibly motivate the world to act. As Andrew Stobo Sniderman and Mark Hanis from the Genocide Intervention Network have argued with respect to atrocities committed by the government of Bashir al-Assad in Syria:

> Imagine if we could watch in high definition with a bird's eye view. A drone would let us count demonstrators, gun barrels and pools of blood. And the evidence could be broadcast for a global audience, including diplomats at the United Nations and prosecutors at the International Criminal Court . . . The better the evidence, the clearer the crimes, the higher likelihood that the world would become as outraged as it should be.[87]

While it remains unclear whether drone footage of atrocities would be sufficient to motivate states to act, it would at the minimum supply some video evidence that might be usable at the International Criminal Court (ICC). It is possible to imagine that video evidence from drones will be an important complement to other forms of evidence used to hold the perpetrators of human rights abuses to account.

One of the key questions regarding drone activism will be who owns the drones and the pictures and footage coming from them. Some drone activists are hopeful that the United States and other powerful governments with drone assets will be willing to use the pictures and video that they produce to either publicize abuses or provide evidence in court cases against human rights abusers. However, it is not always clear that the US military and others would do so. At a minimum, turning over drone footage would reveal their ability to watch and record events on the ground, which the United States generally tries to protect as secret. But also, admitting that a government has footage of human rights abuse may trigger some international legal obligations to investigate it and may, in some circumstances, yield legal responsibilities to act for the state recording the footage.[88] For example, if the United States detected evidence of genocide with a drone overflight, it would be positively obliged to act under international law to prevent the genocide from continuing. Given this possibility, it is likely that human rights activists will find that governments vary in their transparency and willingness to cooperate regarding the detection of human rights abuses with their drones.

For this reason, a number of human rights advocacy organizations have begun to experiment with using drones of their own to collect footage and shame governments into acting. For example, Human Rights Watch used a Sensefly drone in 2017 to fly over waste sites in Lebanon to uncover evidence of the burning of household waste and prove allegations that the government was failing to curb this unhealthy practice.[89] Yet human rights organizations using drones will face some of the same problems that governments do when protecting their intelligence sources and methods. Almost all forms of digital or video recording—whether from a cell phone video or drone—contain a trace signature that can lead governments back to the person who recorded it. In other words, human rights activism can shed light on abuses, but it can also yield information about the person publicizing the abuses and possibly put them at risk. As the human rights organization Witness has noted:

It is clear that new technologies, particularly the mobile phone, have made it simpler for human rights defenders and others to record and report violations, but harder for them to do so securely. The ease of copying, tagging and circulating images over a variety of platforms adds a layer of risk beyond an individual user's control. All content and communications, including visual media, leave personal digital traces that third parties can harvest, link and exploit. Hostile governments, in particular, can use photo and video data—particularly linked with social networking data—to identify, track and target activists within their own countries, facilitated by the growth of automatic face detection and recognition software.[90]

Human rights activist organizations are now faced with an additional set of challenges if they are to include drones in their arsenal to expose human rights abuses. They must develop protocols to "scrub" video footage of personal data and secure means of holding and transmitting video evidence collected from drones in order to protect the activists themselves.[91] They will also have to find some means of authenticating the video and ensuring that it shows what its authors say that it does. This feat is especially difficult in chaotic wartime environments, but particularly important if drones are to be a trusted mechanism for collecting usable evidence of human rights abuses and atrocities.

The End of Anonymous Dissent

From targeted killings to surveillance, drones enable all sorts of actors—good, bad, and indifferent—to pursue goals and take risks that they may have otherwise found too costly, difficult, or politically controversial. But drones are rarely neutral in terms of power. In the hands of human rights activists, they are a tool which can be deployed against the secrecy of the state to reveal evidence of abuse, government malfeasance, or negligence. In the hands of the government, they can be deployed against those same activists under the guise of law enforcement surveillance. In this case, drones can be folded into the coercive power of the state, amplifying its capacity to watch its citizens, regulate their behavior, and punish those responsible for challenging them.

In the United States, there is some evidence that drones are extending the surveillance capabilities of the government, though their diffusion has been uneven across the federal government. Aside from the expansion of the use of drones by CBP and DEA, other federal agencies have begun to experiment with camera-equipped drones. For example, the Bureau of Alcohol, Tobacco and Firearms (ATF) spent $600,000 on six drones in 2015, but found that they were largely inoperable for most major tasks.[92] The FBI invested in thirty-four drones at a cost of $3 million, but also found that some were ill-suited to their needs, and struggled to train and retain qualified pilots.[93] The chief problem with much of the expansion of drones across the federal government is that it has been conducted without proper oversight or any transparency, as many drone purchases and use were undertaken with no public notice.[94] This is partly because many members in Congress encourage drone sales—sometimes to drive business back to their districts—and allow federal agencies to spend on drones with little direct oversight or accountability.[95] This loosening of the rules for the sale and use of drones by federal agencies is also supported by pro-drone caucuses, such as the Congressional Caucus on Unmanned Systems (CCUS), and industry groups, such as the Association of Unmanned Vehicle Systems International (AUVSI). Industry groups in particular have used campaign contributions to encourage members of Congress to authorize spending on drones at the federal level. Only the fact that federal spending on drones is so decentralized and often wasteful has saved the US government from developing an even stronger capability to surveil its own population.

The military has also conducted drone overflights over US territory, though the Pentagon insisted that the drone flights were "rare and lawful."[96] There were fewer than twenty instances of the US military conducting drone missions in US airspace between 2006 and 2015.[97] In 2018, the US military flew eleven missions in US airspace, mostly for training and emergency response:[98] Reaper and Predator drones were used to monitor wildfires and flooding and also in search and rescue.[99] The US military is also exploring the use of drones for dropping parcels of food for hurricane and other disaster relief.[100] The Pentagon insists that there is a high bar for approving these missions, based largely on necessity and operational need, and that the Secretary of Defense or someone authorized by him or her must be responsible for approving any proposed mission. At present, the Pentagon does not permit armed drones for anything other than training purposes and does not permit surveillance of US citizens unless authorized by law and approved by the Secretary of Defense.[101] But the law and guidelines governing drone use remained somewhat vague and the Pentagon admitted that there were no regulations for how to respond if there was a request from a state or local official (for example, a governor) to "borrow" US military drones in a crisis.

This is a problem because on balance there has been more interest in surveillance at the state and local level. A number of police organizations have been publicly weighing policies to use drones to monitor crowds in protests and at major sporting and entertainment events like the Super Bowl or the Oscars. If this became standard practice, it would effectively make surveillance without a warrant a routine procedure at events with big crowds. But no effort to make this a standard practice has yet succeeded. In 2013, the Los Angeles Police Department was reported to be examining the use of drones to monitor the crowds at celebrity-filled events like the Oscars.[102] By 2018, however, they had yet to use drones for this purpose due to sustained opposition by civil liberties groups.[103] In May 2018, a bill in the Illinois state legislature would have allowed police to use drones to take photos and videos of protests, which in turn could be used to develop lists of those participating.[104] With intense opposition from the ACLU and other privacy advocates, it was ultimately defeated.

One reason why drone surveillance has been so controversial in the United States is that its political implications may be very different. Until recently, surveillance largely occurred with the knowledge or even complicity of those surveilled. The most famous example of extreme surveillance, the panopticon of philosopher Jeremy Bentham, presumed that all involved knew they were being watched and in fact changed their behavior accordingly.[105] Even weaker and more limited forms of surveillance in prisons, schools, and factories tended to assume that those being watched knew they were being watched; this fact, and the implied power behind it, had a disciplinary effect on people's behavior within those spaces.[106] The same dynamic applied in public spaces. For decades, individuals in airports and train stations have known that their actions would be captured on CCTV cameras and adjusted their public behavior accordingly.

At the same time, everyone also knew that there were spaces—perhaps at home, perhaps in marginal or neglected parts of each town or city—where no cameras would be present. It was in these unmonitored spaces, often in cafes, theaters, and schools, where political activism and dissent typically flourished. Even in democracies, dissent did not generally flourish in well-monitored spaces where the oppressive glare of surveillance could reveal who was mobilizing and why. Anonymity was an essential precondition to challenging the government. This is why anti-government protesters tended to adopt masks such as the famous Guy Fawkes mask in public protests, even in democracies. In autocracies like the Soviet Union, the preservation of these scarce private spaces—sometimes in university classrooms or cafés—was particularly crucial for providing the anonymity needed to challenge the regime. But today the combined effects of multiple new technologies—from the monitoring of internet traffic and metadata to facial recognition technology and now drones hovering over public spaces—is eliminating anonymity and crippling effective dissent. This explains in part why protesters in the United States so often cite George Orwell's "Big Brother" when law enforcement propose expanding drone surveillance over protests and other political events. There is a sense that freedom is possible only when some kind of anonymity from surveillance, however limited, is also possible and that many types of modern technology are effectively erasing the possibility of anonymity.

The spread of drone surveillance is dangerous from this vantage point for two reasons. First, drone surveillance may strengthen the hand of government or law enforcement officials against protesters, tilting the balance even more toward those who already have power. Second, drone surveillance can eliminate the anonymity needed to effectively challenge the state and reduce or eliminate those areas in the public and even private sphere when an individual could safely assume they were unmonitored. With drones that can fly and record nearly everywhere with little trace or notice, the area where anyone is truly free from being watched is diminishing by the day. Although the US Department of Justice insists that it will not monitor political protests, many state and local law enforcement officials have not made such a pledge and can use drones, along with other new technologies, to do so.

At present, the degree to which surveillance can identify political protesters and dissenters is partially limited by the technology itself. Despite vast improvements in their cameras, drones cannot see everything on the ground and have a limited perspective on what they do see. But facial recognition technology is now growing by leaps and bounds and will eventually be linked to the cameras with which drones are equipped. Pioneered by the military in places like Afghanistan and Iraq, facial recognition technology is now being deployed by a number of law enforcement organizations to identify criminals and build a database of faces that can be located in a crowd. One project, the Biometric Optical Surveillance System (BOSS), is designed to use algorithms to find faces in a crowd and match them against databases of known criminals.[107] But although BOSS was able to make identifications quickly, it struggled with accuracy and issued too many false results, especially with photos taken at an angle. Other programs have been tried by local police departments on an experimental basis, but the overall success rate remains low. The FBI is currently developing a Next- Generation Identification program that will collect fingerprints, iris scans, and facial recognition photographs and allow cross-checking against driver's licenses and other photos.[108] This will eventually be handed over to all 18,000 or so law enforcement organizations in the United States. As of 2016, an independent analysis of facial recognition programs found that photographs of 117 million adults in the United States are stored in facial recognition databases. Even though facial recognition software has a high rate of error, particularly against minority groups such as African Americans, relatively few legal and political barriers appear to be standing against its use.[109] Safeguards against data misuse are also weak, especially inside the government. In 2019, it was revealed that Immigration and Customs Enforcement (ICE) used facial recognition technology on millions of photos kept in state driver's licenses databases.[110] Over time, facial recognition software will be connected to feeds coming from drones and will make the possibility of identifying the faces of protesters in a crowd a reality.

While protesters in the United States and Europe are reacting to the possibility of Orwellian big brother surveillance of this kind, in other countries it is already occurring. In 2012, Russia began to use drones to monitor pro-democracy protesters.[111] Today, the Russia Guard (which is similar to the national guard) uses drone technology stockpiled by the Interior Ministry to watch political protests.[112] In China, drones are used to monitor border regions, such as Xinjiang, while the central government is using facial recognition technology to scan people going in to tourist sites and compare them to a centralized criminal database, ironically dubbed Skynet, as in the *Terminator* films.[113] Especially in regions where the persecuted Muslim minority Uighurs live, the Chinese police use a range of biometric data and other means—for example, GPS trackers in vehicles and government spyware loaded into cell phones—to track individuals who may be considered suspect. Citizens are even given a numeric score, which indicates whether they are considered safe or suspect and determines the rights and opportunities they are provided by the government.[114] Although drones are only one part of the vast surveillance apparatus that has been constructed by the Chinese state, they have played an important role in diminishing anonymity and limiting the possibilities for dissent in some of the most restive regions of the country. In 2017, only 1,000 drones were deployed for tasks such as tracking police suspects, monitoring traffic, and locating opium farms, but more will come in the future as the intense surveillance of Uighur population increases.[115]

Another potential danger of using drones to augment the surveillance and coercive capacity of the state is that they may make monitoring, and even repressing, suspect populations easier. This can be seen in the way modern technology has led to the building of internment camps in Xinjiang, but it can also be seen elsewhere. Pakistan has used drones for targeted killings in the FATA. Israel has been using drones for surveillance of Palestinian populations in the West Bank and Gaza for some time but has recently shifted to policing these areas from the air. For example, in March 2018, the Israeli Defense Force (IDF) used drones to deploy tear gas and disperse protesters in Gaza[116] and again in May 2018 when mass protests broke out over the decision by the United States to open a new embassy in Jerusalem.[117] Against these drones dispersing tear gas, Palestinians flew kites, some of which were equipped with crude explosives. It was the first clear example of what might be described as imperial policing by drones, where a superior government force uses drones to keep a restive population away from their main population centers. As the attacks continued, Israel responded with drones that could knock down the flaming kites and even drop tear gas along the border.[118] For the Israeli government, drones are helping to make a morally intolerable situation more politically acceptable by reducing casualties at home and keeping the conflict out of sight of most of the population.

This practice is spreading. As of 2019, both Pakistan and Nigeria have conducted drone strikes on secessionist regions, suggesting other governments will look to drones to manage problems that would otherwise demand a political solution. The risks of such an approach are growing as drone technology becomes cheaper and more widely available. Although Israel used its own drones for its missions, a number of private companies are now offering crowd control drones at cheap rates on the international market. One drone, the Skunk riot control copter, is equipped with tear gas grenades, paint and pepper balls, and a set of loudspeakers to communicate with the crowd.[119] It is being sold by Desert Wolf, a South African defense company, and has found an eager market among some police and private companies. Other similar, models, equipped with cameras and riot control agents like tear gas and smoke bombs, are now being developed or sold by Chinese, Israeli, and Austrian companies.

Conclusion

It is commonly assumed that a world of drone surveillance will resemble some kind of science fiction nightmare, with governments using drones to convict people of crimes they have not yet committed, as in the film *Minority Report*. Instead, the dangers of the spread of surveillance drones are perhaps less obvious but just as worrying. The use of surveillance drones is growing, even if unevenly, across a range of law enforcement and government actors whose appetites for knowing what their populations are doing is near insatiable. This is happening in an unchecked, non-transparent way with little government direction or oversight. From the adoption of the PSS by Baltimore to the spread of small surveillance drones to dozens of small state and local police forces, surveillance is increasingly becoming normalized across the United States and Europe. A range of actors—state and local law enforcement, but also private companies—are eager to deploy drones and see what they can see. Ordinary legal protections of privacy have proven ineffective against the spread of the surveillance state, both through drones and other technology like CCTV. Given their flexibility and ability to reach less covered spaces, there is a real danger that the diffusion of surveillance drones may reduce the sphere of privacy needed for effective political mobilization and dissent. If drones are ultimately put to watching and identifying political protesters using facial recognition technology, as currently proposed, they will have significantly amplified the coercive power of the state against citizens and also corroded the democratic character of those governments.

Outside the United States, drones are being added to the arsenals of authoritarian governments like Russia and China. Here the danger is greater: drones will eventually become an essential component of the growth of their modern surveillance state, recording protesters in public areas and keeping watch on previously private areas as well. In this way, drones will tilt the balance of power toward the government and make protest and dissent by citizens riskier than it is today. One consequence of the expansion of drone technology may be to make the strong even stronger, thus shifting the balance of power decisively toward those who wield the coercive instruments of power and against those who dare to challenge them. Further, if drones become a tool of imperial policing by states like Israel, there are additional dangers. As the technology develops, we can expect to see more drones equipped with tear gas, smoke bombs, and even small explosives that will be able to disperse crowds and punish those who challenge the existing political order. If drones become a tool of political repression and allow governments to police their domestic enemies with ruthless efficiency, governments with secessionist movements might be less willing to negotiate and grant concessions. The result might be that such conflicts are contained, but not resolved, while citizens in developed states grow increasingly indifferent to the suffering of those making secessionist or even national liberation claims, against them, even when these claims are just. If the gaze of surveillance drones increases the pool of information gathered about suspect populations but not the level of understanding or sympathy for them, it will be hard to see drones as a net positive for human society, especially when concentrated in the hands of the strong and privileged.

7

Dull, Dirty, and Dangerous

AT AROUND NOON ON April 25, 2015, the buildings began to shake in Kathmandu. A magnitude 7.8 earthquake had struck the Gorkha region of Nepal, approximately 48 miles northwest of the city.[1] Within an hour, two additional aftershocks, measuring 6.6 and 6.7 on the Richter scale, struck the region, adding to the destruction and panic that followed. The earthquake was felt as far away as India and China and was the largest recorded in eighty years.[2] The worst affected areas were in Kathmandu and the surrounding regions, but the damage was hardly confined there. The United Nations estimated that the earthquakes affected eight million people, nearly a quarter of Nepal's population.[3] The death toll was 9,000, but could have been significantly higher if the earthquake had not occurred at lunchtime, when many people were away from their homes. A further 16,800 people suffered injuries and 2.8 million were displaced by the earthquake across the country. Perhaps more devastating was the fact that 600,000 homes were destroyed that day, with another 300,000 partially damaged.[4] More than 90% of the homes near the epicenter were entirely flattened.[5] Approximately 20% of Kathmandu's most famous landmarks—including centuries-old temples of cultural and religious significance—were fully destroyed, with nearly all of the rest partially damaged.[6] The earthquakes also caused a massive avalanche on Mount Everest, killing at least nineteen mountaineers and injuring sixty-one, while sending others fleeing for their lives.[7] One experienced expedition leader later described a devasted base camp as looking like, "it had been flattened by a bomb."[8]

In the affected regions, the problems seemed only to multiply. Thousands of those whose homes had been destroyed moved into the streets and lived in tent encampments. The days following the earthquake remained nerve-wracking for many displaced persons who ventured to their former homes to reclaim what was left behind. But no one was sure which structures were stable and which were only minutes away from collapsing. The misery was compounded by a pounding rain, which flooded the tent encampments and made the delivery of relief even more difficult. Aid workers who rushed to the scene recognized that the concentration of people in a wet, dirty, and rubble-strewn environment increased the risks of poor sanitation and disease.[9] But the obstacles in the environment were considerable. The sprawl of tent cities holding displaced people and debris from collapsed buildings rendered some roads inside the city impassable. Outside Kathmandu, rural regions badly affected by the earthquake were cut off by landslides over key roads, while bad weather hampered helicopter access by relief agencies.[10] Dozens of smaller aftershocks also afflicted the region and terrified the displaced residents, culminating in another magnitude 7.3 earthquake that struck the region in mid-May 2015.[11] That aftershock killed at least sixty-five people and injured another 1,900 more.[12] The urgency of the relief effort was underscored by the imminent arrival of monsoon season, which promised even more flooding and landslides.

In the immediate aftermath of the earthquake, Prime Minister Sushil Koirala declared that the Nepalese government would be on a "war footing" while it dealt with the relief efforts.[13] But its actual record was significantly less vigorous. The government was overwhelmed by the difficulties of operating in such an environment and lacked enough manned helicopters for effective delivery of relief.[14] Many of the displaced began to complain that they were being neglected by their government. The United Nations quickly established a flash appeal for Nepal earthquake relief, valued at an estimated $415 million, though the aid ultimately fell short of that total.[15] A number of countries—including India, Bangladesh, China, the United Kingdom, and the United States—provided funds, food, and medicine and even military assets to relieve suffering and get the country back on its feet. The Indian military also played a crucial role in airlifting food and medicine into the country and coordinating the relief process. But the relief effort remained uncoordinated, with multiple governments, international relief agencies, and NGOs sweeping into the country, each with their own priorities. Even the airports were clogged with traffic from incoming relief planes and tourists fleeing the country.[16]

Among the people who arrived for the relief effort were a new class of "digital humanitarians" committed to finding ways to harness modern technology to alleviate the suffering of others in natural disasters.[17] Some of the technology that they proposed using was fairly basic and widely accessible to people in developing countries. For example, almost everyone in Nepal had a mobile phone, and texts from those phones could identify where trapped people were, which roads were blocked by landslides, and where relief was needed most. If the affected population was connected to the Internet, even more opportunities for relief efforts based on what became known as "big data" became possible. For example, social media such as Twitter could be used to glean information about the damage and, if properly geo-located, create a more textured picture of the regional impact of the disaster. But the digital humanitarians were also committed to using drones to help identify which buildings had been damaged and which roads were blocked and holding up relief efforts. A number of international agencies and NGOs soon arrived in Nepal equipped with small but capable drones to aid in relief efforts. These drones were used to identify victims of the earthquake for search and rescue missions, to evaluate ruined heritage sites, and also to take images of damaged buildings so that they could be assessed to see how likely they were to collapse.[18] Aerial imagery could be used to create 3-D models of buildings and provide a more accurate picture of which buildings were likely to collapse given the structural damage they had suffered in the earthquake.[19] In this new "digital humanitarian" model, data could be aggregated from multiple drones and other sources and collectively mapped using a software application such as Micromapper or OpenStreetMap. This would allow effective "crowdsourcing" from multiple UAVs and rapid delivery of the data to aid agencies and other relief organizations.

The advantages of this approach were clear. Digital humanitarians were able to provide better, more detailed images of crumbling buildings and blocked roads than satellites, and this additional level of detail saved time and effort for relief organizations operating in a difficult, time-sensitive environment. But these operations in Nepal were hardly an unqualified success. Nepal had effectively no regulations on drones at the time of the disaster, and as a result many operators were confused about the rules for flying them. Were drones nothing more than hobbyist devices that could be used freely, or did they require the approval of the civil aviation authorities? While most organizations behaved responsibly, some flew drones without checking with local leaders or governments. Others flew too close to military installations and historic sites, while others inadvertently buzzed over and alarmed groups of traumatized people. The Nepalese government, already overwhelmed with the scale of the disaster, found some of these "disaster drones" an added worry rather than a relief. Yadav Koirala, an official with the Home Ministry of Nepal, said later that "drones were an added burden to us. There was no restriction on them . . . They flew everywhere but offered no feedback to the government . . . It was difficult to control them."[20]Some Nepalese officials worried that broadcasting images of the destruction wrought by the earthquake would damage the country's valuable tourism industry.[21] Other agencies worried about a collision between a drone and a manned aircraft around Kathmandu's airport. The Nepalese military and intelligence establishment became alarmed about the prospect of spies using the humanitarian drones as a cover, a fear particularly aggravated when a Chinese national was caught filming army headquarters with a drone.[22] In the middle of the humanitarian effort, Nepal banned drones from flying for more than fifteen minutes and issued no-fly zones for security buildings, houses, and religious and cultural landmarks. The regulations also stipulated that drones could be flown no higher than 100 meters and no more than 300 meters from the pilot.[23]

The Nepalese earthquake had become a testing ground for digital humanitarians to use drones for good purposes, but the effort had run aground because of ambition, lack of coordination, and carelessness. Yet it also raised a series of important questions about the nature of humanitarian work in the drone age. The digital humanitarianism movement seeks to take advantage of the capabilities of drones to do the kind of work—often described as dull, dirty, or dangerous—that others are incapable of or unwilling to do. Their efforts are designed to show that drones are empowering international and local actors in humanitarian disasters in ways never before imagined. Some advocates point to the spread of drones as democratizing data collection—in other words, stripping it from the hands of often selfish or parochial governments—and making the world "a better, safer place."[24] Yet despite good intentions, some elements of the digital humanitarian movement are suffused with a technological utopianism that conflates what can be done with what should be done. Drones have unleashed new capabilities for those who wish to help, but they have also complicated the humanitarian space by introducing new actors whose incentives may not align with the local government or population. As the Nepal earthquake disaster revealed, there is a risk that the needs of the local population would be swept aside by organizations more eager to demonstrate the proof of concept for humanitarian drones than to realistically address the needs of those suffering. All of those advocating for drones as a force for good—digital humanitarians, activists, and even UN peacekeepers—need to be careful that they do not lose sight of their goals or become more risk-taking as a result of access to the technology. For "disaster drones" to be successful, they will need to prove that new technology like drones are necessary for achieving humanitarian objectives, rather than just an opportunity to test the capabilities of new technology amid human suffering.

Disaster Drones

The Nepal earthquake was not the first time that disaster drones were deployed in a natural disaster or emergency. As with most things related to drones, this began with the military. Immediately following a massive earthquake in Haiti in 2010, the United States provided overflights by Predator and eventually Global Hawk drones and released satellite imagery which aided the relief efforts.[25] While this assistance was useful, the government of Haiti remained uneasy about the unfettered use of small drones over the disaster zone. Its initial reaction was to ban small UAVs for fear they would collide with civilian air traffic moving through the airspace for the relief effort. Later, in 2012, the International Organization for Migration (IOM) received approval to use smaller drones to assess the damage done to Haiti after its deadly earthquake. In cooperation with the United Nations, the IOM used drones to identify debris hazards, new construction, water drainage capabilities, and temporary housing in Port-au-Prince and the surrounding region.[26] The IOM used drones to assess destroyed structures, to conduct surveys of public buildings, and to monitor internally displaced people (IDPs) and camps.[27] Images supplied by drones constituted an improvement over satellite imagery, which by some estimates tended to underestimate the scale of the damage in Port-au-Prince.[28] The results, according to the UN project leader, were "incredibly satisfying" and also notable because the Haitian displaced population was curious rather than hostile when seeing drones float by overhead.[29]

In the Philippines, drones were also used to assess the damage caused by Typhoon Haiyan in 2013.[30] In this instance, one of the biggest problems was that the satellite imagery was outdated and some communities were relying on hand-drawn maps.[31] To address this situation, a number of organizations sponsored projects and volunteered their time and resources to improve the quality of images available on the ground. Medair and Catholic Relief Services partnered to use drones to collect aerial imagery in the reconstruction efforts after the disaster.[32] A comparatively new organization, the Humanitarian OpenStreetMap Team (HOT), used satellite imagery donated by a range of actors such as Yahoo!, Microsoft, Digital Globe, and the US State Department to create a picture of where the hardest hit regions were and which roads were considered impassable. They were also helped by the fact that the cost of satellite imagery had plummeted over the last decade and as a result was now used by a greater number of organizations and not just governments.[33] By one estimate, over 1,500 volunteers in eighty-two countries participated in the effort to collect and map recent satellite imagery onto the terrain.[34] With this data as a base, they were able to add UAV imagery from local non-governmental organizations (NGOs) and create a detailed, street-by-street picture of the impact of the disaster. The chief obstacle that NGOs faced was a lack of co-ordination; many of these organizations did not know each other and were not aware of each other's activities, which in turn undermined the effectiveness of the collective relief effort.[35] Yet despite these limitations the response to Typhoon Haiyan suggested that crisis mapping by volunteers coordinated largely over the internet could pay dividends for relief efforts. One assessment of drone use in the response to Typhoon

Haiyan concluded that drones were able to speed up relief efforts, eliminate wasted time and effort, and improve the accuracy of the delivery of food and medicine to those suffering.[36]

For many in the digital humanitarian community, Haiti and the Philippines served as proofs of concept for the value of crisis mapping using drones. Similar crisis mapping projects were launched in the aftermath of Cyclone Pam in Vanuatu in 2015 and a series of earthquakes that rocked Ecuador in 2016.[37] There, drones were used to monitor the stability of roads and even assess the risk of landslides.[38]n 2015, the World Bank funded a series of drones used by the NGO Drone Adventures to map flood-prone areas of Dar-es-Salaam in Tanzania in order to improve crisis response.[39] In 2015–2016, the European Union began to experiment with drones for search and rescue missions. Many of these projects were collaborative and drew technology and insight from private companies, NGOs, and universities, with funding from major corporate donors and international institutions like the World Bank. At the same time, networks of roboticists and development professionals emerged to coordinate the use of UAVs for humanitarian purposes. Among the most notable of these is the UAViators, the Humanitarian UAV Network, which seeks to disseminate regulations about the operating environment, best practices, and lessons learned from these missions. In 2017, this network boasted over 500 pilots in seventy countries and served as a hub where calls for humanitarian drone pilots could be made in the wake of a natural disaster.[40]

The success of these disaster drones hinges on both technical and human requirements. At a minimum, it is critical to deploy the right kind of drone—a number of different types of drones have been deployed for these types of operations. Multirotor drones, such as the popular DJI Phantom quadcopter, are simple to launch and do not require much space for take-off; they can be bought cheaply and used easily in urban areas. Their chief limitation is a limited battery life; they can only remain in the air for a short time.[41] Fixed-wing drones can carry a heavier load and fly for a longer period, but they need a launching strip, which can be difficult to find in dense urban or mountainous regions.[42] One particularly promising fixed-wing model, the Sensefly eBee, has a longer range, high levels of automation, and can remain in the air for a longer flying time than other drones. It has been used in Haiti and Nepal.[43] But it is not as simple as throwing a drone into the sky. For effective crisis mapping, drones need to be equipped with cameras with powerful resolution to get the kind of detailed images that are needed to depict on-the-ground damage. In general, drones flying at high altitude often get lower levels of resolution for their images, but those flown at lower altitude need to have a higher resolution camera and can generally cover less ground. The resolution of the camera can vary depending on size and cost. Visible light or RGB (red green blue) cameras are often better for crisis mapping because their pictures are similar to satellite images and can be more easily interpreted by aid workers.[44] These drones must also have a gimbal to stabilize the camera during flight. Aside from cameras, drones can also carry infrared sensors which can detect vegetation and water.[45] For drones to be viable in humanitarian crises, drones typically need some professional-grade features, which raises the costs involved and rules out some of the popular off-the-shelf commercial models.

The decision by a government or NGO to use a drone is partly dependent on the alternatives available. In some circumstances crisis mapping could be more effectively or cheaply done with satellite images or even pictures taken from helicopters. The reduced price of satellite imagery on the commercial market, from vendors like Google Earth, gives a clear advantage to it over drones. This is particularly so if satellites or military-grade drones can cover more ground. For example, the US military's Global Hawk drone can take wide area shots, akin to shining a set of floodlights on a sports stadium. By contrast, most disaster drones cover far less ground and are closer to shining a spotlight on that same sports stadium. One estimate suggests that on average disaster drones can cover approximately 3 sq. km in a day, or 10 sq. km a week, for data collection and processing.[46] However, one of the most important advantages disaster drones provide over satellites is that they can capture images below clouds. Some commercial satellites are not equipped with cameras that can peer beneath cloud cover and see the ground below. This is particularly crucial for disaster relief because major storms like hurricanes and cyclones come with dense cloud cover that can obstruct satellite imagery for days or even weeks.

For disaster drones, what matters most is the quality of images and their correspondence with the real world so that they highlight areas where relief efforts are needed. For this to work, images taken from drones need to be spatially accurate and related to real-world maps. Most disaster drones are equipped with lightweight GPS units that are capable of locating exactly where an image was taken. These images can be combined with widely available satellite images and geometrically corrected to become uniform so that they adhere to common geographic coordinates and "knit together" on a single map.[47] The creation of what is known as a georectified orthomosaic—essentially a distance-corrected map that corresponds to what is actually on the ground—sounds more complicated than it is. There are many free or cheap software packages that can help crisis mappers create these maps on their laptops. It is particularly important to make sure that normal two-dimensional pictures do not distort the distance between two points—such as towns or buildings—on a map. It is not hard to imagine how important this is for time-sensitive relief efforts that depend on traversing large distances accurately to deliver food, medicine, or other supplies to needy populations. Disaster drones can also produce other useful images. For example, some drones are equipped with software that allows them to create 3-D models based on a calculation of the volume of structures appearing in images, which are useful for assessing whether buildings are damaged or structurally unsound and should be evacuated. Thermal maps drawn from drones are also useful in detecting structural damage to roads or the source of groundwater discharges.[48]

This imagery, once collected, must still be processed quickly and delivered to the relief agencies to be useful. This is one of the chief limitations of disaster drones. At present, in many disaster zones, the infrastructure and the drones themselves are not present on the ground and must be brought in by external parties. One estimate suggests that drones take 6.5 days to arrive to the scene of the disaster, well beyond the crucial seventy-two-hour window when most relief is needed.[49] To be most effective, drones need to be in place before the disaster or delivered rapidly immediately thereafter. This is often not the case in the disaster-prone parts of the developing world. Most of the drone labs employed by digital humanitarians are in major hubs of the developed world, specifically in North America, Europe, and parts of Asia. The teams of volunteers and the drones themselves must get to the disaster zone quickly, which is difficult if commercial air travel is limited or expensive. Even when the teams of volunteers and drones are in place, processing the imagery quickly enough to be useful presents a challenge. Large-scale data collection of thousands of images, or the collection of particularly high resolution images, can slow the processing time considerably. There may be days or even weeks between when the images are taken and when they are developed and returned to the disaster zone. This time delay is less important if the drone project is itself not urgent. For example, some drone humanitarian projects are proactive, mapping areas of a region that might be susceptible to flooding or landslides. In these cases, the delay is unfortunate but does not detract from the value of the project. But for immediate disaster relief, when lives are at stake and the risks multiply as the days pass, getting disaster drones to the right location and analyzing images and delivering them to relief workers is a time-sensitive proposition. In these circumstances, the arrival of disaster drones can sometimes be too little, too late.

Aware of this fact, some organizations have rushed into disaster zones with drone projects designed to shorten the time that people must wait for relief. This was the case in the Nepal earthquake, where frequent media coverage of the disaster drew a number of NGOs and private organizations to the country and led to drones flying unchecked everywhere. This rush to the disaster zone is in part a function of the technology itself: cheap and readily available drones empower small groups of individuals who might otherwise have no direct way to act in a crisis. Many of these teams are well meaning, but such a rush to a disaster zone provides a space for opportunists and those less interested in helping people than experimenting with the technology to arrive. Moreover, even the best and most skilled drone teams are not always coordinated, either with the government or with each other. In Nepal, this produced a strong backlash and made some people increasingly suspicious of drone humanitarianism. Patrick Meier, the founder of the Humanitarian UAV Network, called the rogue drone pilots "cowboys" and estimated that a lack of cultural sensitivity and respect for the local population set the humanitarian UAV movement back at least six months.[50] In response to the problems seen in Nepal, Meier and his colleagues generated a database of UAV regulations by country and produced a series of guidelines for drone pilots to ensure that their drones were used in a way that respected the preferences of the local population. This development of a code of conduct, as well as the training and skill-building associated with it, will go a long way toward ensuring that disaster drones are used more responsibly in the future. But at the most fundamental level these codes of conduct remain voluntary, and there remains a danger with drone humanitarianism that some organizations will ignore the guidelines and laws and run roughshod over governments that are otherwise distracted and incapable of enforcing their laws on UAV use.

This is particularly important because disaster drones are not used in a normal environment where their presence might be otherwise expected. In some parts of the developing world, drones are not commonly seen and their sudden appearance can alarm the population. This is particularly the case in environments where high-end technology like drones typically belongs to the military. Especially in countries in which the military is not widely trusted, drones flying overhead might be seen as an attempt to conduct surveillance on the population for unknown ends. When dealing with a traumatized population which has recently experienced a natural disaster, these concerns over the ultimate purpose of drones are magnified. The last thing that humanitarian UAV users would want to do is to alarm or even terrify a population which has been already coping with a natural disaster. One way that this might be remedied is to paint the drones a specific color to signal their humanitarian purpose and to distinguish them from military drones. But such a measure might not be enough to put some lingering unease about the technology to rest. While more than a majority of aid workers (60%) surveyed in a 2016 study had a broadly positive view of drones in humanitarian work, a minority raised concerns that drones might be perceived as "inhumane" or too distant from the population to be useful. Others expressed concerns that their use will contribute to the trend of delivering aid at a distance and removing needed aid workers from dangerous environments.[51]

Many of the advocates of humanitarian UAVs recognized that it is crucial that drones not be seen by the local population as a foreign object, alien or otherwise hostile, but instead see them as part of an effort that has a human face. One way that this is being addressed is by investing in partnerships with local NGOs and universities so that drone humanitarian projects come from the societies that they are helping. WeRobotics, an NGO based in Switzerland and the United States, builds partnerships with universities, governments, NGOs, and other community organizations in the developing world to identify places where drones and other robotics might be useful and to develop projects to meet local needs.[52] The goal is to create "Flying Labs" where local organizations can develop hubs to train local people to use the technology and teach others. WeRobotics currently has projects underway in Nepal and Tanzania and plans for as many as fifty more.[53] In 2019, in a partnership with the Centers for Disease Control and Prevention (CDC), they were also engaging in training and capacity building for cargo drone use in Papua New Guinea.[54] By undertaking these projects, WeRobotics addresses some of the concerns about the sustainability of these drone projects and builds capacity and buy-in by the local population and its representatives. This is a crucial step: to turn people in the developing world from objects of the drone's gaze into agents in charge of what the technology does next. Such local ownership and buy-in will address cultural and social issues surrounding drone use and make humanitarian relief through drones more effective and sustainable.

There is no doubt that disaster drones have empowered new actors to take images of crisis zones, but it is still unclear whether their use will produce a sustained and large scale benefit in humanitarian disasters. Most of the cases in which disaster drones have been used could be described as pilot projects where the intended scale of the impact was small. For this reason, there is a risk that drone humanitarianism may become an example of tokenism—that is, when organizations invest in a project that does only a small amount of good and is directed more at proving the value of the technology than alleviating human suffering. Even if this does not happen, it is unclear that disaster drones can be scaled upward in a cost-efficient way to produce the kind of sustained or systematic change in disaster zones that their most optimistic advocates promise. It is also unclear whether all drone humanitarian projects will adopt the goals of sustainability and local buy-in that WeRobotics and others endorse. But the promise of disaster drones has only recently come into view. While there is no conclusive proof that providing more information and improved maps will vastly improve the response of governments, IOs, and NGOs to humanitarian disasters, there is more than enough evidence to support modest investment in drone projects that will complement existing capabilities (through satellites and other means of learning what is on the ground) and allow for a precise and carefully tailored response to the needs of those affected by disaster.

Search and Rescue

Another function of drones in humanitarian emergencies is search and rescue. The advantages of drones for this function are clear: they can cover large swaths of ground, operate beneath the cloud cover that typically obstructs satellites, and with the right cameras they can see through some obstacles (like tree cover) to find missing or endangered people. It is not hard to imagine the usefulness of drones in finding missing children, for example, or lost hikers in a vast national park. In humanitarian crises, the same technology could eventually be put to work in finding those stranded or endangered by floods and other disasters. The problem is largely one of access. Today, most NGOs and IOs operating with drone humanitarian teams do not yet have the capacity to conduct effective search and rescue. In part, this is due to timing: it takes too long, on average for a drone humanitarian team to arrive in a crisis zone for them to be useful in most search and rescue missions. In the aftermath of an earthquake or flood, the chance of survival for those lost and trapped declines precipitously as each day passes. In the little less than a week it takes for most drone humanitarians to arrive in a crisis zone, those in most need of help may have already been found or even have died. Seen from another light, the problem is also one of location. Some of the world's worst disaster zones—places where earthquakes and flooding, for instance, are recurring—are located in the developing world where the infrastructure both for delivering relief and for supporting humanitarian drones is embryonic. If organizations like WeRobotics succeed in boosting local capacity for drone use in the disaster zones of the developing world, search and rescue will become more feasible than it is today.

At present, most of the sustained efforts for using drones in search and rescue have been pilot programs by police and civil defense forces in the developed world. In the United States, for example, drones have been purchased by a large number of police forces responsible for extensive rural areas like Texas and the Grand Canyon. In the United Kingdom, the Greater Manchester Fire and Rescue Service has flown Skyranger drones since July 2015 and have a dedicated Aerial Imagery Reconnaissance Unit (AIR Unit) to provide overhead imagery of incidents and their ensuing rescue operations.[55] In Europe, the European Emergency Number Association (EENA) signed a partnership with the company DJI to train first responders to use Phantom and Inspire drones as a first response following a disaster and to find missing people.[56] The European Commission has recently devoted €17.5 million to the Integrated Components for Assisted Rescue and Unmanned Search (ICARUS) program, which equips first responders with drones and other robotics to respond to emergencies.[57] The ICARUS program also promotes the development of drones and other tools and integrates them into the standard operating procedures of first responders across Europe.

These efforts at using drones for search and rescue are being tested by the refugee crisis that has engulfed southern Europe since 2012. Thousands of refugees, fleeing growing chaos across the Middle East and North Africa, arrive annually in Europe from Syria, Iraq, Libya, and Afghanistan.[58] In 2015, according to FRONTEX, the European Union border and coastguard agency responsible for monitoring refugee movements, 1.8 million refugees tried to enter EU member states, often via water routes across the Aegean or the Mediterranean Sea.[59] For many European governments, the influx of war refugees is potentially destabilizing, especially if the refugees are able to claim legal right to remain in their countries due to the threat of harm if they were returned. For this reason, the European Union has invested in interdiction of refugees at seas, often with patrol boats in the Mediterranean, which are designed to stop traffickers moving people into Europe.

Leaving the moral and legal issues aside, the process of interdiction at sea is challenging from a technological vantage point: patrol boats must find small vessels and dinghies that move under the cover of darkness toward entry points into Europe. Satellite images are ineffective because it takes too long to produce and update them—by which time vessels detected have moved on and are lost. From a search and rescue vantage point, timing is crucial. Some of the ships carrying refugees capsize in rough weather, leaving them drowning at sea and creating an immediate need for emergency response. The UN High Commission for Refugees (UNHCR) estimated that 5,000 refugees died crossing the Mediterranean Sea in 2016 alone, making it one of the deadliest years for refugees in history.[60] To deal with this crisis, the European Commission proposed a plan, worth €22 million, to set up an EU drone fleet equipped with video feeds, sensors, and chemical "sniffers." By September 2018, FRONTEX was beginning to test drones in Greece, Italy, and Portugal to monitor the EU's borders. The purpose of the EU drone fleet is not exclusively humanitarian; the "sniffers" are included to catch boats that are violating EU emissions standards.[61] But the drones are also designed to find smaller, makeshift crafts during day and nighttime hours.[62] The EU drones are equipped with thermal sensors, which make spotting refugees at night easier, and can pick up the heat signatures of refugees swept overboard at night and at risk of drowning.[63]

Although FRONTEX remains the chief organization for managing the European Union's approach to the refugee crisis, another agency—the European Maritime Safety Agency (EMSA), based in Lisbon, Portugal—has been tasked with arranging for drones for this purpose. EMSA has a long record in enforcing maritime safety and environmental regulations, but the humanitarian dimension of its new mission is relatively novel.[64] Initial efforts to turn this proposal for humanitarian search and rescue drones into reality has run up against a host of obstacles, not the least of which is developing coherent EU-wide regulations to integrate drones into the civilian airspace.[65] Across the European Union, there is a reluctance to embrace "good" drones given the negative connotations that the technology carries due to its use for military operations and targeted killings by the United States. But perhaps most importantly, the effort to use drones for humanitarian purposes has run up against questions of cost: large military drones such as the Reaper or Global Hawk are too large and expensive to be viable for these tasks, while small commercial drones are too limited in capacity to be useful. In effect, organizations like EMSA need a drone that has the endurance of some of the military drones, but with fewer capacities for holding payloads and a lower price tag.[66] EMSA is confident that it will be eventually able to find a supplier to provide these drones and service the ground stations needed to run them, thus allowing for cost-efficient rescue operations at sea, which will compliment rather than replace its existing operations.

The advantages of drones for this particular search and rescue mission are clear. Drones can cover a wide area, detect vessels and people that are hard to see with satellites and ordinary cameras, and communicate almost instantaneously with ground control stations. Drones can also approach human trafficking vessels while making very little noise, which is important because the sound of the approach of manned aircraft often makes the traffickers flee to avoid detection and capture at sea. Drones can also last longer at sea and fly for hours without the need to allow crew to rest. But the ambitions of those planning to use drones for search and rescue are not limited to detection by sight alone. Leendert Bal, director of search and rescue operations for EMSA, envisions a future in which drones may be able to detect mobile phone signatures or SIM cards of refugees at sea.[67] Sensors that were able to detect and locate mobile phone signatures would reveal where people are even in low visibility environments, such as rescue operations at night or in bad weather. Another longer-term objective would be for drones to drop life vests or even small rubber boats to help refugees whose boats have capsized.[68]

While the European Union's efforts at integrating drones into its search and rescue operations are embryonic, they are an important indicator for the new capabilities this technology provides to humanitarian actors and changes the actions they will consider as appropriate. For most actors, drones will be just another tool in the toolbox for conducting effective search and rescue operations. But in the case of the European Union, the technology is moving the boundaries of what is considered legal and appropriate as well. While these search and rescue drones in the Mediterranean have been cast by their defenders as the quintessential example of humanitarian drone use, they are used in the service of a refugee policy that is widely criticized as inhumane by organizations like Human Rights Watch and others.[69] The European Union's efforts to deploy drones to maximize situational awareness could be read as an attempt to deny the refugees' right to asylum on the continent and, in some cases, to forcibly return them to dangerous environments.[70] Although it is described as "search and rescue," it is part and parcel of the European Union's efforts to police its border regions against refugee claims. This illustrates that the technology is itself not value neutral, as it enables a policy which may have some legal basis but is considered ethically questionable. At a minimum, drone technology cannot sort those who might have a right to stay from those who have no such rights, as this is a legal question. Instead, the deployment of drones offers the misleading appearance of control over a refugee situation that is rapidly worsening with every year that passes.[71]While the camera on the drone itself yields precise images, it enables a humanitarian policy conducted at a remove from those who suffer the most.

Delivery Drones

The next stage in the evolution of the humanitarian drone will involve delivering parcels of food and medicine to populations in need. At present, most of the discussion of delivery drones has revolved around the plans of large companies like Amazon to deliver consumer packages to residences in the United States and Europe. At present, these plans are unlikely to revolutionize commerce in the developing world as some of their advocates claim. Delivery of ordinary lightweight packages by drones may be less cost-efficient than delivering products by ordinary truck in many densely populated areas. In part, it depends on the cost of drone deliveries. At present, the cost estimates are highly speculative ($1–6 per package), but unless the cost curve changes in favor of drones—and it might, as one analyst suggests Amazon can get the cost down to 0.88 cents per package—it will be hard to compete with trucks that can deliver hundreds of packages per day.[72] This is even a bigger problem for heavier packages (over 5 pounds) which are more challenging and expensive to move. Drones will also need to be equipped with "see and avoid" technology and insured against accidents or theft in many residential environments, a factor which may alter the calculation of cost-efficiency for commercial drone delivery. There is also a thicket of safety regulations and laws that delivery drones must contend with in most locations in the developed world. Although start-up companies like Flirtey are convinced that there is potential in commercial drone delivery in the United States and other places, it will take regulatory change and considerable investment by major commercial players like Amazon or Walmart and a reduction in costs of drone deliveries to turn this dream into a reality.[73]

At present, the greatest potential of delivery drones does not lie in getting packages or novelty goods to relatively well-off consumers in US cities. Rather, it lies in the efficient delivery of medicine, blood, and other essential goods to populations who lack effective road networks or for whom the delivery of goods by truck is cost inefficient. Consider, for example, how expensive it is to deliver an Amazon package to a customer in a city compared to a customer in a rural area. A truck might make dozens of package deliveries in the same few blocks in New York City, thus making the costs involved—for example, the labor of the driver and the costs of running the truck—worth it from the company's perspective. But driving a single package to a customer a hundred miles or more from the delivery center is expensive, costing the company much more for the hours of labor of the driver and gas for the truck than they might get back in shipping charges. In these cases, there is a compelling cost-efficiency argument for using a drone rather than sending a driver out to spend several hours delivering one package.

Such remote populations are often in rural areas in the developing world, but they can also be located in geographically isolated parts of countries in the developed world. Delivering to remote populations is often called "last mile" delivery because it takes the package for the last section of a supply route. In parts of the developing world, this "last mile" delivery may mean the difference between life and death. There, many routine diseases become life-threatening either because vaccines and medicines are not available or slow to reach those in need, or because medical test results take so long to be delivered that treatment starts late, if at all. In these cases, delivery drones could be successful if they shorten the time periods between the emergence of symptoms, their diagnosis, and their treatment. By some estimates, one third of the world's population lacks access to essential medicine, in part due to poor roads and the failure of "last mile" supply networks.[74] As a result, approximately 40% of vaccines supplied to people in the developing world expire before they are delivered and ready for use.[75] If the supply chain could be tightened by the use of drones, medicine and test results could be delivered rapidly and the gap between diagnosis and treatment would become smaller.

The technology behind mass-scale drone delivery is not quite ready, but it shows promise for specific kinds of payloads. Most delivery drones can hold only a few kilograms of weight and fly a relatively short distance, often less than 150 km.[76] To be effective, experts estimate that delivery drones should be sent out on short-range missions (typically 40 km or less) to help in areas with enough population density to make the trip worthwhile.[77] But one of the chief constraints facing such drone use is that the infrastructure for large-scale delivery drones simply does not exist at the commercial level. While the US military has made large deliveries of up to 2,700 kg with K-Max or Snowgoose large cargo drones, this movement of bulk cargo is not possible or cost-efficient for many NGOs and other organizations with less budgetary largesse.[78] Instead, most off-the-shelf drones are best used for high-value, low-weight products (blood samples, vaccines, tests) and some food parcels, water, mobile phones, and other lightweight, highly useful goods.[79] As a general rule, the heavier the payload, the harder it is to have a cost-efficient and reliable delivery to underserved parts of the developing world. While this might change in the future, adding additional payload to a delivery drone certainly increases the costs associated with its operation.

Other technological limitations make drone deliveries harder than simply flying and dropping the cargo. Many of the existing delivery drones are operated by rechargeable batteries, which must sustain the entire return flight or be designed so that they can be recharged at a waystation somewhere along the flight route. If the journey is too long, pilots must pre-determine a recharging station along the route and have enough people and infrastructure there to recharge and relaunch the drone. This means that even medium-range drone deliveries require a sophisticated operational plan, as well as plans for what could go wrong, before payload can be effectively moved. But perhaps the biggest issues that drone delivery advocates face is take-off and landing. Most fixed-wing drones require a runway for take-off and landing, which does not always exist in the locations that need the delivery. When these runways are available, fixed-wing drones are a preferred option because they can fly for longer and carry more payload than comparable quadcopters. When runways are not available on either end of a proposed delivery, pilots must be satisfied with quadcopters that are capable of vertical take-off and landing, but can carry fewer kilograms of payload and fly a shorter distance. In both cases, one of the key questions is whether the drone should drop its cargo in mid-flight or land to deliver it directly. The latter is safer and more reliable, but once on the ground drones can be stolen, hacked, or destroyed rather than returned safely. Dropping cargo from the sky is riskier and may miss the target but does not carry the same risk of deliberate or malicious disruption.

Efforts to use delivery drones in rural regions of the developing world have been underway for some time, though relatively few projects have moved beyond the proof of concept stage. A Silicon Valley start-up called Matternet has been deeply involved in some test cases using its M2 quadcopter drone, which has substantial battery life and intuitive handling. In 2014, Médecins Sans Frontières (MSF) worked with Matternet to experiment with the use of small, payload-bearing drones to transport the test results of those suspected of contracting tuberculosis in Papua New Guinea.[80] A similar effort was made by the World Health Organization and Matternet to use drones to connect hospitals in the urban and rural regions of Bhutan.[81] In 2016, a UNICEF-Matternet joint project in Malawi delivered laboratory samples for early infant diagnosis of HIV faster than motorbikes over distances up to 10 km.[82] In early 2016, a South Korean company called Angel Wing delivered medicine, needles, and vaccine vials to rural health clinics in Nepal.[83] Most of these projects were small in scale, testing the concept of operations rather than making a sustained impact, and with modest operational ranges. Yet the operations themselves pointed to greater potential in scaling up these projects for a much wider impact.

One way that this will happen is by investing in local partners who are able to assist in getting a project off the ground. In 2015, the United Nations Population Fund (UNFPA) worked with Drones for Development, an Amsterdam-based start-up, to use Dr. One, a hybrid drone designed to deliver contraceptives to rural clinics in Ghana. One advantage of this program was that it was done in partnership with Ghana's health service. The Dr. One drone is designed to use lightweight material and open source electronics to allow local operators to use and maintain the drones at low cost. The project resulted in savings of about €3,000 compared to similar operations with motorbikes. There are still obstacles to scaling this project up to the national level—for example, a central mechanism needs to be established to coordinate and run the drone flights and the supply chain on both ends must be made more efficient for different types of payloads—but the Dr. One model showed that it could work.[84] It also showed that some drone models could be made in a flexible way to accommodate multiple types of payloads and could be made sustainable so long as local governments and firms were involved in the operations.

Other companies are experimenting with different private-public partnership models to build local capacity for drone use and maintenance. The underlying idea is that these projects should not only entail Western countries flying drones over the developing world, using them as almost a laboratory for the technology, but rather they should produce a lasting benefit to the country itself through local partnerships and capacity building. Zipline, a Silicon Valley start-up, has developed a fixed-wing drone that can fly a radius of 80 km (49.7 miles) from its ground control station through all weather conditions (fig. 7.1). Equipped with GPS and an autopilot, Zipline's drones can fly 150 km round trip powered by its lithium ion batteries without needing to be

FIGURE 7.1 Zipline drone launch in a California testing facility, March 2018

recharged.[85] They are relatively fast, flying at 100 km per hour, and can deliver parcels in as little as thirty minutes.[986] The drones drop packages using small paper parachutes rather than landing, thus obviating the need for runways at the drop point and limiting risk to the drones themselves. They can also take off and land with a catapult launch system that limits their physical footprint on the ground.

The Zipline drone's expected payload is small—1.8 kg, or roughly 3.96 pounds—but the intention is to deliver blood and emergency medical supplies to rural regions on a "just-in time" basis. Demand for certain medicines is intermittent in many rural clinics and keeping supplies of every medicine that might be needed in each rural health clinic is costly and inefficient.[87] Some medicines—for example, poison antidotes or snake-bite anti-venoms—may not be needed all of the time, but when needed, are urgently needed. Many medicines also have a short shelf life and can expire in storage before they are needed and used. It is obviously wasteful to simply hold them in stock at every location only to discover later that they cannot be used. To deal with these logistics problems, Zipline shortens the supply chain by allowing centralized health services to rapidly deliver medicine and blood on an as-needed basis. In 2019, Zipline has expanded its repertoire of delivered goods to include 170 different products, including anti-venoms, vaccines, and rabies treatments.[88]

The first country that agreed to work with Zipline was Rwanda, which has a strong central government and a good health service, but like many countries around the world, has problems with delivering blood and other medicines in time to meet the needs of urgent cases. Most rural delivery in Rwanda is done by motorbike over dirt roads, which is expensive and cumbersome given the state of the roads in the mountainous country. Another advantage to working in Rwanda is its regulatory environment: the Rwandan government is aiming to make itself a technology hub of East Africa and has relatively few regulations on flying drones compared to more regulated areas like the United States and Europe.[89] Unlike many developed countries, Rwanda's airspace is uncrowded and easy to use for drone test flights. Rwanda also has high levels of mobile phone usage in the country—one estimate suggests that 70%–80% of Rwandans have access to a cell phone—so this allows for rapid communication of needs from rural regions.[90]

Zipline's approach requires local doctors to send a text with details on the needed blood or medicine to a distribution center outside Kigali, which then prepares the package for the drone and deploys it on a preprogrammed flight to the specific location where it is needed.[91] In 2016, Zipline began service by making 50–150 daily deliveries to twenty-one of Rwanda's district hospitals, covering approximately half of the area of the country.[92] By December 2018, Zipline was making deliveries to thirty district hospitals and 473 public health clinics around Rwanda and estimated that it was delivering more than 65% of all blood transfusions in the country outside the capital city of Kigali.[93] Its experience in Rwanda suggested that the local population was not hostile to the use of drones, in part because the Rwandan military had not used drones domestically and therefore the technology had few of the negative connotations it has elsewhere.

Zipline is clear that it is not a charity, but rather a for-profit health logistics company whose services are purchased by the local government.[94] This is very different from the traditional humanitarian approach in which drones and services are simply donated to countries in the developing world. This suggests that Zipline and other similar drone start-ups should not be cast in purely humanitarian terms, but rather as commercial endeavors designed to generate profits for their creators. There is nothing unusual or wrong with this; many other service providers in the humanitarian aid and post-conflict reconstruction worlds are also for-profit companies. But it does impose some additional burdens on the governments who make these bargains with commercial companies. One is to prove that the measure is cost-efficient: is this the best way for the Rwanda to deliver essential medical supplies? Or would it be more efficient to invest in building sustainable roads, rather than a high-tech solution? In part, the answer to this question hinges on how expensive it would be to build roads and other infrastructure in the developing world. It is not uncommon for countries in the developing world to have substantial deferred maintenance on existing roads and to struggle to afford that along with new infrastructure. For this reason, their governments may find that they get a bigger, immediate "bang for their buck" and cover more people with drone deliveries than they might with long-term, costly road improvements.

One country that has made this cost calculation in favor of drones is Ghana. In 2019, Zipline announced that it was partnering with the Ghanaian government to expand drone delivery operations on a larger scale than in Rwanda. Zipline aims to deliver 148 different medicines, vaccines, and other products to as many as 2,000 different health facilities across the country[95] and to build four distribution centers, each equipped with thirty drones, to enable this service. It estimates that it will make 500 deliveries a day from each distribution center, serving 12 million people.[96] Although the majority of its work is in Rwanda and Ghana, Zipline is not exclusively focused on Africa and plans to extend its operations to Latin America, North America, and other parts of the world. In each of these expansions, the rationale behind drone deliveries will hinge on complex calculations of cost and effectiveness against alternative options.

In the developing world, a particular challenge for drone deliveries is ensuring that these operations are sustainable over the long run. There is a danger with all of these projects that for-profit companies will simply use the airspace of the developing world as a laboratory, fine-tuning the technology for more lucrative markets like the United States while ultimately leaving the population there with nothing. Zipline has sought to address these concerns by working with Rwanda's Ministry of Health and by hiring locals to do repairs, to operate the drone's take-off and landing, and to process orders from local health clinics. If this investment continues in other locations and with other companies, it suggests that the deployment of delivery drones in a humanitarian context can be sustainable provided there is a commitment to develop the capacity of the local population to assume control over many, if not all, of the dimensions of a drone-based project.[97] If this capacity is built, it can also lead to entirely indigenous and self-directed drone delivery projects, as recently happened when Nepalese entrepreneurs and technicians developed a homegrown drone delivery service for medical supplies in the Himalayas.[98]

Another project which straddles the humanitarian and commercial dimensions of delivery drones are the Redline cargo drones, hosted at the École Polytechnique Fédérale De Lausanne (EPFL) as part of their Afrotech program. As envisioned by former *Economist* journalist, novelist, and founder of Redline Jonathan Ledgard, fixed-wing drones would radically expand the delivery of cargo across the African continent, carrying packages of emergency medicine but also ordinary commerce weighing as much as 10 kg (22 pounds).[99] Unlike Zipline, which aims to be cost-competitive with, and possibly replace, ground transport, Redline drones would be considered a supplementary system to the existing road and ground transportation networks. As such, they would not operate in a purely "last mile" context, but would rather fly from fixed, futuristic-looking drone ports that would be built throughout a particular country and would sustain the transport of medical and other commercial goods over a wider region. These drone ports, designed by the noted British architect Lord Norman Foster at the cost of $70,000 each, would become hubs of emergency supplies and later commercial goods, as well as refueling and repair stations for the drones themselves. The current Redline plan is to construct eighteen drone ports in Rwanda but eventually to move on to other, larger countries in Africa.[100] At present, both types of drones remain in the prototype stage and the drone ports have not yet been built.

Redline drones would carry medical supplies, but its successor—the Blueline drone, capable of carrying a payload of 100 kg—would focus on commercial deliveries across underserved parts of Africa. Ledgard sees an enormous untapped potential in consumer spending in Africa among those who have cash but lack access to high-quality goods in major market towns because of problems with infrastructure and the postal delivery service.[101] The ultimate goal of the Blueline drones would be to connect towns of between 20,000 and 50,000 people and to build a web of air cargo pathways that could facilitate the movement of goods and maximize the choices available to consumers. In his conceptualization, the humanitarian delivery drone (Redline) serves as a gateway for a wider commercial application of delivery drones (Blueline) that will stimulate economic growth across Africa. In the long run, Ledgard hopes that the drone ports themselves would become central hubs for drone and non-drone commerce in a particular region.

The Redline/Blueline model is an explicit effort to humanize drones and to recast them as something other than a purely military instrument. Drones have an inherently negative connotation because, as Ledgard notes, "there is a stickiness around the 'drone' word. If people were as scared of drones as they were a year ago, when there was a muddling in the public mind between military and civilian drones, there would be a problem. But people are now accepting that there are military and civilian drones." At the same time, "the bristling word 'drone' which always conveys the idea of an insect, or of a worker doing a mindless task, is a problem. Intellectually, it's better to think of this as flying robotics. But it's hard in English to find a good word for this." He has experimented with a number of other descriptions for cargo drones, including "flying donkeys," in order to convince the local population that they are a normal, almost routine, part of commerce, rather than something to be feared. If the Redline/Blueline model succeeds, it will show the local population that medicine and consumer goods are just as likely to come from drones as are bombs. As Ledgard notes, the success of the Redline/Blueline model could be ultimately described as a "postmodern version of the idea of converting swords into ploughshares."[102]In cases like Rwanda, in which the post-genocide government has not only survived but emerged as a strong regional player, it is possible to see the investment in delivery drones in exactly these "swords and ploughshares" terms. But in cases where swords are still present—that is, in cases where countries have fallen into the abyss of civil war—delivery drones struggle to get off of the ground. To some extent, this is because delivery drones are hardest to make work in environments where they are most needed. As the Zipline example shows, some delivery drones projects presuppose a government that can work with aid givers, establish workable regulations, and exercise some control to de-conflict the local airspace. It is no accident that both Zipline and Redline have been tested in Rwanda, one of the strongest and most capable states in sub-Saharan Africa. These missions also presuppose trained staff, community engagement, and pre-mission planning, which often involves people on the ground in the area where the drops will happen.[103]

In cases where political order has utterly collapsed due to civil war or other calamities, none of these preconditions are present, and as a result delivering humanitarian relief via drones becomes much harder. In 2014, Mark Jacobsen, a former US Air Force C-17 pilot, founded the Syria Airlift project to deliver medical supplies and water purifiers to besieged populations in the Syrian civil war, in which hundreds of thousands have died since 2011. His hope was that one team with ten fixed-wing aircraft could deliver 400 pounds of goods, in small packages, each night.[104] The project had some successful test runs, but ultimately failed at the pilot stage. The Syrian airlift project fell apart for a number of reasons, not the least of which was that requirements for safety and reliability for drones flying in such ungoverned, dangerous airspace increased the complexity and costs of the project. Few traditional backers, such as governments, were willing to assume the liability for funding delivery drones flying in airspace populated by US and Russian aircraft or controlled by the dictator Bashir al-Assad. Equally, for-profit companies could not be convinced relieving suffering through drone deliveries was somehow profitable.[105] The risks of delivery drones in these cases—that the packages might go awry, fall into the wrong hands, or be mistaken as bombs—were simply too high. With today's technology, delivery drones are possible, but to work they need a favorable regulatory and political environment, a local population not immediately suspicious of the technology, and the alignment of the incentives of the project leaders and backers. It is unlikely that delivery drones will be a solution for reaching vulnerable populations in civil wars without significant improvements in the technology, high levels of confidence in their precision and reliability, and a significant drop in cost.

Peacekeeping Drones

While humanitarian organizations have been reluctant to deploy drones into conflict zones, the United Nations has been far more willing to countenance the use of drones in violent environments to keep the peace, although it is a relative latecomer to drone technology, only beginning to explore its potential in a systematic way in 2013. But the first use of a surveillance drone by UN forces was much earlier, in 2006, when European peacekeepers deployed to the Democratic Republic of the Congo (DRC) as part of the United Nations Organization Mission in the Democratic Republic of Congo (MONUC),[106] designed to monitor a cease-fire between various state-backed militias that had plunged the country into a bloody civil war in the late 1990s. The observation mission was composed of Belgian peacekeepers and a specially constituted European Force (EUFOR), both of which operated under a UN authority. The first drones came not from the United Nations itself, but from the national governments supplying the forces in the Congo. Both EUFOR and the Belgian forces used surveillance drones to interdict illegal arms shipments and had some success confronting the perpetrators with images of the shipments. The experiment ended after one drone was shot down and another was crashed in Kinshasa, killing one woman and injuring at least two others.[107] This crash also happened at a sensitive time, only a few days before a crucial election ratifying the post-war peace.[108]

In 2009–2010, Irish troops deployed as part of a peacekeeping mission in Chad brought drones with them, which were used for surveillance. The situation was similar to that of the DRC: the drones came as part of the equipment belonging to a national contingent, as opposed to the United Nations itself, but were eventually put under the UN's legal mandate for aerial surveillance.[109] The drones were used to monitor the movement of armed groups along the borders with Sudan and the Central African Republic and to improve the protection of refugees, internally displaced persons, and others.[110] They were particularly useful in detecting incursions and threats to the civilian population when domestic opposition forces came across the border from Darfur in spring 2009.[111] The drones were widely considered successful, especially for making sure convoy routes were clear of bombs and other threats, but they had some limits when operating at night.[112] In Chad, the drones were capable of picking up images and relaying them back to the UN command but the absence of drones and helicopters capable of night flight slowed the response to threats like banditry and attacks on civilians.[113]

Despite these successes, the idea that the United Nations might use drones and other means to boost its surveillance capacity in peacekeeping made a number of governments particularly uneasy. Governments are typically jealous of their sovereignty and are reluctant to let anyone, especially from a foreign country, operate in their airspace without approval. In the DRC, Rwanda objected to the use of drones in part due to fears that UN drones would be hijacked by intelligence services of Western governments and used to expand surveillance in Africa and elsewhere. This is particularly important because information gleaned from drones could convey advantage in the tense geopolitics of the Central African region. Oliver Nduhungirehe, a Rwandan diplomat, argued that "Africa must not become a laboratory for intelligence devices from overseas . . . We don't know whether these drones are going to be used to gather intelligence from Kigali, Kampala, Bujumbura or the entire region."[114] Some of the objections draw from the latent anti-colonialism that often distinguishes African politics and makes governments, like Rwandan government, suspicious of initiatives coming from Washington, London, and Paris. But some speculated that Rwanda, Uganda, and other countries were concerned that drones might reveal clearly the degree of their support for predatory militias in the DRC.[115] Others were concerned that drones might eventually put peacekeepers out of work. For some countries in Africa and Asia, being a paid peacekeeper for the United Nations is a lucrative business, which they are loathe to lose to a new technology. But the reluctance to using drones in peacekeeping is not confined to Africa. Other countries, such as China, Guatemala, Pakistan, and Russia, also raised concerns over the use of drones in UN missions in Africa.[116] Proposals to use UN surveillance drones in South Sudan fell afoul of similar concerns regarding information gathering and were outright rejected by its government.

Even with these concerns and objections, the momentum toward using drones in peacekeeping was difficult to stop, in part due to advocacy from within the United Nations to harness the technology for good. The first serious effort by the United Nations to support the use of drones was in February 2013, when the newly expanded United Nations Organization Stabilization Mission in the Democratic Republic of the Congo (MONUSCO) was granted the ability to buy and fly its own drone fleet. The first calls to equip MONUSCO with drones were made in 2008, when UN officials called for a surveillance capacity to boost their force protection and to allow for constant monitoring of the movements of government troops and rebel groups, as well as route clearance.[117] But for years the debate was stalled over questions about the appropriateness of deploying drones in peacekeeping operations and the ownership of the images and intelligence coming from the drones. The obvious shadow hanging over this debate was the United States' use of drones for targeted killing under the Obama administration. UN rapporteurs and human rights councils had traditionally been hostile to the Obama administration's expansive policy on the use of drones, especially in countries like Pakistan and Yemen, where the United States was not officially at war. That lingering unease translated itself into delays and some restrictions on the nature of the United Nations' drone use. The approval for drone use in the DRC happened in January 2013, but UN officials noted that it was contingent on the consent of the host government and that this authorization was "without prejudice to the ongoing consideration by relevant United Nations bodies of legal, financial and technical implications of the use of unmanned aerial systems."[118] The degree of unease around the United Nations' approval of drones in the DRC shows how much the technology itself has been associated with its use in armed conflict and targeted killings. As a consequence of this, advocates of "good drones" in humanitarian relief and peacekeeping missions are often forced to engage in a sustained public relations campaign to sell drone use to the public, especially in war-torn environments and those where drones had been used previously for military advantage.

Under the leadership of Hervé Ladsous, the head of the United Nations Department of Peacekeeping Operations (DPKO), the UN Security Council approved the deployment of drones as part of a new and stronger mandate for MONUSCO's in the DRC. Perhaps most important was the fact that the new mandate to use drones extended beyond protecting UN forces from attack and authorized the United Nations to use drones for the protection of civilians.[119] Implicit in this mandate was the view that drones would be effective in deterring rebel groups from straying across borders or threatening civilians.[120] The United Nations initially planned to send three drones to the DRC for deployment in the northern Kivu province, but eventually pushed that number up to five. It selected the Falco fixed-wing drone, operated by the Italian company Finmeccanica, for the mission.[121] Nearly half the size of a Predator, the Falco drone is capable of flying for ten to twelve hours at a time and at a distance of 150 miles. Operating at up to 18,000 feet, the drones are capable of seeing a wide swath of territory and capturing images of moving groups of people or trucks. Falco drones were equipped with synthetic aperture radar and a camera capable of taking nighttime and thermal images, as well as providing both still photos and real-time video.[122] The capability to take good nighttime photos is crucial because it provides some degree of coverage for periods of time when no UN peacekeepers are normally available. A lack of infrastructure and poor roads in the DRC meant that UN peacekeepers generally patrolled in a limited radius of 10 to 15 miles from their home base and returned home at sunset. As a result, most militia activity, especially from particularly aggressive and deadly groups like the M-23 rebel movement, happened at night.[123] For the United Nations, providing some coverage at night was crucial because MONUSCO had come under fierce criticism for failing to stop or respond to night banditry.

The United Nations' investment in drones in the DRC constituted approximately 1% of MONUSCO's budget.[124] The original agreement paid a company called Selex $13 million per year to run and service its drones.[125] Each Falco drone flew between one and two missions per day of about five hours in duration. By some estimates, the drones were vastly more efficient than manned aircraft given that one UAV could cover about the same ground as fifteen to nineteen manned helicopters.[126] The use of drones in the DRC also freed up the helicopter fleet for other tasks and allowed UN peacekeepers to reposition themselves against greater threats.[127] There were also unsubstantiated claims that some rebel groups, such as the M23 and FDLR, who were backed by Rwanda, were more likely to surrender because of the pressure placed on them by drones. In DRC operations, drones were also able to spot capsized boats and illegal smuggling operations.[128] But rebel groups hidden under a dense canopy of jungle were less likely to be affected by drones.[129]

Despite these accomplishments, drones are not an unabashed success in the DRC. They are not in the air constantly nor are they capable of seeing a lot of rural areas where rebel group activity might be threatening the welfare of civilians. With often only one drone in the sky at a particular time, large parts of the DRC—itself a country the size of Western Europe—go unmonitored.[130] Even for those areas monitored, the imagery was returned too late to be considered useful, thus undermining the claim by UN Secretary General Ban Ki-Moon that drones would improve "timely decision-making."[131] Some critics suggested that the record of drones were overblown and that they had made "no discernible difference" to most ordinary people. At present, there is no systematic evidence to suggest that drone peacekeeping is

FIGURE 7.2 Technicians examine an unarmed UN drone in Goma, North Kivu province, Democratic Republic of Congo, 2013.

effective in reducing violence. It is particularly difficult to assess the validity of claims that drones have been effective in deterring rebel groups from crossing borders or preying on the civilian population. Some UN officials are convinced that the mere appearance of drones is enough to worry rebel groups and to get them to rethink some of their actions, much like a thief would be deterred by a CCTV camera in a store.[132] UN officials have admitted to flying Falco drones at low altitudes and exploiting the sounds they make to deter rebel groups on the ground.[133] They have also admitted to having drones hover over militant camps, as a way of saying "we know where you are, surrender."[134] One UN official reports that that rebel groups have moved away from areas where drones are in the air.[135] If this is correct, drones may be altering the dynamics on the ground, but it remains to be seen if they are reducing violence. For example, the presence of drones in daytime might deter rebel groups from launching attacks when drones are watching, but this might only make nighttime attacks more likely. In other words, altering the dynamics on the ground does not necessarily imply a reduction of violence, only a redistribution of it. Actually measuring the effect of deterrence—that is, whether an overflight stopped an attack—is notoriously difficult to do and has not been attempted by the United Nations.[136] The idea that drones will deter attacks is intuitively attractive, but it is hard to know for certain whether these attacks have been truly stopped or merely re-directed toward times when the drones are not watching.

Even if the United Nations did get the images generated by the drones, it is not self-evident that they would respond quickly enough to make deterrence work.[137] The time that it takes to capture and relay images is a substantial problem, particularly for large geographic areas like the DRC. The images must be conveyed to a ground station and analyzed before they can be useful. Traditionally, the United Nations has been unable to do intelligence analysis in a systematic way because of internal reluctance to do something so explicitly "military". It also struggles because national governments tend to be unwilling to share their intelligence with the United Nations or loan assets like drones to its forces. But the United Nations is moving in fits and starts toward developing a capacity for collecting and analyzing intelligence and in December 2014 published a report called *Performance Peacekeeping* that examined ways that the DPKO could use new and emerging technologies to enhance their ability to do peacekeeping missions.[138] Aside from recommending that drones be used for peacekeeping, policing, and border control, the report also recommended that the DPKO and other relevant bodies boost its capacity to collect data and images from a number of sources (satellites, sensors, CCTV cameras, and drones) to make peacekeeping a much more nuanced and information-rich enterprise. At the moment, this remains aspirational, as the UN mission in the DRC tended to have few analysts combing through the images, collecting and converting them into operational recommendations.[139] Only $13 million of the MONUSCO's overall budget of $1 billion is currently spent on surveillance, and only a fraction of this is devoted to analysis.[140] By contrast, the UN mission in Mali has been able to use drones and to develop a more sophisticated joint intelligence center, led in part by European troop contributors with their own assets, than was possible in the DRC.[141] But national governments who use drones are more likely to collect intelligence and share it with a secure circle of close military allies, as happened in Mali, than with the United Nations itself. Until the United Nations invests enough to develop the surveillance and analysis capability that national governments have, it is unlikely that their drones will be as comparably effective.

A more general problem with drone peacekeeping is that the presence of the drones themselves raises expectations in the civilian population beyond what the United Nations can currently achieve. To populations traumatized by war, the sudden appearance of drones can look like a technological miracle and suggest that the United Nations will be constantly watching and protecting them. The reality is very different: most drones currently only cover a limited area during a short period of time, and as such they can be evaded by adaptive rebel groups. Drones cannot fly effectively in bad weather or dense cloud cover and will crash or fail in sandstorms and other extreme weather. Even if drones spot incursions, it does not necessarily follow that the United Nations can or will respond with force to protect a civilian population. As the US record in Rwanda demonstrated, knowledge of an atrocity as severe as genocide does not imply that a government or organization will be moved to respond.[142] Drones are a particular problem for setting expectations because as a sleek, futuristic technology they appear to promise too much. Some Congolese, for example, puzzled over why the drones were not armed and why they did not simply kill the rebel groups threatening them.[143] Setting expectations among the population about what drones can and cannot do—and by extension, what the organization possessing them is willing to countenance—is at least as important as the technology in the missions.

There is a further irony with expectations inside the United Nations itself. While raising the expectations of the population, drones can also increase pressure on UN commanders by raising expectations of what they can and should do in time-sensitive environments. A UN mission equipped with drones can see more and should make fewer mistakes than one in which information was received from other sources. For UN commanders, the flow of rich images from drones may add to the time pressure of their decision-making and raise expectations that they will make fewer, if any, mistakes. One danger pointed out by John Karlsrud and Frederik Rosén is that "radical improvements in knowledge accessibility will also lead to heightened requirements for the application of this knowledge when using force, as well as an increase in the demand for that force."[144] UN commanders may find themselves being asked to do more to respond to images of atrocities coming from drones while at the same time to make fewer mistakes than ever before. The presence of video from drones means that they will be held accountable for mistakes in ways that would have been impossible before. In other words, drones may modernize peacekeeping, but also make it more demanding than ever.

Risks

The use of drones may be different in a number of ways for humanitarian operations and for peacekeeping, but it presents three similar risks. The first is privacy. From crisis mapping to watching borders in conflict zones, surveillance drones can capture images of people not just in public settings but also in homes, hospitals, and other locations where they have a reasonable expectation of privacy. Some of these drones will take only still images, but others will take live video and draw on other sources of data, including eventually SIM card data from cellphones, from targets on the ground. In emergencies like Nepal's earthquake, it is obviously not possible to seek the consent of those filmed from above with drones. Equally, it is hard to control how the footage will be used. Especially following disasters in which new, often unregulated, actors appear on the scene with drones, there is relatively little information about who is filming individuals or what they will ultimately do with the footage. Similarly, when drones are handled by contractors, it is unclear what these private, for-profit organizations owe the local population in terms of privacy expectations.[145] As chapter 6 discusses, drones are scrambling the conventional expectations of privacy in the developed world, but they are also doing so in humanitarian crises in the developing world where the population is even more vulnerable. Some organizations are beginning to recognize this challenge and respond with guidelines. In January 2017, the Harvard Humanitarian Initiative published the Signal Code, which attempts to ground the protection of information in humanitarian crises as a human right and offers some guidelines for how responsible organizations should behave.[146] Other organizations like UAViators have also begun to develop codes of practice which lay out the expectations of privacy that individuals on the ground should have in humanitarian operations. The problem remains one of enforcement: if an organization violates the privacy of an individual during a humanitarian or peacekeeping mission, who can punish them effectively? How can we protect the privacy of people in desperate situations whose names we may not even know?

Another set of risks concerns who owns the data and images that are produced as a result of drone overflights. The guidelines and procedures for retaining and sharing information and images in crisis zones remain unclear and somewhat contradictory. Many of the "digital humanitarians" flying private drones over places like Nepal and Haiti have argued that data and images should be shared widely and even uploaded to common software like OpenStreetMap in order to make relief efforts easier. Although strictly speaking the images are owned by the organization that flies the drones, the cultural emphasis of the digital humanitarian movement is toward open source sharing so that as many people as possible can see and help those in need. But an absence of strong regulations on sharing images and data from drones presents a problem because freely shared images could be misused by some governments to hunt enemies, attack dissidents, and put pressure on refugees to leave. In other words, the digital humanitarian movement needs to be careful about simply throwing images up online with the assumption that only people who want to help will see them. As responsible organizations like UAViators acknowledge, this movement needs instead to develop data and image protection policies that acknowledge how these resources could be misused for malicious purposes. The United Nations is slightly different: it owns the data and images drawn from the drones and is keenly aware of the potential risks of misuse. As a result, the United Nations does not always share what it collects even with the host government or partners in its operations. This can cause friction with some host governments who have claimed that the United Nations is collecting intelligence on security threats which it refuses to share. But the United Nations has its own problems in terms of data ownership. As an institution, the United Nations has traditionally lacked a clear policy on privacy and has weakly enforced the handling of classified data and images inside some missions.[147] The absence of clear regulations and guidelines for the use of data and images from UN drones will only multiply the policy dilemmas that commanders face on the ground. For example, it remains unclear whether the United Nations can or should share data with local police forces or whether evidence that it collects could be turned over to the International Criminal Court.[148] Transparency about what is collected, and who may see it, remains a serious concern for UN member states.[149]

Finally, the humanitarian use of drones has itself been complicated by UN efforts to use drones for peacekeeping purposes. Those deploying drones in humanitarian emergencies for mapping or search and rescue have always framed their drone deployments as consistent with the principle of neutrality. Such deployments are committed to helping all and delivering information and aid to everyone in need rather than only one side in the conflict.[150] This principle of neutrality is well established in humanitarian practice and enshrined in UN resolutions and the practice of the International Committee of the Red Cross (ICRC). For many digital humanitarians, it is an article of faith and a principle by which the legitimacy of their operations is conferred. By contrast, a number of recent UN peace enforcement missions are explicitly non-neutral and involve the United Nations taking a side and punishing the parties seen as most threatening to the peace. In the DRC, for example, UN forces fought with government forces against the M-23, FDLR, and other rebels seen as dangerous to the peace; in Mali, UN actions were directed against Islamist militias threatening the government. In these cases of peace enforcement, UN surveillance drones may not be armed but they are certainly working for one side in the conflict. For this reason, a number of humanitarian organizations are concerned that the use of drones by the United Nations for peacekeeping might mean that "the lines are blurred between help and harm," or between neutral and non-neutral actors in a conflict.[151] One consequence of this might be that rebel groups will fail to distinguish between neutral humanitarian drones and non-neutral UN drones and attack the former. Others worry that humanitarian organizations on the ground will be targeted in reprisal for military activities that the drones of the United Nations or national governments are involved in. In a world where a number of actors are deploying drones in conflict zones for different reasons, drawing a recognizable distinction between humanitarian drones, UN drones, and even local military drones is increasingly important. One simple potential remedy for this is to paint drones in different colors to ensure that they are distinguished as humanitarian, as opposed to military or UN vehicles. But even this will require the local population understanding the difference and tailoring their responses to the drones accordingly.

Conclusion

The use of drones for "dull, dirty, and dangerous" missions has been both lauded and criticized for providing a positive face for a technology that is more closely associated with military use and targeted killings.[152] There is certainly evidence that drones can do good in humanitarian and peacekeeping missions and bring capabilities for surveillance and data collection honed in military operations to bear in innovative ways. At their best, they may be able to "lift the fog of war" that makes delivering humanitarian relief and establishing a workable peace so difficult.[153] But it is important to measure the risks of their use in these environments, rather than just succumb to techno-utopianism about their potential effects. Drones will not replace aid workers or revolutionize aid delivery in the short term. Even the strongest advocates of humanitarian drones acknowledge that the technology must improve and the costs of drone use must plummet in order for the impact of drones will be known.[154] It is also important that the organizations' goals for drones remain limited: it is possible, in an environment of overwhelming need, for goals to expand and for drones to be attached to all manner of problems, even when they are not the best tool to use.

Perhaps a greater risk of drones in this space is that of virtual humanitarianism. The movement toward disaster drones has produced a rush of new actors into crisis areas, but many are start-ups whose interest in the country extends only for a short period of time. Their surveillance and airdrop activities, however useful in the short term, are not always certain to produce a long-term benefit for the population, especially as global attention shifts to the next crisis. The danger is that humanitarian crises and peacekeeping missions could become little more than laboratories to test drone technology on a population in stark need but with little agency about how it is treated. Drone humanitarians must guard against the belief that they can simply provide data or airlift parcels of food and medicine by drones and be done with a crisis. Such an approach might be a natural temptation with unmanned aircraft, but if left unchecked it will undercut the genuine humanitarian purpose of the mission. For both humanitarian crises and peacekeeping, it is important that drones not be directed simply toward managing the effects of a crisis, but rather be embedded in a comprehensive approach that is directed at ending political instability and alleviating human suffering over the long term.

8

The New Race

ON SEPTEMBER 9, 2013, the Japanese Air Self Defense Force (JASDF) detected an unknown aircraft hovering approximately 200 km north-west of the Senkaku islands in the East China Sea.[1] The ownership of these uninhabited rocky islands, only 2.7 square miles in size, has been long disputed by China and Japan. The islands—known as the Diaoyu islands to China—have been under Japanese administrative control since 1972, although China had never relinquished its legal claim that the islands were part of its sovereign territory.[2] The scramble for control of these desolate islands had intensified as China's booming economy increased its appetite for oil and gas supplies worldwide. Reports of vast deposits of oil and gas in the seabed below the islands had made seizing control of them and asserting ownership rights over the surrounding territorial waters a high priority for China. Accordingly, its efforts to challenge Japanese control over the islands had become more insistent. China Maritime Surveillance (CMS) vessels sailed near the islands and manned Chinese aircraft edged around the margins of the island's airspace.[3] But until that September afternoon, China had never attempted a drone incursion over the Japanese-controlled airspace over the islands.

Within minutes of the incursion, the JASDF scrambled F-15 fighter jets to confront the aircraft. They were shocked at what they found: a sophisticated Chinese surveillance drone, believed to be a Harbin BZK-005, a high-altitude, long-range drone employed by the Chinese navy. The Japanese fighter jets shadowed the Chinese drone until it eventually departed the islands' airspace. While China had deployed drones in the South China Sea before, it never had done so to test Japanese control of the Senkaku Islands. Japan was alarmed and demanded an immediate cessation of drone flights over the islands. China's response was blunt: "China enjoys freedom of overflight in relevant waters . . . The Chinese military will organize similar routine activities in the future."[4] Under pressure, Japanese Prime Minister Shinzo Abe granted approval to the Ministry of Defense to shoot down any drone that ignored warnings to leave Japanese airspace. An unnamed Japanese defense official remarked that the Chinese drone incursion "was unexpected . . . I fear we are in the position of being one step behind."[5] Weeks later, Japan approved new guidelines for the JASDF to shoot down drones that enter Japan's airspace if warnings to leave were ignored.[6] Geng Yansheng, a spokesperson for China's Ministry of Defense, said that such an act by Japan would constitute "an act of war" and lead China to strike back with "resolute measures."[7]

Alarmed at the escalating rhetoric, the United States jumped into the fray to show its support for Japan. By early November 2013, the Obama administration had agreed to deploy its Global Hawk drones on joint missions with Japanese E-2C early warning aircraft in the airspace over the islands.[8] Japan also revised its defense spending plans and ordered the purchase of several US drones for its own fleet. On November 23, 2013, China responded by unilaterally extending its Air Defense Identification Zone (ADIZ) over much of the East China Sea, including the Senkaku islands. This expansion of China's claims rattled the United States, Japan, Taiwan, and South Korea and increased pressure on Washington for a tougher response. Chinese foreign ministry spokesperson Hong Lei remarked that due to the ADIZ extension, the islands were "an inherent territory of China. Japan's seizure and occupation of the islands are illegal and invalid."[9] The drone incursion became a forerunner of an expanded Chinese claim over territorial waters in the East China Sea, while the islands themselves had now become "a focal point for Sino-US competition and potential conflict."[10]

The Senkaku islands controversy sharpened the lines of division between China and Japan and led both to take measures to prepare themselves for a future war. For both countries, improving their ability to signal, threaten, and fight with drones became a high priority. In 2012, China began the construction of two drone bases to boost its capacity for surveillance over the Senkaku islands.[11] By 2014, China was allegedly planning eleven coastal drone bases to allow it to conduct surveillance over an even larger area.[12] In June 2015, a leaked People's Liberation Army (PLA) document revealed that China was considering deploying its Yilong surveillance drones to the airspace over the Senkaku islands.[13] Japan took notice of China's growing interest in drones and responded in kind. In 2014, the Japanese government announced plans to increase its spending on drones by 300%, a three-million-yen investment that would give it substantial surveillance capacity by 2020.[14] By 2015, Japan was debating shooting down Chinese drones that intruded into the airspace, but acknowledged that doing so could provoke an international crisis.[15] Incursions around the Senkaku islands have continued with Japan reporting another Chinese drone hovering over its territorial waters in May 2017.[16] In 2019, the Trump administration decided to up the ante by selling surveillance drones to Malaysia, Indonesia, Philippines, and Vietnam as a way of signaling that it would no longer "tiptoe" around Chinese aggression in the South China Sea.[17]

In time, the standoff over the Senkaku islands may be remembered as one of the first of the drone age. Today, a growing number of states are using drones as China did: to test the nerves and strategic commitments of their rivals, to chip away at relationships of deterrence, and to gain leverage in crisis bargaining scenarios. While most states are using drones for routine surveillance, emergency response, and other ordinary tasks, governments are introducing military grade drones into long-simmering regional conflicts to gain advantages over their rivals and to see what response, if any, they provoke. This matters because the unmanned nature of drones can change the underlying calculations of risk that lie beneath deterrence and coercion. In most circumstances, when using manned aircraft states are naturally sensitive to the risks of losing crucial military personnel and assets and tend to tread carefully in crisis situations. States also fear that downing another state's aircraft can damage their reputation and even invite more provocations. As both the United States and Soviet Union acknowledged during the Cold War, attacks on manned aircraft that kill a pilot or crew can precipitate a crisis with a momentum that proves hard to reverse with ordinary diplomacy.

With today's unmanned aircraft, the calculation is not so straightforward. Almost all of the underlying assumptions about risk, credibility, and the potential for crisis escalation are scrambled when a human being is no longer in the cockpit. With no lives at stake, governments may see drones as cheaper or more expendable and take risks that they would not do with manned aircraft. For example, they may use drones to rattle their opponents and to test the limits of deterrence, under the assumption that an unmanned aircraft will not provoke the same military response that that manned ones do. They may further gamble that shooting down a drone will not precipitate an irreversible international crisis, as killing a pilot in a manned aircraft probably would. Aside from increasing risk-taking in deterrence, drones may also lead to an expansion of a state's goals and lead them to try to coerce enemies in new and potentially destabilizing ways. In a world awash with drones, strategic moves once thought too dangerous become possible, even attractive, if they can be conducted without physical risk to human pilots. But this perception of limited, controllable risks may be wrong; there is no guarantee that one's enemies will "see" drone incursions in exactly the same way as their opponent and respond with restraint. A world full of risk-taking with drones, but marked by uncertainty about how strategic moves involving them will be interpreted, is one in which crises may escalate in surprising ways.

Global Diffusion

The risk of new strategic interactions between states with drones is growing precisely because the technology has spread so quickly, but unevenly, around the world. Less than two decades ago, only a select number of governments with highly capable militaries, such as the United States, Russia, China, Israel, and the United Kingdom, had even small drone fleets. Even these governments tended to see drones as unreliable and expensive and cut those programs first to save other defense programs in times of austerity. As the air wars in Afghanistan and Iraq ramped up, the United States vastly increased the size of its drone fleet to meet reconnaissance and combat demands by ground commanders and to target al Qaeda and associated forces. By contrast, most of the other countries waited until the technology matured and costs decreased to build up their drone fleets in a significant way. When the technology became reliable and the cost curve flattened, drones began to spread to more countries worldwide, depending in part on their technological capacity to develop and run them.[18] At present, at least eighty countries have unarmed drones and twenty-four have armed drones.[19]

Although drone diffusion is a fact of life, not all the world's drone programs are the same; they range in sophistication from relatively small boutique programs to growing multipurpose drone programs for combat, surveillance, and other uses. For many states, indigenous development of drone technology has moved slowly and remains largely imitative, copying popular US and Israeli models and adapting them to local needs. In some cases, the rationale for developing a local capacity for building drones is essentially one of prestige: developing a token drone industry is a signal that the state is a serious military player in the world. As a status symbol, a drone industry has become a "must have" no matter how strapped defense budgets may otherwise be.[20] For this reason, many governments now sponsor small or moderately sized research and development (R&D) programs for drone technology, which are in turn supported by smaller commercial drone firms hoping to corner the local market. This effectively makes the global diffusion of drones look greater than it is. The drone programs in many states have small ambitions. Many are designed only to be effective inside their own borders, and do not aim to match the level of sophistication of US drones like the Reaper or the Global Hawk. But other countries have set their sights higher. Through government-led and private investment, countries such as Russia, China, Turkey, India, Pakistan, and Iran have produced increasingly complex drone models, sometimes by harnessing innovations from their commercial drone industries or by copying US and Israeli technology.

Skeptics have argued that global drone diffusion will run into its natural limits due to the fact that the technology requires a costly organizational infrastructure, as well as technical base of expertise, to support its development and use.[21] It is certainly true that military-grade drones are not an easy thing to construct from whole cloth. Designing drones takes a significant base of technological expertise and a defense manufacturing capacity; using them as the engine of a targeted killing campaign based on information dominance, as the United States has done, takes vastly more resources and is an option unavailable to all other governments at present. [22] Even to deploy drones on a regional basis, as China does, takes a network of military satellites, intelligence capabilities, and data bandwidth capacity that few countries have available. But more to the point, only a very small number of states have a compelling reason to develop drones with that kind of reach. Only governments facing substantial security threats, such as terrorist groups like al Qaeda or regional rivals, have a reason to invest deeply in drone technology in the way that the United States, Israel, and now China have done.[23] For the rest of the world, living with relatively peaceful, stable borders, there is no incentive to construct an equivalent of the Reaper. Instead, they will be satisfied with smaller range, less expensive drones that can monitor borders, stop illegal smuggling in drugs and other banned goods, and conduct routine military tasks like target-spotting for their ground forces. This suggests that indigenous drone development will stratify into two tiers: a select number of states who have the need and the organizational and infrastructure base to build high-quality military drones, and a wider array of states engaging in the production of drones which may be only a cut above the commercial models available but are sufficient for routine tasks.

To a degree, this stratification of the world of drones is complicated by two cross-cutting factors. The first is the explosion of the commercial drone market. An estimate by the Teal Group in 2013 suggested that drones will remain a global growth industry for the foreseeable future, with annual sales expected to double from $5.2 billion to $11.6 billion by 2023.[24] The Teal Group also estimated that approximately $89 billion would be spent in research and procurement on drones over the next decade.[25] In 2011, there were 680 active drone development programs run by governments, companies and research institutions, compared to only 195 in 2005.[26] This market has only grown. In 2016, PricewaterhouseCoopers LLP projected that the global market in drones would ultimately be worth $127 billion by 2020.[27] According to an estimate by Goldman Sachs, between 2017 and 2021, the majority of spending on drones will be concentrated in a few countries with highly developed programs (United States, Russia, China, United Kingdom,

Australia) but smaller programs in countries beyond this set will never-theless constitute $8 billion in spending.[28] Even where the commercial drone industry is small, it can yield benefits for smaller militaries be-cause they can use homegrown drones for routine tasks like reconnais-sance and monitoring the perimeters of a base. The global expansion of companies like DJI also allows some cash-strapped militaries to buy "off-the-shelf" drones for these tasks. Beyond the fact that a booming commercial market means more drones are available to buy, expertise also travels between the commercial and military R&D world, espe-cially in countries where there is a permeable border for employment between government and private industry. In such cases, even a small local drone industry will have a spillover benefit in improving the quality of drones available to that country's military.

The second complication for this picture is the military export market, which is putting more advanced technology into the hands of governments that could not otherwise develop it. The countries that export military-grade drones can be divided into two camps. The first camp consists of the first movers in drone technology, the United States and Israel, who have years of experience developing and using military grade drones and have a sophisticated defense-industrial base with deep expertise and extensive commercial links. Unsurprisingly, the United States and Israel are best positioned to capitalize on the drone export market and have been the leading global exporters so far. The second camp consists of relative newcomers to the drone industry—such as China, Russia, India, South Korea, and Japan—who have poured re-sources into drone development and are poised to catch up, even imper-fectly, with the United States and Israel in the export market. According to an estimate by IHS Janes produced in 2016, these five countries are due to account for $3.4 billion in drone export sales between 2015 and 2024.[29] Another new potential exporter is Turkey, which has used its defense manufacturing base and its experience in a long-running war against the Kurds to hone indigenous models and prepare them for sale.[30] Of these six countries, China is the best positioned and most eager to capitalize on the export market. This suggests that the export market in the future may be dominated by three large exporters (the United States, Israel, and China), with the other exporters fulfilling highly specialized demands for drones. In part due to the actions of these three actors, the export market will drive global drone diffusion, produce competition that lowers costs, and put sophisticated drones in the hands of governments that could not otherwise develop them.

United States

The vast R&D base for military and commercial drones present in the United States puts it in a strong market position for exports, although its actual record of global exports still fall below those of Israel. According to the Teal Group, the majority of global drone spending—65% of the R&D and 51% of the procurement—will come from the United States between 2013 and 2023. Its long-running investment in drone technology has given the United States a qualitative advantage, and it often develops models years ahead in quality compared to others. This lead in R&D is due in part to deep links between government and private industry. The US commercial drone market—already a world leader, but estimated by the FAA to grow tenfold between 2017 and 2021—is a source of expertise and ideas for innovation in the military drone sector.[31] The United States has also developed an array of private and university-based drone labs which are partially supported by Department of Defense funding. This funding allows commercial labs to experiment with drone prototypes and to develop capacities for autonomous flying, swarming, and other types of innovations.

Given its lead in advanced drone technology, the United States is in the best position to take advantage of the opportunity to sell drones on the international market. Yet its actual record of exports is relatively modest. The total direct military commercial sales for US drones was $240 million during the period 2005–2011, with an additional $144 million transferred as part of the foreign military sales program over nearly the same period.[32] Most of the early US drone sales went to NATO allies, such as the United Kingdom and Italy, and US officials initially rebuffed calls to sell sophisticated models such as the Reaper or Global Hawk to other long-standing partners. This reluctance was partially due to legal agreements on technology sales and transfer under US law. For the United States, all international arms sales are governed by the Arms Export Control Act (1976), which sets standards for designating the "friendly countries" that the United States can export to, and all US companies are bound by the International Traffic in Arms Regulation (ITAR) law which requires State Department approval in advance of a sale.[33] Drone exports were particularly restricted by the Missile Technology Control Regime (MTCR), a voluntary agreement among thirty-four states which limits the sales and export of heavy payload weapons, and by the Wassenaar Agreement, which requires participants to share information on deliveries, especially for dual-use technology.[34] The Obama administration proposed changes to the MTCR to enable more drone exports, but was unable to win agreement on the proposed changes from other signatories.[35] Further concerns among US officials over the ability of the United States to monitor the uses of drones by other countries also led to caution in implementing an aggressive export policy.

This caution over drone exports incurred the wrath of powerful drone manufacturers who believed that the United States was foolishly sacrificing a chance to become the world's most prolific drone exporter. With jobs and profits on the line, major companies like General Atomics, Northrup Grumman, DJI, and others sought allies on Capitol Hill and formed unmanned aircraft caucuses in the House and Senate, in part to push for reforms to export policy.[36] Drone lobbyists worked to reclassify some drones to exempt them from MCTR and Wassenaar restrictions and to allow manufacturers to tap growing markets in Asia, Europe, and Latin America.[37] A 2012 Congressional Research Service report put the pro-export argument succinctly: "Much new business is likely to be generated in the market, and if US companies fail to capture this market share, European, Russian, Israeli, Chinese or South African companies will."[38] Worried that these companies will lose market share to foreign companies, the Obama administration began to consider expanding the range of states that are pre-approved for drone sales.[39] The initial draft guidelines proposed by the Pentagon in 2012 suggested that the United States would be permitted to export at least some drones to sixty-six countries around the world.[40] But developing a policy ultimately took two years of intense behind-the-scenes negotiations between a number of government agencies that had a stake in drone exports.

What emerged was a cautious, hedged export policy that shied away from blanket approval of drone sales, particularly of the armed variety.

In February 2015, the State Department announced a final policy for the export of military and commercial drones. According to their guidelines, the United States could export armed military drones to other governments under strict conditions, among which was that the recipient government must agree to end-use assurances to ensure US technology was not deployed for illegal purposes, such as repression of the local population. The new export policy did not have a formal list of countries approved for drone exports, as some in the Pentagon originally envisioned, but rather processed each request for a sale on a case-by-case basis.[41] The policy had a "strong presumption of denial" for licenses to export lethal drones and required four specific commitments from those that received US drones: (1) that the drone system would be used in accordance with international law, including international humanitarian and international human rights law; (2) that the drones would be deployed "only when there is a lawful basis for the use of force under international law, such as national self-defense"; (3) that the drones would not be used to "conduct unlawful surveillance or use unlawful force against their domestic population"; and (4) that drone users would have to agree to require technical and doctrinal training to "reduce the risk of unintended injury or damage."[42] These standards are relatively high and, as one critic noted, it is not clear that the United States would meet them for some of its own military operations, especially targeted killings.[43] The Obama administration's policy was designed to be consistent with the MTCR and to restrict the transfer of high-payload drones.[44] Under the new export policy, medium- or high-altitude drones would rarely be exported and then only to NATO allies who could use them for counterterrorism and other US supported missions. In many of these cases, they would be used for operations in which the United States shared a military command with its NATO allies, such as airstrikes in Libya and Iraq. Even with these restrictions, US officials nevertheless argued that this export policy would vastly improve the capability of partner nations and improve interoperability with US military forces, especially for intelligence, reconnaissance, and surveillance missions.[45]

The Obama administration's export policy on drones was designed to resolve some of the uneasiness around unleashing drone sales worldwide, but it left a number of questions unanswered. The unclassified parts of the export policy did not make clear what substantive penalties, if any, a state would incur for violating any of these end-use assurances. The United States could use its existing Blue Lantern end-use monitoring system, administered by the State Department, to check compliance with these conditions. In general, a failure to meet end-use assurances or issues involving misuse or misdirection of US technology leads to a revocation of a license and a ban on future sales. Such a program would presumably allow the United States to use its market power to punish those who misuse US technology, but monitoring compliance in this way would be limited by the high levels of secrecy that many countries apply to their drone programs. It is also not clear if excluding countries from future US drone sales would be an effective deterrent, given the growing number of other exporters in the current drone market. To do so would require the United States to induce its allies who are drone exporters, such as Israel, to refuse to sell to those who misuse drone technology. Even if Israel and NATO allies like the United Kingdom were cooperative, the United States has relatively little leverage to force other less friendly drone exporters like Russia or China to do the same; on the contrary, in a world awash with drones, every instance of US restraint opens up a business opportunity for someone else.

This new policy enabled the United States to export non-lethal drones to a larger range of non-NATO allies, including Panama, Burundi, Lebanon, Egypt, and Pakistan.[46] As the export policy was being shaped, the Obama administration also sought to build an international consensus on how exported drones might be used. In October 2016, with the new presidential election looming, the Obama administration issued a joint declaration on the export and use of armed drones.[47] The policy, supported by forty governments, called for compliance with international law, transparency, and a voluntary code of standards for the sale and use of drones. While some powerful NATO drone users signed up to this agreement, many US rivals for the export market—Israel, China, Russia, among others—did not agree to the declaration for both financial and political reasons. At a minimum, such a code of conduct would forbid the sales of armed drones by those states to some lucrative markets and push them into direct competition with the United States for the few "approved" markets left. But some also noted that the call for voluntary standards was ironic because the Obama administration could not even confirm that its own drone usage, particularly for targeted killings, met these standards. There were few political reasons to say "yes" to a lame duck Obama administration and potential costs to incur for doing so if the Republicans won. Beyond that, this proposal seemed like slamming the stable door shut after the horses had already bolted.

The administration of President Donald Trump was initially slow to move on drone exports but eventually broke with the caution that characterized the approach of the Obama administration. In April 2018, the Trump administration announced a series of policy changes designed to loosen the restrictions on drone sales.[48] It changed the process by which sales had been approved and eliminated some of the bureaucratic barriers to direct sales by companies to foreign governments. It also reclassified drones with strike-enabled technology, like laser target designators, as unarmed, thus making them easier to sell. The Trump administration also signaled a willingness to renegotiate the MCTR to exclude drones and some of their related components. Part of the administration's objection to the MCTR was that the United States was losing not just customers but control, as its allies were increasingly buying drones from China. When this happened, the United States essentially lost the ability to influence how those drones were used.[49] The goal of the Trump administration was to increase the speed and volume of US exports and allow US manufacturers to compete more effectively with Israel and China for business with non-NATO countries. But this was not an overwhelming success: slow approval processes for foreign exports inside the Department of Defense delayed the sales and the United States still found itself losing business, especially in the Middle East, to China.[50]

Israel

One of the most notable states that rejected the Obama administration's call for voluntary export standards was Israel. Israel plays a complex role in the drone export game, producing and selling more than virtually any other player in the export market, but calibrating its export policy in a way to avoid a direct confrontation with the United States. For Israel, the commercial imperative to sell drones is balanced against the geopolitical imperative of maintaining its alliance with the United States. In many cases, however, the commercial imperative prevails. Israel's powerful defense manufacturers and commercial aviation industry have pushed the government to capitalize on the advantages that the country possesses in the drone market. A number of Israeli companies, such as Israel Aerospace Industries, Elbit Systems, Rafael, and Aeronautics Defense Systems, are world leaders in drone technology R&D. Historically, Israel has been the world's most significant drone exporter, selling 60.7% of all drones sold worldwide between 1985 and 2015. By some estimates, Israel has sent drone technology to at least fifty countries since it began exporting in the mid-1980s.[51] One report on Israeli drone production and sales concludes that "the reality is that if you scratch any military drone and you will likely find Israeli technology underneath."[52]

In recent years, Israeli drone exports have increased in volume. Between 2005 and 2013, Israeli drones exports have totaled $4.6 billion in sales to a wide variety of countries, including Australia, Singapore, Turkey, Thailand, Azerbaijan, Nigeria, Uganda, and Indonesia.[53] According to the Stockholm International Peace Research Institute, Israel accounted for 41% of all drone exports between 2001 and 2011.[54] By 2014, drones accounted for 10% of all Israeli military exports.[55] Israel is projected to export $500 million per year of drone-related technology (including both anti-drone technology, sensors, etc.) with an annual increase of 5%–10% between 2015 and 2020.[56] Although Israel sells to a wide range of countries, its biggest clients are countries with sizable militaries, such as India, Brazil, and the United Kingdom. Israel's export destinations include Europe (50.2%), Asia Pacific (33.3%), South America (11.2%), and North America (3.9%).[57]

Israel has managed to make drones attractive to a wide range of countries with very different needs. For example, its Heron model, widely used in Gaza, has become a default surveillance drone for many states without indigenous production capabilities. One reason that Israel has been successful is that it has emphasized the production of drones which are adaptable both to a wide-range of purposes and to very specific niche needs. Such adaptability enables Israeli companies to aggressively advertise their products to a wide range of potential consumers inside and outside the military.[58] To some extent, Israeli drones also benefit from US restraint. Since they are not subject to the same foreign military sales legal restrictions that the United States imposes, Israeli companies are free to sell to a greater range of countries than their US competitors.[59] In general, Israel is also less concerned about selling to countries that have a long history of human rights abuses.[60]

Israel's export strategy is also tailored to boosting its ability to sell. Israeli companies sometimes set up a network of subsidiary companies in countries where there is a strong preference for domestic production.[61] This allows them to sidestep regulations that insist on domestic production for military assets. Israel also sometimes sells through third country suppliers.[62] As a result of these moves, Israel has now become the main supplier of drones for many states in Europe, Asia, South America, and Africa. The chief limitation that Israel faces in exploiting the drone market lies in the hostility that it experiences in the Middle East. Most neighboring Arab states will not purchase from Israel, which in turn leaves an opening in that regional market that other countries like China and Russia will seek to exploit. The sales of Chinese drones to the United Arab Emirates, Jordan, and other states show how this gap in the market might be filled in the future.

While eager to exploit the drone export market, Israel is also generally careful to pay attention to US sensitivities when approving exports to Washington's enemies and less trusted friends. In a few cases, Israel has brokered co-production or service agreements, rather than outright sales, in an effort to mollify the United States. Nevertheless, Israel has shown itself willing to break with the United States over drone sales to big markets like China and Russia. In the 1990s, Israel sold China 100 Harpy armed drones.[63] When Israel considered selling an anti-radar attack drone to China in 2005, the Pentagon temporarily shut Israel out of the F-35 aircraft program.[64] Similarly, in 2009, the United States objected to an Israeli sale of drones to Russia, according to documents released by WikiLeaks.[65] With Russia, Israel continues to do a particularly delicate dance. It has sold drones to Russia and to its enemies, including Georgia, leading to the unusual situations of both combatants in the same war using Israeli technology.[66] The Israeli Foreign Ministry has been willing to use drone sales as a bargaining chip to get Russia not to sell weapons systems to Iran, although it has to steer clear of transferring to Russia any technology jointly developed by US manufacturers for fear of violating US law and intellectual property standards.[67] Israeli manufacturers have also courted controversy in other ways. In 2017, one Israeli drone manufacturer, Aeronautics Defense Systems, was accused by the Israeli government of using the Orbiter 1K drone in a kamikaze attack on Armenian soldiers in an attempt to sell Azerbaijan the technology.[68]

China

Of the newcomers to the export market, China is in the best position for selling in the years ahead. The Chinese drone program began in the early 1960s, when the Soviet Union transferred its La-17 target drone to the PLA for target practice. This drone was reverse engineered and rebranded as the Chang Kong 1 target drone. Similarly, the PLA's recovery of a US Firebee drone provided the underlying technology for what became the Chinese Wu-Zhen 5.[69] In the 1980s, China began to invest in indigenous production and to develop a manufacturing base across government, private industry, and academic centers.[70] While many of China's drones still lag in quality behind the top US and Israeli models, the scale of the investment across these sectors in China suggests that it will soon become a global competitor with the United States and Israel. A 2012 assessment from the US Defense Study Board concluded that China could "easily outmatch or outpace US spending on unmanned systems, rapidly closing the technology gaps and become a formidable global competitor in unmanned systems."[71]

While many drone newcomers build relatively small or specialized drone programs, China has done the opposite, investing heavily in an array of R&D centers around the country and diversifying its production of drone models. Some of its leading universities, such as Beijing University and Nanjing University, have already yielded drone models which have been deployed by PLA units.[72] China has strong existing capabilities in avionics, propulsion, and flight control systems that can be rapidly adapted to drones.[73] In addition, virtually every major Chinese arms manufacturer has a research center on drones, and dozens of different models are in development.[74] The Aviation Industry Corporation of China (AVIC), state-owned and one of the leading commercial aviation companies in the country, also has a number of subsidiary companies which are heavily involved in drone R&D. Chinese intelligence is also interested in hacking US companies for information on their programs.[75] Many of China's drones are designed for target practice, ground support, and reconnaissance, but China is beginning to invest in long-range drones that could extend its ability to engage in surveillance.[76] China is also investing heavily in unmanned combat aerial vehicles (UCAVs) and mini- and micro-drones that might be used to overwhelm opponents. By 2020, China was expected to match overall US spending on drones.[77] While China's drones are not generally equivalent in quality to US models, they are often seen as good enough in quality and cheaper, thus drawing interest from countries that are barred from buying US-made drones.

China's drone program is also designed for a different type of battle than much of the US fleet. Most of China's drones are flexible platforms, used for surveillance, reconnaissance, and combat operations, and some of these are designed to counter US power projection capabilities in places like the South China Sea. A 2012 report from the Defense Science Board noted that China's move into unmanned systems was "alarming," especially given China's evident interest in using drones to swarm US vessels in the event of a naval conflict.[78] An estimate of China's drone program from Project 2049 noted that unlike the United States China's goal was to have drones that were successful in denied or contested airspace.[79] China has also been pursuing small, agile, and autonomous drones that could disable radar on a naval vessel, overwhelm it, and enable other, more conventional attacks. Another option that China has explored is the use of small, disposable drones that could fly directly into targets. Along these lines, China has been investing heavily in electronic warfare packages on its drones to jam GPS and other communication devices, to disable early warning systems and some defense systems, and to confuse the detection of incoming threats to US vessels.[80]

As of 2012, approximately 93% of China's produced drones were tactical in orientation.[81] But China is not content to leave the export of medium- and high-altitude drones to the United States alone. China developed its own iterations of the medium-altitude Reaper model, known as the Wing Loong, which can conduct reconnaissance missions but also fire missiles at targets. It has also developed a sophisticated high-altitude stealth drone, called the Lijian (or Sharp Sword),[82] a UCAV specifically designed for reconnaissance and combat operations in contested environments. Broadly similar to the US X-47B drone, the Sharp Sword has two internal bomb bays, equipped for holding a payload of 4,400 pounds of explosives and is designed to obscure its exhaust fumes to avoid detection by enemy radar.[83] While the United States abandoned the X-47B due to cost concerns, China pursued the model by mimicking US technology and now has the lead in the export market for those looking for stealth combat drones.[84] China has also developed the Divine Eagle drone, a highly secretive, high-altitude drone designed to provide a comprehensive picture of a specified battlespace, including tracking targets like US warships and enabling attacks through other means.[85] In contrast to the low- and medium-altitude drones that China intends to sell, China keeps these high-altitude drones, designed for a quiet but deadly serious geopolitical competition with the United States, cloaked in secrecy.

This penchant for secrecy regarding drones has extended to China's export policy as well. The desire to tap the export market has been among China's most important motivations for rapidly catching up in drone technology. Unencumbered by the MTCR and the export restrictions faced by the United States, China sees an extraordinary opening for exporting surveillance and attack drones to governments to which the United States cannot sell. Zhang Qiaoliang, a representative of the Chengdu Aircraft Design and Research Institute, remarked that, "The United States doesn't export many attack drones, so we're taking advantage of that hole in the market. The main reason is the amazing demand in the market for drones after 9/11."[86]

In 2011, China began offering for export the Wing Loong drone, which is capable of holding laser guided missiles and is comparable to the Predator in flight range. Unlike the Predator, it costs only $1–2 million.[87] In September 2013, the *China Daily* reported that China was making substantial gains into the drones market, and was negotiating with at least three governments on Wing Loong sales.[88] China eventually exported the Wing Loong I and II models to a range of countries, including Egypt, Saudi Arabia, Nigeria, United Arab Emirates, and Kazakhstan. In 2016, China reported making its "biggest overseas purchase order" in the sale of Wing Loong II drones. The details were kept quiet, but most reports suggested that the buyer was Saudi Arabia.[89] China also allegedly struck a co-production agreement with Saudi Arabia. The sale to Saudi Arabia—rumored to be for as many as 300 drones—suggests that the Middle East offers a significant opportunity for China's drones due to the export restrictions facing both the United States and Israel in that region. China has also sold drones to Iraq and the United Arab Emirates, which in both cases, have been used for strikes against militants.[90]

But China has also capitalized on selling capable drones that can be purchased by countries without deep pockets for defense spending. One of its most successful lines has been the Caihong (Rainbow) drone models. The Rainbow models, originally designed for reconnaissance and short-range surveillance, have now been equipped with air-to-ground missiles and rockets that can be used to attack targets at short range. Among the most popular is the Caihong-4 model, which can be equipped with laser-guided missiles capable of reliably finding targets and even piercing through 40 inches of armor.[91] While there are persistent doubts among experts about the technological capabilities and reliability of the Caihong models, the Caihong system is widely seen as good enough by a growing number of governments. Although the Caihong-4 can only carry two missiles and remain in the air for six hours, it is sufficient for fighting enemies at close range, especially if those enemies are relatively poorly armed.[92] As one expert remarked, these drones are "fast becoming the Kalashnikovs of the drone world—entry-level alternatives for countries eager to achieve a basic unmanned strike capability quickly and cheaply."[93] The Caihong drones have been allegedly purchased by Iraq, Nigeria, Pakistan, Egypt, the United Arab Emirates, and Saudi Arabia. There is already evidence that Pakistan, Iraq, and perhaps Nigeria have used them to kill militants in their territories. At the moment, the range of these drones is limited, but future iterations of the Caihong model will be capable of operating at a distance of over 4,000 miles.[94] The base model Caihong drones are also significantly cheaper than comparable US or Israeli models and come with none of the concerns over end-use assurances that US exports do. For this reason, sales are swifter and less encumbered by burdensome regulations. By some estimates, China's drones cost as little as a quarter of the comparable US models.[95]But the chief attraction is that China does not disclose details of its sales. In the words of Ian Easton of the Project 2049 Institute, China's drone export strategy revolves around a three-pronged offer of "price, privacy, and product."[96] This ask-no-questions approach is allowing China to capture a significant portion of the market for small- and medium-sized drones, especially among countries that are not authorized to buy from the United States and Israel. Over time, China could be a powerful force for the global proliferation of drones, as it has little compunction about selling drones to countries with poor human rights records and can exploit the market in selling to governments in the Middle East and Africa that are forbidden from accessing US technology.[97]

Other Exporters

Aside from the three main exporters, there are a number of latecomers to drone development and export that could play an important role in the future. What distinguishes the latecomers is their capacity: all have a substantial military infrastructure and robust commercial defense sector that could be used to exploit the drone market with sufficient investment. One of the most important latecomers is Russia. Long one of the world's biggest arms exporters, Russia has lagged behind the United States, Israel, and China in the production of drones and is now racing to catch up to their capabilities. In 2013, the Russian government announced a $13 billion, multi-year effort to develop military drones by the end of the decade.[98] It has also closed deals with Israel and the United Arab Emirates for the purchase of drones, while in August 2013 the Russian Defense Ministry ordered that the speed of drone development must be doubled over the next decade to capture its share of the export market.[99] For years, Russia relied on Israeli-provided technology, but it has found its access to some Israeli models blocked by US pressure. It discovered to its chagrin that it lacked enough drones to dominate the battlefield in Georgia in 2008 and was determined to ensure that this did not happen again.[100] But by 2015, Russia was estimated to have approximately 800 drones, with the greatest emphasis placed on small tactical drones for use by its own armed forces.[101] In 2016, it deployed up to sixteen different "low, small" drones in eastern Ukraine and has used these to achieve air superiority, to identify targets, and to rattle its opponents.[102] It has also used drones in Syria to spot targets, to conduct reconnaissance of the battlefield, and even disseminate propaganda.

But Russia is not content to ignore medium- and high-altitude drones or new technology. After years of relying on Israeli medium-range drone models, Russia has recently begun to spearhead the production of a new long- and medium-range attack drones similar to the US-made Reaper drone.[103] Some of these models were expected to be available in the 2018–2020 time frame.[104] It is also developing heavy strike UAVs, such as the Korsar and the Sukhoi's Okhotnik UCAV, which also bears some similarities to the cancelled US X-47B.[105] Russia is investing in swarming drone technology, with semi-autonomous function, with a view toward using this in future battles.[106] One of its planned exports will be an exploding suicide drone produced by the Kalashnikov Group—the same company that produced the ubiquitous, eponymous rifle. As with the rifle, the aim is for the new Kalashnikov drone to be sold widely and cheaply across the world.[107]

Russia has a number of obstacles before it develops a strong drone export capacity. Its R&D base still lags behind that of the United States, Israel, and China, and it has not developed attack drones to match its ambitions. What the Russian industry has developed does not match the country's needs for ISR, particularly for patrolling its vast territory. Yet once its drone technology catches up, Russia will be in a good position to take advantage of the growing global demand for drones and the gap in the Middle Eastern market. It is already working on joint projects with Iran.[108] Like China, Russia is less encumbered by human rights considerations over its sales of drones and has an opportunity to sell to states (such as Iran, Syria, and others) that are not authorized for drone sales by the United States or Israel. It is not yet a leading global exporter, but if the record of Russia's arms sales are anything to go by it is just a matter of time before it achieves this status.

Three other latecomers also have the capacity to play an important role in the export market. In Asia, both Japan and South Korea have the organizational and defense industrial capacity to become major players in the drone export market. While Japan has lagged behind China in domestic production of drones, it has invested heavily in building its domestic capacity for drones and may reach its desired fleet size even earlier than the projected date of 2023. By some estimates, it now has the fastest growing drone fleet in the world.[109] Using both US Global Hawks and its own models, Japan envisions a future with expanded ISR capabilities in its airspace and immediate region, including the South China Sea and the Korean peninsula.[110] Similarly, South Korea is developing its own set of strategic UAVs designed to provide deterrence against North Korea and to expand its own ISR capabilities. Like Japan, South Korea is temporarily relying on US drone technology, but plans to develop an indigenous medium-altitude, long-endurance drone for ISR missions over North Korea by 2020.[111] Over time, given their strong technological capacities, Japan and South Korea could become important exporters of drones. Finally, Turkey has also been exploring the export market and is now trying to sell unarmed and armed drones to Saudi Arabia and Qatar, while also exploring a joint production agreement with an Indonesian company.[112]

Finally, a number of European countries are possible, future drone exporters. Despite having very sophisticated, powerful defense industries, many European countries have lagged far behind in drone domestic production. While European companies are pouring millions into R&D, the hard fact is that their governments are mainly takers, rather than makers, of drones.[113] As a result they are ceding the export market to others. As the European Commission noted in 2012:

> [UAS] technologies are a source of important spin-off to civil aviation and a key element of the future aeronautics sector. Presently, the US and Israel dominate the sector although also other non-European countries show great potential to become strong competitors. The European aeronautics industry is still lagging behind and must quickly catch up to be able to compete on this global emerging market.[114]

The degree to which Europe has fallen behind in drone production means that even countries with significant defense export industries, like the United Kingdom, Germany, and Italy, have become reliant on US and Israeli drones. One estimate by the *Guardian* newspaper in 2015 suggested that the United Kingdom is the world's biggest importer of (largely US) drones.[115] Others, like Germany, have been importing

Israeli Heron drones and US Global Hawk drones for tactical and strategic purposes.[116] Under the auspices of NATO, a number of European countries are also pooling contributions to field five Global Hawk drones as part of the Alliance Ground Surveillance (AGS) system,[117] designed to provide NATO allies with ISR assets. It also illustrates how dependent these countries are on US drone technology, at least until their domestic UAV industries are further developed.

There have been a number of multilateral efforts by European countries to pool resources in order to build drones which might be competitive with popular US and Israeli models. Among these is the Future Combat Air System, led by France and Germany, which aims to produce a mix of manned aircraft and UCAVs that should be able to conduct surveillance, target selection, intelligence gathering, and targeted strikes in combat situations.[118] It is due to be rolled out for demonstration by 2025 and be the basis for a new concept of operations by 2030. Another joint endeavor between France, Germany, Spain, and Italy is the European MALE Remotely Piloted Air System, nicknamed the Eurodrone. This is a medium-altitude drone broadly comparable to the Reaper.[119] This project, in the works with Airbus and other European companies for a decade, carries the promise of delivering a share of the global export market for medium-altitude drones to a consortium of European companies. It aims to be in the air by the mid-2020s. Finally, there is the multinational effort, led by Dassault Aviation, behind the nEUROn, a UCAV designed to be stealthy and work in contested environments. All of these models are in the testing stage, but may eventually go into production and possibly export in the coming decades.

What Impact Will Drones Have?

The combination of the expansion of domestic drone production around the world, the rapid development of commercial drones that can be retrofitted for a military purpose, and the booming export market together suggest that drones will soon be in the hands of most militaries worldwide. But will this matter? For governments blessed with peaceful neighborhoods, few military rivals, and stable borders, drones are likely to remain just another useful tool in the arsenal, an efficient way to do routine tasks more cheaply or with less risk to personnel. For the majority of new users, it is unlikely that drones will alter their strategic position in any fundamental way. But there is a subset of cases in which the introduction of drones may produce a very different effect. Today, drones are being inserted into a series of active conflict zones where they have not previously been deployed. This marks an important shift. For the last two decades, drones have been used over battlefields in the undergoverned spaces of the world (for example, the tribal areas in Pakistan and Yemen, or Iraq and Syria) in which the United States and its allies were pitted against non-state actors like al Qaeda or the Islamic State. In these circumstances, the United States typically has full air superiority at high altitudes, meaning that it has destroyed enemy air forces (if they exist) and faces no real risk that its aircraft will be shot down. These conflicts can be best described as asymmetric, pitting a well-armed state against a weaker non-state actor in an environment where the drone user effectively "owned" the skies. In these conflicts, drones were not required to be nimble or to rapidly respond to threats in their environment, but could instead loiter safely to identify their targets and strike them at will. Even Israel's drone use follows this broad asymmetric template. Israel deploys its loitering drones against Palestinian militant groups in Gaza and the West Bank for surveillance and aerial policing, and faces no real air-to-air risk from drones owned by Hamas and other militant groups. With medium-altitude drones concentrated exclusively in the hands of the United States, Israel, and a select few NATO allies and deployed in asymmetric contexts, few states have had to imagine how things might change if the principal enemy were another state with its own equally capable drones.

Today, drones are being introduced into that exact situation; they are now in play in interstate rivalries in which long-standing enemies eye each other warily and monitor each new technological advance to see if it will confer any advantage. In these circumstances, the introduction of drones may give one side an advantage or, at a minimum, sharpen the conflict between them. The degree to which drones can change the equation can be seen in the US and Chinese drone flights in the South China Sea, and in other locations as well. Both North Korea and South Korea are deploying drones along the demilitarized zone (DMZ), and India and Pakistan are flying drones over each other's territory along the Line of Control (LoC) in Kashmir. In the war in Iraq and Syria, US and Russian drones glide through the same airspace on opposing sides, a situation that would have been considered almost unthinkable less than a decade ago. In all of these cases, governments are deploying drones in the context of what the scholar Thomas Schelling once called "dirty bargaining": a contentious negotiating relationship in which the threat of violence is ever present.[120] This in itself is not a new development. States have always intermixed threats of violence and diplomatic pressure in crisis zones, and nuclear deterrence during the Cold War often depended upon careful aerial surveillance such as is seen today with drones. But drones may alter the underlying calculation behind "dirty bargaining" because, as unmanned platforms, they so dramatically change the risks and costs involved.

This raises the question as to whether and how drones will be used in dirty bargaining. Some experts have argued that drones themselves do not fundamentally change the strategic logic of bargaining in a serious way.[121] According to this line of thinking, current generation drones are a tool, like any other, and are not as effective as manned aircraft for deterrence or coercion because of their technological limits. It is certainly true that current generation drones like the Reapers or Global Hawks are poorly suited for contested airspace and would be more vulnerable than manned aircraft to being shot down. For example, unlike manned aircraft, most medium-altitude drones are not quickly adaptable and have few, if any, air-to-air defenses.[122] Drones are also subject to other potential disruptions, such as jamming, hacking, or spoofing, which can lead to data links being disrupted and drones crashing or falling into an enemy's hands. This is why one US Air Force general concluded in 2013 that, "Predators and Reapers are useless in a contested environment."[123]

Since they are slower and less agile, current generation drones like the Reaper might then not be the best choice for a government hoping to deter or coerce another government in a crisis situation. Even if that were not the case, skeptics argue, it is not clear that the introduction of current generation drones into conflict zones would necessarily make crisis points more unstable. Surveillance drones may have the beneficial effect of reinforcing deterrent relationships between states by improving the flow of information and reducing the risk of miscalculation.[124] If, for example, a state is able to track the movement of troops along a border and determine that the movement is a training exercise rather than an invasion, this may reduce the risk of accidental war. For example, India is deploying forty-nine hand-held mini-drones along its borders with China and Pakistan to watch troop movements in high-altitude regions.[125] Such an effort, while fallible, might provide better intelligence to Indian commanders and reduce the risk of misperception and conflict. Similarly, the United States is conducting drone surveillance operations over Iran from its bases in Qatar and the United Arab Emirates, thus providing fine-grained details on Iran's nuclear programs that may help verify its claims about the purpose of its facilities.[126] The use of surveillance drones, and the quality of the information that they provide on a rival's activities, may reduce uncertainty, enhance decision-making, and make accidental war less likely.[127]

On balance, more information is a good thing, and no state would want to know less about a crisis that it faces. Yet there are three reasons to be less sanguine that the improved quality of information from drones will always be a force for stability in crisis zones. First, it is not clear that more information will always be a net positive in crisis bargaining situations. Information alone is not useful unless the receiving government is able to collect and analyze the data efficiently. Even the United States has struggled on this point, with senior Pentagon officials noting that they cannot analyze all of the data that they collect from their vast drone surveillance apparatus.[128] Whether a government can actually use the new information provided by drones is likely to vary based on its organizational characteristics and intelligence capabilities. Some militaries that lack the deep pockets and extensive intelligence collection and analysis capacity of the United States, China, and Israel may find that additional imagery produced by drones will not be processed quickly or efficiently enough to be useful in crisis-management scenarios. Given that most new drone users will fall into that category, it is not clear that the rapid provision of rich, detailed information will always be as beneficial to others in crisis situations as it might be to the United States.

Second, the provision of vivid imagery from drones may not always be a net positive for decision-making. Vivid information has been shown to draw some types of decision makers away from other salient or more useful information.[129] In a crisis situation, vivid drone imagery of a potential incursion could exaggerate cognitive, affective, or psychological biases that decision makers already have. For example, such imagery may heighten their sense of urgency, exacerbate pathologies in the decision-making process, or truncate their deliberations by forcing a decision too quickly. In crisis bargaining situations, some decision makers will profit from the ability to visually track an enemy's movements and reassure themselves of its non-threatening nature; others may panic and rush to decisions based on the same drone imagery.

Information can also cut in different ways depending on whether it is released into the public domain or used as leverage. The danger of vivid aerial imagery inflaming a crisis, charging public opinion, and giving a new character to routine or transactional violence was clear in a crisis around the LoC in Kashmir in 2016. In September of that year, India claimed to have conducted a series of "surgical strikes" on militants in Kashmir.[130] Although Pakistan denied that this happened and claimed the strikes were an "illusion," India essentially backed the Pakistani leadership into a corner by letting it be known that the entire operation was filmed with drones and hinting that footage could be publicly released.[131] The result was an even stronger public condemnation and retaliatory actions by Pakistan, which then shot down an Indian drone on November 20.[132] The prospect of drone imagery and the public reaction to it may have produced more pressure on Islamabad to respond more forcefully to an incursion or violation of deterrence than it otherwise would have and further ratcheted up tensions between the countries.

Third, it is unlikely that drones will continue forever to be able to provide the kind of information that improves decision-making in crisis situations in contested environments. This is largely because states in those situations have strong incentives to take countermeasures to knock drones out of the sky and to reduce what they can see. As drone technology improves the flow of information about what other actors are doing, there will be pressure for more aggressive (and risky) countermeasures designed to block the gaze of drones, as well as calls for more aggressive operational security about controversial activities, such as building missile sites. Aside from investing heavily in counter-drone technology, states are considering new steps that they can take to shoot drones from the sky if they are detected in their airspace. Russia has developed a new model of its Tor M2 surface-to-air missile defense system, designed deliberately to shoot down drones.[133] Another response in the future might be the development of counter-UAV operations in which drones are developed specifically for the purpose of detecting and destroying other drones.[134] This may lead to a dangerous sequence of UAV operations and counter-UAV strikes by adversaries, thus setting the stage for conflict spirals and dangerous accidents.

Even if a war of drone versus drone does not develop, states will not accept the widespread use of surveillance drones, and the corresponding loss of secrecy, without a response. One way that they may respond is to go further to ground to conceal their activities and to adopt stronger countermeasures to block their detection. Russia has recently fielded a Krasukha 4 radar system designed to block surveillance of ground targets and emphasized that the system was capable of blocking both Global Hawk and Reaper drones.[135] China has also developed a home-made laser defense system capable of shooting down US drones.[136] It is just a matter of time before states in regional conflict rivalries—North and South Korea, or India and Pakistan—also begin to invest in anti-drone technology in this way. They may also likely build even more underground nuclear and military facilities to avoid drones' gaze. In the long run, the spread of surveillance drones—and the corresponding conclusion that one must assume everything is being watched from the skies—may paradoxically lead these states to become more opaque, not less, due to aggressive countermeasures and improved operational security over military bases and sensitive locations.

Aside from these arguments, there is a more fundamental reason to believe that crisis zones populated with drones may be more dangerous than those with manned aircraft. If those using drones are more risk-taking because a pilot is not in danger, a world full of drones might be one in which crises are more, rather than less, likely to escalate into a war. The core of this claim—that risk-taking in war grows in dangerous ways if decision makers and warfighters are insulated from its costs—has a long intellectual lineage from Immanuel Kant to the early twentieth-century anti-war movement.[137] Although many forms of military technology reduce or even remove elements of risk during wartime, drones go further than most by abolishing entirely any physical risk to personnel on their own side: their uniqueness comes from their ability to "allow you to project power without projecting vulnerability," in the words of Lt. General David Deptula.[138] As David Hastings Dunn has pointed out, drones appear as a disembodied threat, which appears to "enable their use with domestic political impunity, minimal international response and low political risk and cost."[139] This does matter if the absence of risk does not affect strategic choice, but evidence from the US targeted killing program suggests that it does. Even President Barack Obama acknowledged this, saying that his use of unmanned aircraft had edged closer to being a "cure all" for terrorism and eroded some of the decision-making barriers on the use of force during his term in office.[140] The central characteristic of drones—that they are unmanned and therefore wholly detached from the risks associated with losing personnel—may induce leaders and even the public to roll the dice and use force in more risk-taking ways than they might with manned aircraft.[141] Although there is not enough evidence to say that drones alter strategic choice in every case, there is some early evidence that operational patterns and norms evolve when drones are put into play. An analysis by the Royal United Services Institute (RUSI) in London found that three of six drone users in the Middle East altered their operating norms, and engaged in more extra-territorial strikes, as a result of having access to the technology.[142]

Deterrence

If drones change governments' calculation of risk and their ensuing behavior, they will begin to affect relationships of deterrence between long-standing rivals. One way that this might happen is that states begin to use drones to test the nerves and strategic commitments of their enemies. States may calculate that small drones will be able to evade detection in contested space or at the least be able to evade being shot down. Drones are particularly conducive to what Alexander L. George referred to as "salami tactics"—that is, small steps used to probe or test deterrent relationships by violating commitments in a small but measurable way.[143] The crisis in the Senkaku islands in September 2013 provides an obvious example, but there are others in Kashmir and the Korean peninsula. In August 2013, India flew a drone from Kashmir into Pakistani airspace. Pakistan scrambled its aircraft in response and demanded an explanation of the violation.[144] In July 2015, Pakistan shot down an Indian drone taking aerial photographs near Jammu and Kashmir, but India's appetite for drones was undimmed, and it continued to fly even more drones near the LoC.[145] India also pressed the United States to sell it Predator drones for surveillance, while Pakistan increased its efforts to boost domestic production of surveillance and armed drones.[146] By 2019, Pakistan was allegedly flying its own surveillance and armed drones over the LoC in an attempt to rattle India, while at the same time seeking to buy Wing Loong II drones from China.[147]

The same dynamic is emerging on the Korean peninsula. In spring 2014, three crude drones allegedly from North Korea crashed in different locations in South Korea, one near the city of Peju, another close to the DMZ, and another on Baengnyeong island in the Yellow Sea.[148] They were reported to have been equipped with cameras that contained images of the Blue House, the South Korea's presidential residence. In January 2016, it was estimated that at least six North Korean drones had crashed in South Korea, and at least one of them contained images of South Korea's nuclear reactors, presumably to allow them to be sabotaged.[149] While the drones themselves were poor quality compared to those South Korea has, this carried its own advantages. As Van Jackson put it, "it's the low performance qualities of North Korea's drones that enable them to evade South Korea's defenses, which are optimized for more traditional threats from bigger, faster high-altitude aircraft."[150] These efforts culminated in South Korean forces shooting at a North Korean drone over the DMZ in January 2016.[151] Yet another North Korean drone escalated a missile crisis in May 2017, with South Korean forces firing ninety shots at it before it disappeared.[152] If North Korea develops more capable armed drones, the tempo and nature of these tests of South Korea's deterrence posture may also change.[153] Although the details of North Korea's drone arsenal are hard to confirm due to the regime's secrecy, a South Korean think tank estimated that North Korea had approximately 1,000 drones which it could deploy for surveillance and for terror attacks with crude chemical and biological weapons.[154]

There is also some evidence that even powerful states can find their posture of deterrence challenged by weaker states with drones. In June

2019, Iran shot down a US Global Hawk drone over the Strait of Hormuz in the midst of a crisis that had been escalating since the Trump administration pulled out of the Joint Comprehensive Plan of Action—the international deal to review and contain Iran's nuclear program. There was some precedent for Iran's gambit, as it had hijacked an RQ-170 Sentinel drone by "spoofing" it in 2011.[155] But this was different. The drone was shot down, not hijacked, the Global Hawk was a much more expensive, potent symbol of US power, and the shoot-down occurred in the middle of a crisis. Would Iran have been just as willing to test the United States by shooting down a manned aircraft, especially in the middle of a crisis? On balance, Iran has tolerated the limits of US deterrence in the Middle East and refrained from open attacks, although it has conducted a covert campaign of subversion and support for terrorism to harm US interests. But with drones, the calculation appeared to change. Iran gambled that the United States would be more willing to let them destroy a drone without a military response. This calculation was proven correct: President Trump said that he decided against a US military response to the destruction of the Global Hawk at the last minute because the response would cost 150 Iranian lives, thus making it disproportionate.[156] This crude moral calculation—that one drone was not worth that many lives—ultimately rebounded to Iran's benefit. But the entire events are marked by uncertainty: Iran had no guarantee that the Trump administration would respond as it did or that the limits of US-Iran rivalry in the Middle East would remain intact. Yet it was still willing to take a shot against a much more powerful opponent because, in the end, the target was only a drone. This interaction may have emboldened Iran even further. In July 2019, one or more Iranian drones approached the USS *Boxer* in the Strait of Hormuz and were disabled by the United States, with no Iranian response.[157] Although this crisis appears to have been averted, there is also no guarantee that a future drone interaction between the United States and Iran—or indeed between any two other states—would remain limited in this way. Much depends on the nature of the drone incursions and even the whims and personalities of the leaders involved.

The response might also depend on whether the details around the drone incursion are clear and widely accepted, as in the drone attack on Saudi Aramco oil facilities in Abqaiq, Saudi Arabia, in September 2019.

Initial reports suggested that Houthi rebels in Yemen launched a coordinated attack with at least ten drones on the oil-processing facilities as revenge for aerial bombings conducted by the Saudi military in Yemen's civil war.[158] But the United States, Saudi Arabia, and others soon doubted that the Houthis were responsible and pointed a finger at Iran, a longtime backer of the Houthis. The government of Saudi Arabia presented evidence of Iranian cruise missiles in the wreckage and suggested that some of the damage was due to missiles rather than drones.[159] While the Houthis continued to claim responsibility, Iran denied any involvement in the attack and cast doubt on the evidence that Saudi Arabia presented. Although US and European leaders eventually agreed that Iran was responsible, confusion over what caused the damage and who was responsible undercut momentum for a response, at least over the short term. The lesson was that violations of deterrence with drones might be possible, even attractive, if the incursion itself can be cloaked in uncertainty.

As these incidents show, among the dangers of using drones to test deterrence is that their low cost might make such violations of deterrence more routine, even normal, and sustainable over the long run. This is in part because low-cost violations of deterrence are more politically sustainable than those in which pilots are at risk. Democracies are naturally cost sensitive, and their publics—as well as their leaders—will resort to options which minimize costs, especially for attacks that are going to be frequent. As Amy Zegart has argued:

> Precisely because drones are lower-cost options to fulfill a threat, they are more likely to be initiated, sustained and supported by a domestic public. With drones, the low-cost threat becomes credible for the first time: "I can send drones at you, all day long, with no risk to me" becomes plausible, sustainable, attractive, and true. Indeed, the drone threat could become more credible than higher-cost signals to put boots on the ground. Why? Because high-cost signals may show a willingness to initiate a course of action but not to sustain it. In the long run, domestic audiences are more likely to maintain support for options that do not run risk of American lives; presidents, legislators, and adversaries should know this at the outset. [160]

By making these threats to deterrence not just possible but routine, drones may begin to shake the agreed limits of previously stable deterrent relationships and make them bearable for cost-sensitive governments, effectively giving them a free hand to test deterrence far more often than they might with manned aircraft. Given how cost effective drone attacks can be—the Saudi Aramco attack knocked out 5% of the world's oil supply at the cost of only a few drones and cruise missiles—it is not hard to see how tempting it might be for other governments to test deterrence with drone incursions.

When drone incursions happen, the responding state faces a number of important questions. Is a drone incursion a sign of serious intent to harm, or is it just a trivial foray which suggests nothing has changed? Does a drone incursion present a credible threat, or can it be dismissed? Answering these questions is crucial for deciding what kind of response is appropriate. Judging this accurately is hard because the rules of engagement for violations of deterrence with drones are unclear and may evolve in very different ways from manned aircraft.[161] It is possible that there will be an implicit understanding between states that destroying drones is not a big deal, or that it is not worth paying for that destruction in human life. If that is the case, a drone can be shot down, or the responding state may respond with its own drones in a tit-for-tat exchange. But it is equally possible that states will have varying interpretations of what is and is not appropriate to do to an unmanned aircraft, producing disagreements about whether there was a violation of deterrence and what the appropriate response may be. This is particularly problematic in deterrence relationships because they rely on what Schelling referred to as a "tacit bargain"—that is, an unspoken agreement about what is and is not permissible between two actors.[162] No one yet knows what the tacit bargains surrounding drones will be. The absence of an agreed interpretation of a destroyed drone—that is, whether it should be treated as an aircraft and interpreted as a major provocation or dismissed as easily as a downed weather balloon—may not be a problem in some cases, but it may suddenly become a serious one in others. Although some experts are confident that all countries appear to know that their opponents will value drones differently than manned aircraft, there is not yet enough evidence to accept that conclusion given the limited number of drone-to-drone interactions in crisis zones so far.[163]

Coercion

Armed drones might also be used to coerce other states into changing their behavior. In theory, drones are attractive for use in acts of coercion because they radically lower the costs of coercion for the attacker, effectively allowing power projection at lower levels of risk.[164] Because drones are also generally cheap, drone coercion could be sustained for a long time, especially against a comparatively weaker enemy.[165] As Zegart has argued, armed drones can keep the weaker party trapped in a "constant state of ambush," fearing attack if they defy the stronger party.[166] Especially in cases of stark asymmetry, drones can increase the certainty of punishment for the weaker party and shift more vulnerability onto them, as the coercer can just keep promising that more drones will keep coming at them until they comply. Following this logic, it is possible to imagine a scenario in which governments would threaten their enemies with repeated drone attacks in an attempt to coerce them over an issue of importance. A government might threaten important military assets and personnel, or even other heads of government and high-ranking officials, to get them to change their behavior. In a different form, coercion is what the United States has been doing with non-state actors like al Qaeda and the Islamic State. This model—called "targeted hurting"—may be the way in which coercion by drone occurs in the future.[167]

Understanding whether coercion would actually work with drones—an often cheap, risk-limiting technology which presumably conveys a weaker signal of resolve—is hard to do at present. It is worth noting that the United States has only been able to coerce with drones given clear air superiority and a much weaker non-state opponent. Coercion might not work the same way between rival states. For example, if North Korea threatened repeated drone attacks against South Korea to coerce it for some purpose, it is not clear that South Korea would be sufficiently alarmed to do what was demanded. It might choose to simply bear the cost of North Korea's attacks, to respond in kind with drones, or even to develop its anti-drone defenses in order to nullify the potential threat. There is not enough evidence of state practice with drones to know definitively how coercion would operate when there is a rough strategic balance between the two government rivals. Because current generation drones are largely designed for uncontested airspace, few governments in this situation have used them for coercion for fear that they would be quickly shot down. But as drones become more capable of flying in contested environments and using stealth to cloak their activities, more governments will have an incentive to take the risk of using drones for coercion, if only to see how effective such a low-cost tool would be at changing an enemy's behavior.[168] This incentive is likely to increase as next generation UCAVs, capable of moving fast, using stealth, and striking deep into an enemy's territory despite air defenses, come into the hands of more governments worldwide.[169]

How coercion would play out when one state attacks another to force it to scale back its support of non-state actors is also an open question. For example, Pakistan has long supported militant groups in Kashmir to contest Indian control over the territory. India has responded with strikes against militants in Kashmir, but both India and Pakistan have observed tacit bargains in their responses to ensure that the situation in Kashmir does not generate an irreversible momentum for war. Drones, however, may be changing this calculation. In 2018, one Indian general announced that, "the Indian Army is capable of using drones to attack hostile targets inside Jammu and Kashmir and across the Line of Control, and sees no problem in using them provided the nation is willing to accept mistakes and collateral damage." The chief concern was not whether Pakistan would respond but whether the "international community [would] get after us" over civilian casualties from the strikes.[170] This suggests that at least some in India's military believe that the risks associated with a coercive drone strike against Pakistani-backed militants across the LoC are manageable and that restraint would prevail in Islamabad. But it is not clear that the Pakistani government would show such restraint in the face of an armed drone strike from India. Would Pakistan simply accept it, perhaps with some protest as it does with the United States, or would the fact that its bitter regional rival India launched the attack change its calculations? Pakistan might calculate that the reputational costs of being coerced by a drone attack from India are high and might respond with equal or greater force. This calculation might also change depending on whether the strikes are a one-off event or a semi-regular occurrence. Drone coercion is a seductive idea, but the chief contrast surrounding it—that it is so costless for the attacker, but implies so much about the weakness of the attacked if they give in—suggests that the underlying logic might be different than it is with other forms of coercion.

The biggest problem with using drones for coercion concerns credibility. It is not self-evident that drones will convey the same kind of message that manned aircraft do. The absence of risk to a pilot suggests that their use may be seen as a less credible threat by opponents and hence less useful to states as a signal than a manned aircraft. For coercion to work, a state must be able to send a message that it is serious about the issue and willing to pay the costs to get an enemy to change its behavior. If all a state is willing to send is a drone, it may not be taken seriously. Moreover, if routine incursions by drones become almost normalized between enemies over time, it will undercut the urgency of a threat and render achieving goals through coercion more difficult.[171] The result of making drone incursions routine is that the underlying calculations behind coercion—for example, how many drones does one send to change an enemy's behavior, and what response will follow—will be harder to make. If governments decide to try coercion with drones, they may need to be prepared to escalate the level of threat beyond what would be ordinarily needed with manned aircraft in order to convey their resolve. It may not be sufficient to send a single drone; instead, a government might have to send dozens to make it look like a real threat. Because drones are detached from risk to pilots, their natural lack of credibility as a threat may have the perverse effect of making governments more likely to ramp up the intensity or scale of the threatened violence.

Accidents and Spirals

A final reason to be concerned about the growing drone arms race is the danger of accidents and the conflict spirals that can subsequently arise. While drones are becoming more sophisticated, they are still prone to frequent accidents. According to an estimate in 2010, the United States has experienced at least seventy-nine drone accidents costing at least $1 million each, as well thirty-eight Predator and Reaper drone crashes during combat missions in Afghanistan and Iraq.[172] Unmanned platforms such as the Pioneer and Shadow drones have even higher rates of accidents.[173] A later estimate in 2014 put the total number of major drone crashes at over 400 since 2001.[174] Although it is estimated that some of these accidents are caused by human error, and that the accident rate is declining for military aircraft, these rates are still higher than comparable manned aircraft.[175] Among the problems noted in military drone crashes were a drone's limited ability to detect other aircraft or bad weather, pilot error, mechanical defects, and unreliable communication links.[176] It is also possible that less sophisticated models sold by China and other new suppliers will have a higher rate of accident than the more robust US models. Simply as a matter of probability, drone accidents will become more commonplace as more military drones take to the skies in the future.

Some military drone accidents may have political costs that are hard to estimate, especially if they cause a substantial loss of life or are interpreted as intentional. The possibility of a military drone colliding with a civilian airliner, while improbable, is not impossible. In 2004, a German UAV nearly crashed into an Ariana Airlines Airbus A300 carrying 100 people in the skies over Kabul.[177] Over the last fifteen years, drones have been equipped with anti-collision software designed to avert such crashes, but dangers remain. One estimate in 2012 found that at least seven US Predator or Reaper drones have crashed overseas in the vicinity of civilian airports.[178] In September 2013, the United States was forced to move its drone operations from Camp Lemonnier in Djibouti due to concerns that drones would crash into passenger planes from a nearby airport.[179] Whether a state views an accident as negligent or intentional will matter. If a US drone struck an Iranian passenger airliner, for example, it is not hard to imagine the incident causing a serious international crisis, along the same lines as following the accidental US downing of Iran Air Flight 655 in 1988. The risk of a conflict spiral from a drone accident between India and Pakistan, or Israel and one of its neighbors, should not be ignored. Similarly, a collision between a Chinese drone and a Japanese civilian aircraft in the East China Sea could produce disastrous consequences. It is further possible that governments or insurgents will try to hijack drones and cause international incidents with crashes or accidents. Hijacked drones are a particularly attractive way to test an enemy as they are hard to trace and can shield the perpetrator with a degree of plausible deniability. As drones wind up in the hands of more unscrupulous actors, and less reliable drones sold by China and others flood the market, governments around the world will face a vastly increased risk of a conflict spiral from drone misuse, hijacking, or collision with a civilian aircraft.

Conclusion

The world will soon be full of drones. Given the explosive growth in the government-led, commercial, and export markets, it is impossible to stop the spread of drone technology or to convince states to not use them. For most countries, there is no reason to do so, especially if drones are only going to be put to routine purposes. But for a select number of states, embedded in deeply contested rivalries, the introduction of drones will have important strategic consequences. From the Strait of Hormuz to the South China Sea, we are moving into an age in which drones may tilt the strategic balance between adversaries in some of today's most dangerous flash points. In these, drones will not be a mere substitute for manned aircraft, replicating the same calculations of risk and opportunity that have governed how these crises unfolded in the past. Because they are unmanned, drones are fundamentally different and the strategic calculations around their use will be marked by uncertainty rather than the tacit bargains that have prevailed for decades.

At present, we do not know how this will turn out because only a few years have passed since many of these states have acquired drones. But there is evidence that unmanned platforms will make governments become more risk-taking and expand their ambitions in ways to test their enemies' mettle. This is potentially dangerous. As a technology that promises to make violence antiseptic and carefully controlled, drones may convince governments that a potential violation of deterrence is negligible, or that it can be managed. Alternatively, it could convince them that coercion will be easier or more sustainable than it actually is. It may also convince them to be careless with the risks around drone use and to proceed while unaware of how conflict spirals could emerge even following simple accidents and other misunderstandings. All of this flows from the fact that the unmanned nature of drones detaches risk from the act of flying in another's airspace and inflates the confidence of users to set new, perhaps ambitious goals. This is why it is so crucial for drones to operate with some code of conduct and set of norms, as the Obama administration envisioned, rather than let the technology proliferate without clear legal and moral understanding of how it may be used. Such an effort would be imperfect and marked by defections, but the alternative—a world of drones with no rules—might be worse. In that world, we may find that the biggest strategic effect of drones would be to exaggerate the hubris of the governments who have them and to make the path to war different but no less likely.

9

The Future

THE DRONE AGE EVOKES both dreams and nightmares. For some, the arrival of unmanned aircraft fulfills a dream that began in the earliest days of aviation. As drones have become cheaper and easier to obtain, the skies are now open to those who once saw flying as an unattainable goal. The spread of drones to more people unleashes human potential and holds with it the tremendous promise of advancing scientific knowledge and commerce. From environmental conservation to archaeology, drones can be put to a nearly unlimited number of peaceful purposes. It is also true that drones will allow people to achieve some goals faster and with less risk than ever before. This is not an advantage which can be easily dismissed. As many of today's societies have tried to control risk or even banish it from daily life, the prospect of being able to fly, and even fight, without jeopardizing the life of a pilot seems almost irresistible. And if drones can also help to alleviate human suffering in war zones and crisis areas, as many defenders of the technology suggest, it is hard to see why we should be afraid of a world full of them.

Yet the nightmares about drones continue. Drones tap into a deep vein of fear about technology and its potential misuse that has been part of the public debate over aviation for at least a century. In 1907, the writer H.G. Wells reacted to the emergence of manned aircraft by depicting a future with aerial bombardment of cities and even kamikaze attacks in his novel *War in the Air*. Wells saw manned aircraft as harbingers of doom that inevitably would be misused for our own destruction during wartime. Today, the fact that drones are unmanned makes it even easier for critics to depict them only as uncaring robots and see their spread as our worst fears come to life. These fears play a disproportionately large role in the public debate, as evidenced by the regular connections drawn by the media between drones and George Orwell's "Big Brother" as well as the *Terminator* films and other science fiction accounts of futuristic technology gone wrong. Given the depictions of drones in movies and television, it is not surprising that many people today fear that their arrival will produce a hellish future of dehumanized warfare and remote slaughter.

Drones also evoke concerns about whether the technology itself will render humans irrelevant or obsolete. Some scholars fear that an over-reliance on drones may lead us to become intolerant of human error and diminish the value of human agency, creativity, and spontaneity.[1] If the skeptics are right, a world awash in drones may not necessarily be more dangerous, but it could be duller, less creative, and ultimately less free. This fear underlies some of the biggest questions surrounding drones in both the military and civilian worlds. Will military pilots be displaced from the prestigious positions they hold now? Will they be reduced to mere custodians of largely autonomous machines? And what happens to the rest of us living under the gaze of the drone's camera? It is hard to feel like a human, with all of the freedom and richness of experience that this implies, when reduced to a pixelated dot under the gaze of a drone. It is even harder to believe that freedom at large will thrive if drones are used to watch us and regulate our behavior.

These fears have begun to mobilize new constituencies to call for a ban on "killer robots" and other related technology. In 2017, Elon Musk, founder of the Tesla and Space X companies, joined forces with dozens of artificial intelligence (AI) industry leaders to call for a ban on autonomous weapons. Such a ban would include drones that use sophisticated algorithms to make decisions about whether and when to kill an enemy on the battlefield. For Musk and others, rendering drones and other weapons truly autonomous in this way carried more risks than benefits. They wrote:

Lethal autonomous weapons threaten to become the third revolution in warfare. Once developed, they will permit armed conflict to be fought at a scale greater than ever, and at timescales faster than humans can comprehend. These can be weapons of terror, weapons that despots and terrorists use against innocent populations, and weapons hacked to behave in undesirable ways. We do not have long to act. Once this Pandora's box is opened, it will be hard to close.[2]

The noted Cambridge physicist Stephen Hawking concurred with the call to "ban killer robots" because the use of AI to make drones and other robots autonomous weapons would bring about risks which humans scarcely comprehend. He warned that, "unless we learn how to prepare for, and avoid, the potential risks, AI could be the worst event in the history of our civilization. It brings dangers, like powerful autonomous weapons, or new ways for the few to oppress the many."[3]

How should we understand the dreams and nightmares that drones evoke? At a minimum, it is clear that the extreme versions of both sides of the argument are overstated. Drones are not a panacea for all of the world's problems; in a world full of drones, war, poverty, and environmental degradation will continue. But drones are also not necessarily harbingers of a nightmarish world where technology is entirely out of control. Too much of our debate resembles a form of shadowboxing in which critics of drones attack fears of what might happen instead of what actually is happening. It is not that Musk, Hawking, and their colleagues are wrong, but that they are peering into the future and identifying future threats at the expense of the ones that have become apparent today. As this chapter will show, there are good reasons to be concerned with how future technological developments with drones will play out. But it is equally important not to ignore what the evidence surrounding the use of drones today tells us about how the technology affects the strategic choices of its users.

Drones and Strategic Choice

This book has suggested that drone technology is important because it has a unique array of characteristics that affect how humans use it. There is no single characteristic of drones that produces radical changes in the way that people behave. Rather, the characteristics of drones together produce subtle but noticeable shifts in the strategic choices of its users. Each characteristic plays a different role. That drones are low-cost makes them easy to buy and quick to spread among even resource-constrained actors, such as NGOs and terrorist groups. That drones are adaptable also makes them attractive to actors, such as militaries and peacekeeping forces, which must deliver payloads in highly complex or dangerous environments. That drones can be precise and yield high levels of information about what they see on the ground makes them indispensable for actors who dream of knowing more about the environment in which they operate. And finally that drones are unmanned makes them invaluable for all actors for whom risk is defined almost exclusively by a loss of life of pilots or other personnel. For societies that have put a premium on measuring and avoiding risk in this way, drones are almost ideally designed: they offer the promise of invulnerability while opening the world beneath the camera's gaze subject to precise investigation and action.

An important theme of this book is that drones are rarely introduced into situations as mere tools that can be applied to a pre-existing mission. Instead, they tend to change the boundaries of the mission itself, leading to what is often known as goal displacement. Goal displacement—or as it sometimes called in the military, mission creep—occurs when the objectives of a particular operation drift or expand over time in response to new opportunities and, in this case, new technologies. As this book has demonstrated, drones are particularly susceptible to this dynamic because they provide the appearance of controllable risk. If there is no pilot and hence no risk of casualties, the reasoning goes, why shouldn't we use drones for yet another task? This reasoning explains why the United States, which initially sought drones for military action in declared war zones like Afghanistan, suddenly discovered that the technology has value for targeted killings outside declared armed conflicts. Similarly, as chapter 8 has shown, states that purchased drones originally for surveillance found themselves contemplating using the technology for limited probes of their enemy's airspace. This type of mission creep can also be seen in responses to humanitarian disasters and in peacekeeping missions. For example, NGOs may buy drones to map the terrain and provide information to first responders in disaster zones but soon find themselves tempted to directly provide relief to vulnerable groups through aid drops. Because they are so readily available and flexible, drones are a gateway into new missions and goals that seem to get bigger all of the time.

A second major theme of this book is that the nature of drone technology affects the calculation of risk associated with it, specifically the decisions over what to do and how to do it. Industry defenders of drones have argued from a standpoint of technological neutrality that drones are just a faster or more efficient way of doing something that one always intended to do. They acknowledge that drones can be misused by actors like terrorist groups, but they argue that the fault lies wholly with the person who misused it than with the technology itself. This book has adopted a different argument: that the technology itself structures choices and induces changes in decision-making over time. In other words, drones, like all forms of technology, are not neutral but rather they reorder the calculation of risk and privilege certain types of actions at the expense of others. While the users themselves remain primarily to blame for bad behavior with drones, this perspective emphasizes that their menu of choices has been altered or constrained by drone technology itself.

To understand this point, a comparison with nuclear weapons is useful. Although they are dissimilar from drones in a number of ways, and clearly more consequential for war and peace, nuclear weapons are nevertheless an example of a type of technology that began to structure the choices of those who adopted it. States with nuclear weapons found themselves thinking of crisis situations in different ways, worrying particularly about the risk of escalation and adjusting their behavior accordingly.[4] In some cases, strategic stability brought about by a parity of nuclear weapons yielded mission creep (for example, by leading to deeper involvement in proxy wars like Vietnam and Afghanistan), as well as substantial changes in the risk calculations of the states themselves.[5] Nuclear weapons also produced an antiseptic language to describe its operations (e.g., counterforce versus countervalue, second strike, and so on) much in the same way that drone-based targeted killings have (e.g., disposition matrix, kill box, military-aged males). Nuclear weapons did not entirely change the options available to leaders, or absolve them of ultimate responsibility for their decisions, but they altered the menu of choices available. Although drones are smaller, more diffused, and less lethal than nuclear weapons, they have also altered the choices available to the diverse array of actors who wield them.

One of the reasons why this point has been so contested when considering drones is that there is a crucial but overlooked distinction between political decision-making and tactical or military decision-making. As chapter 3 has demonstrated, many drone pilots deny that their decision-making has been in any way altered by the fact that the technology is now unmanned and insist that they are governed by the same rules as manned aircraft. The evidence suggests that this is largely true, especially for militaries with a high level of oversight and accountability. Despite what some news accounts imply, US drone pilots are not free to simply kill people at will and are held accountable for mistakes on their watch. Even more, the video record from drones provides incontrovertible evidence of wrongdoing and makes avoiding accountability more difficult. As a result of being located in a rule-bound bureaucratic structure, pilots have a limited degree of discretion about how they use drones.

But at higher levels in government, where the decisions about who and where to strike are made, the political decision-making is not as constrained and rule-bound. Although decisions are not subject entirely to discretion, senior decision makers have more latitude in deciding to expand the area of targeted killings and in assigning new targets to the "kill list." This can be seen in the expansion of targeted killings first under Obama and later under Trump. It is here where the changes in risk calculations and goals when using unmanned technology matters. Although there are only a limited number of cases of leaders equipped with robust drone fleets, the record so far suggests that leaders at the political level may become more aggressive or risk-taking when the lives of pilots are not at stake.[6] This dynamic explains the expansion of the targeted killing list, the increasing number of drone fly-bys and probes at the interstate level, and even some of the risk-taking seen when rebel and terrorist organizations bait more powerful enemies like Israel. This does not always translate into operational looseness at the tactical or operational level, but rather a gradual, deliberate expansion of the role, scope, and geographic range of drones once they are already deployed. In practical terms and in the short term, as chapter 4 demonstrates, this leads to relentless pressure on military pilots to fly and do more with drones, even while they remain constrained at the operational level in their decisions about who to strike.

This dynamic also leads to pressure to learn and do more and more about the battlefield. One of the themes of this book is that everyone, from militaries to NGOs to terrorist organizations, seeks to harness the capabilities of drones to know more about the environment in which they operate. The use of drones generally leads to an increase in the quantity, and often quality, of information available to an actor. But this is not a neutral or even always a positive development. As chapter 8 noted, increased levels of information can generally help decision-making, but it can also exaggerate cognitive bias and lead to overconfidence and errors in making decisions. More specifically, it can lead to overconfidence that one can control one's environment because one can see it; it can lead to underestimating the adaptability of the enemy or target because they appear as mere dots beneath the gaze of the drone. The world does not sit still just because it is being watched. Peacekeepers equipped with drones are now confronting the fact that militia groups are changing their behavior and adapting due to the presence of drones over refugee camps. Even the US military was surprised by the degree to which groups like the Islamic State adapted their tactics and strategies to drones and gradually found ways to use drones against them. At its core, technology like drones can elevate the acquisition of information from a means to an end, making acquiring more information the goal of the activity and losing sight of the reasons why that information was collected in the first place.[7] This can be seen in the US military's insatiable desire for surveillance coverage by drones, collecting so much data on the world's activities that even they are struggling to sort through it and use it effectively.

Perhaps the greatest consequence of the emergence of drones for war and peace is their impact on our thinking. Many of the theoretical criticisms of drones have centered on their impact on notions of warrior honor, now that combat is virtual, remote, or depersonalized.[8] Yet this is misleading: many forms of technology, from artillery to manned aircraft to nuclear weapons are designed to protect one side's fighters while killing the others from a distance.[9] Drones are only the next step in a journey to remote, depersonalized warfare that began long ago. What is more consequential is the mode of thinking that becomes dominant and entrenched with the use of drones. This type of thinking, described by Jacques Ellul as "technique," reduces political and moral issues to problems of technical efficiency.[10] As a comprehensive but reductive mode of thinking, it holds up what machines like drones can do as the ultimate standard for measuring a resolution to a problem. This can already be seen with discussions over targeted killing; instead of asking why we are using aircraft for a task in the first place, we tend to debate instead whether the drone is better than the manned alternative. The answer to the latter question is often "yes," which seems to put an end to the debate. But the underlying questions behind our actions—should we be engaged in targeted killing at all? Should we be surveilling these populations at all? To what end will all of this killing be put?—are often elided. As Ellul would suggest, "technique" tells us that we do not have a problem of militancy in the borderlands of Pakistan and Yemen, but rather one of efficiency and precision with our "vast killing machine" of targeted killings. Similarly, we do not have a political problem of countries generating vast flows of refugees, but rather a practical problem of how best to watch and count them with drones once they appear in refugee camps. As this book has shown, this approach dehumanizes such problems and sidelines the essential moral questions that they raise. If "technique" exercises a colonizing effect on decision-making and spreads to other domains, we should expect the types of mechanical thinking surrounding drones—one in which genuine dilemmas become reduced to problems of technical competence and instrumental rationality—to dominate thinking about war and peace in the future. And that will be an even greater problem if technology continues to advance at the relentless pace that it does today.

Future Trends

Although today's technology yields some conclusions about how drones will affect decision-making, the technology of tomorrow brings with it the possibility of even further acceleration of these trends. Drones are unlikely to remain in their present form for long, as today's laboratory technological developments will radically alter their future development. As an analogy, if we consider today's drones as equivalent to the early manned aircraft, the next generation of technology will change at least as much as aircraft did when the jet age began in the late 1940s. Five technological developments, in particular, will intersect with drone technology and change their operations in remarkable ways, with real consequences for the strategic choices surrounding their use.

Artificial Intelligence

Perhaps the greatest potential change in how drones operate will come from the rise of AI. Although there are debates about how it should be defined, AI can be broadly understood as efforts to use the processing power of computers to simulate or anticipate human behavior.[11] AI is widely touted as the next major revolution in technology and indeed human development; some estimates suggest that millions of jobs are at risk of being replaced by computers whose abilities are indistinguishable from those of human beings.[12] The use of AI is hardly new; some types of AI have been around in embryonic form since at least the early 1950s. More recently, rapid advances in computing and processing power, data storage, communications, and connectivity have brought some of the dreams of AI researchers into reach. In particular, researchers have made significant progress in helping computers to learn from their mistakes and think strategically, even to the point where they can defeat humans in games and other tests of reasoning. Aside from evoking fears from science fiction about computers that learn to turn on their human counterparts, the rapid development of AI has led to concerns that it may outstrip human cognition or change the way that we think, perhaps by pushing us even more toward "technique."[13]

The hype around AI is substantial. Today, it is common to hear companies like Microsoft claim that AI has the potential to change the world in fundamental ways. But governments have not been far behind. In 2017, Russian President Vladimir Putin claimed that "artificial intelligence is the future, not only for Russia, but for all humankind . . . It comes with colossal opportunities, but also threats that are difficult to predict. Whoever becomes the leader in this sphere will become the ruler of the world."[14] While many defense industry leaders hail the potential of AI to revolutionize military operations, others have expressed concerns that it could lead to accidental war. Elon Musk, for example, has predicted that AI should be understood as a "civilizational existential risk" and predicted that it could cause World War III.[15]

At present, the risk of an AI-generated war is low. Researchers are working on a number of different ways for AI to "think" and learn as humans do and have made considerable progress over the last decade, but no AI is yet self-aware and making decisions in the way that is depicted in Stanley Kubrick's *2001* or the *Terminator* movies. Most of what AI currently does is mimetic: it mimics human behavior, either by following decision rules or by deducing behavioral pathways, to produce behavior that resembles that of humans. In addition, the greatest advances in AI have come about in labs, in settings far removed from real life, and most AI-enabled drones have been slow and not as adaptive as humans.[16] There are also pitfalls in getting AI-enabled drones to think in strategic ways as humans do. At present, most AI-enabled drones can follow basic commands, but they cannot quickly adjust to threats or prioritize tasks in the service of the larger mission.

What makes AI potentially important, at this stage, is that it is able to complete tasks at a speed that human beings cannot match. If the raw processing speed of AI was attached to a task like surveillance, there is a danger that killing could become so easy as to be almost automatic. It is this danger that alarmed employees at Google, who protested until the company pulled out of plans to help the Pentagon use AI to sift through and analyze its vast collection of drone footage for military uses, among which might include identifying targets.[17] In the end, Google abandoned the controversial Project Maven and issued new guidelines committing it to avoid technology that causes "overall harm," as well as weapons designed to injure or kill, although there are some ambiguities about whether the company would pursue peaceful technologies which could be retrofitted to that purpose.[18] Moreover, even without Google, research into the use of AI for drones will continue unabated, as other companies are eager to secure generous Pentagon contracts for this technology.

In the future, research into AI will become even more important because it raises the possibility of developing drones that operate autonomously. There is an important distinction between automated systems—which respond along preprogrammed logic steps to stimuli—and autonomous systems, which can decide how to achieve goals within certain parameters.[19] Today, many manned and unmanned aircraft are at least partially automated. For example, commercial airliners have multiple back-up systems and redundancies that come into play along a preprogrammed pathway as systems or parts fail on the aircraft. For many of us, the fact that commercial planes automatically adjust to stimuli, like turbulence, is a good thing. There is no reason why drones cannot be automated to respond to similar stimuli. For example, next-generation drones might be preprogrammed to adjust their altitude and speed to accommodate environmental factors like wind or weather. Alternatively, an automated drone might also be preprogrammed to avoid colliding with another drone or a missile fired at it.

Automated drones have four advantages over drones steered by pilots throughout their mission. First, they would no longer require pilots to fly them throughout the entire mission, thus reducing the tedium and associated costs of paying pilots. Today, some drones like the Global Hawk are already highly automated in that they are preprogrammed to fly between fixed waypoints, with pilots only required to watch the video feed and make adjustments as needed. Second, automated drones might also reduce the risks of crashes. Even today, Global Hawk and Reaper drones are controlled by pilots on take-off and landing, when crashes are most likely. Allowing take-off and landing to be even more automated might lower this crash rate. Third, automated responses from drones might prevent humans from overcompensating for environmental factors like wind or weather. Fourth, drones which are highly automated will respond more quickly than those controlled by humans to external threats like incoming missiles. In contests where speed matters, the few seconds gained by automating the response to the missile might be crucial and save the drone. For these reasons, it is almost inevitable that commercial and military drones will have more automated elements in the future, just as manned aircraft gradually were infused with automated systems to make them more reliable and safe.

To make a drone autonomous—that is, capable of partially directing itself, and deciding the steps it takes, toward its goal—requires a much bigger leap in terms of both technology and military doctrine. An autonomous drone could theoretically be programmed as a scout, surveying the landscape ahead of ground forces, or it could locate potential targets along set parameters (for example, military-aged men apparently holding weapons) and strike at will once they are located. The problem lies in the specification of the parameters: how can these be set without producing false positives (i.e., people misidentified as enemies) and accidents? The danger here is that AI might misinterpret the parameters or confuse the evidence determining that the parameters have been met (i.e., for example, mistaking a man holding a shovel for a man holding a weapon). For this reason, the central issue with autonomous weapons is the degree to which humans remain in the loop. A human pilot can be present at any stage in this loop, which is described as the "sense-decide-act" loop by the Pentagon. As Paul Scharre has pointed out, the fact that a human is directly involve in picking a target, or authorizing it, would render weapons systems only "semi-autonomous."[20] For example, a human might only be responsible for supervising target selection by the drone, or alternatively could have some responsibility for pulling the trigger. Only a fully autonomous weapon or drone would cast the human entirely out of the loop, making the decision entirely on its own in each specific circumstance.

Not all autonomous weapons or drones would necessarily be lethal. It is not hard to imagine that the United States might choose to conduct long-range surveillance with nearly autonomous drones. In these circumstances, the worst that could happen would be that the drone might take too many pictures or videos, or perhaps invade someone's privacy with excessive surveillance. The problem grows much bigger when these drones are lethal. For example, Israel has developed an autonomous Harop drone, which operates like a loitering munition and can dive-bomb targets on the battlefield without a human being in its decision loop. The Harop drone, which has stealth capability, uses its own body as a warhead and can destroy targets through kamikaze attacks in minutes. Even more powerful lethal autonomous weapons systems (LAWS) carry greater risks: they might commit fratricide by killing one's own soldiers, or be hacked by an enemy, or subject to an accident.[21] In each of these cases, there are serious questions about who should be held accountable, especially if the weapons system or drone was fully autonomous when it took these actions. For this reason, a number of military ethicists and senior policymakers have argued that LAWS present a serious challenge to the current laws of war. It is not clear that a fully autonomous drone equipped with a missile, for example, would meet the standard of discrimination and necessity that is needed when using deadly force. It is also not clear who could be held responsible in a military court if a fully autonomous weapon goes wrong or makes costly mistakes on the battlefield.

The development of AI will gradually improve the prospects of automated and eventually autonomous drones and make selecting these options in a cost-sensitive budgetary environment hard to resist. At present, the United States is reluctant to accept fully autonomous drones in part due to cultural and organizational preferences to have a pilot in control.[22] Although the United States is investing heavily in autonomous weapons, and believes that these weapons can ultimately be made compliant with international law of armed conflict, it has nevertheless committed that there will always be a human "in the loop" for any decision to take a life.[23] Yet other countries may make a different decision and put pressure on the United States to rethink this commitment.[24] For example, Russia, China, Israel, and South Korea have made investments in autonomous weapons.[25] Of these countries, China is the most serious player because its substantial investment in AI for commercial and surveillance purposes has yielded a deep industrial base that it could harvest for autonomous unmanned systems.[26] In a confrontation with the United States, China might choose to employ autonomous drones that can adapt more rapidly and attack more nimbly than their US counterparts still controlled by human pilots. If wars against adversaries like Russia and China become, as one military officer put it, "extremely lethal and fast," US reservations about autonomous weapons could give way in favor of necessity and operational efficiency.[27] A world of competing, clashing, autonomous drones is not imminent so long as current policy remains in place, but the fragility of states' commitment to keeping a human "in the loop" remains a serious concern.

Swarming

Another potential game-changer for drones is the development of swarming drones. Until now, most of the military drones used by the United States and other great powers have operated either on their own or as part of a small grouping of other drones, largely (though not exclusively) in uncontested airspace. The bulk of communications to and from the drones were directed to the pilot located on the ground. But this is about to change. Recent advances in AI and robotics have offered new opportunities for drones to fly together and coordinate as part of a swarm. The obvious models for swarming drones lie in the animal kingdom: insects, birds, and some pack animals that work together and coordinate either explicitly or with tacit signals to achieve common tasks. Small unmanned drones, some of which are no bigger than birds or insects, now fly in formation and attack an enemy together in a coordinated way. As Scharre has described it:

> Emerging robotic technologies will allow tomorrow's forces to fight as a swarm, with greater mass, coordination, intelligence and speed than today's networked forces. Low cost uninhabited systems can be built in large numbers, "flooding the zone" and overwhelming enemy defenses by their sheer numbers. Networked, cooperative autonomous systems will be capable of true swarming—cooperative behavior among distributed elements that give rise to a coherent, intelligent whole. And automation will enable greater speed in warfare, with humans struggling to keep pace with the faster reaction times of machines.[28]

One example of this is the Gremlin drone. In 2015, DARPA issued a wish list for next generation technology which included disposable swarming drones that could be dropped from the back of a manned aircraft and retrieved in mid-air. By 2018, Dynetics had acquired a $38.6 million contract to develop Gremlins and proved in a series of test flights with a C-130 aircraft that such operations were possible.[29] Gremlin drones fly in small numbers—perhaps no more than twenty at a time, though larger swarms are possible—can be used for reconnaissance and probing an enemy's defenses, and be redeployed up to twenty times.[30] The US Navy is also experimenting with swarming scout drones, known as LOCUST drones, that can harass an enemy and get them to waste anti-air missiles trying to knock them down.[31] The US Air Force is testing Perdix drone swarms: 100 small drones no larger than a robin bird deployed from two FA-18 Super Hornets.[32]

In 2015 alone, the United States flew eighty Perdix missions over the Alaskan border. In the words of the Strategic Capabilities Office of the Pentagon:

> The Perdix are not preprogrammed, synchronized individuals. They are a distributed brain for decision-making and adapt to each other, and the environment, much like swarms in nature. Because every Perdix communicates and collaborates with every other Perdix, the swarm has no leader and can gracefully adapt to changes in drone numbers. This allows this team of small inexpensive drones to perform missions once done by large expensive ones.[33]

The advantages of swarming drones are numerous. They can effectively compensate for a greater enemy mass (e.g., the larger numbers of ships, tanks, or even people held by an enemy) by dispersing and attacking from multiple vantage points. This has the advantage of confusing enemies and making them fight on multiple fronts, exhausting both their manpower and munitions and saturating their defenses. Drones can swarm targets and confuse the enemy about where to fight and might also be able to disable their command and communications systems. They can also disperse if needed and split up when under fire, coming back together again when their attack resumes. Drone swarms are also remarkably resilient; their distributed intelligence allows individual drones to "trade off" being the leader of the swarm when the leading drone is taken out by enemy fire. Even when elements of the swarm are destroyed, they will gradually degrade in combat power, but not suddenly stop fighting on a dime.[34] Some of the networks behind swarming drones are even capable of healing themselves to keep the swarm going.

Drone swarms are still in the experimental stage and may not see battlefield use for some time.[35] More generally, they have some of the same command and control problems of autonomous drones. In particular, drone swarms often move too fast and in too complex a fashion for a single person to completely control their operations. As a result, commanders are forced to develop broad parameters as instructions for the swarm, which can lead to problems if the parameters are misinterpreted. DARPA is currently working on voice and gesture

recognition for swarming drones to make the swarms more respon-
sive to changing battlefield needs.[36] Swarms will also need to be resilient: electronic warfare, like jamming, is a real risk to their viability.[37]
DARPA has conducted tests of how drone swarms function when
their communications and GPS signals were under attack and there
is some evidence that swarms could continue to function despite such ·
countermeasures.[38]

The United States is arguably behind China in efforts to develop
swarming drones. China has already fielded larger drone swarms than
the United States for reconnaissance missions and even managed to run
a swarm of 1,000 drones for an aerial show.[39] China is investing heavily
in military-grade swarming drones in order to be able swarm US radars
and disrupt their powerful command control networks in the event
of a clash in the South China Sea. The state-owned China Electronics
Technology Group (CETC) has been experimenting with swarms of a
hundred or more drones for some time, essentially in anticipation of
a battle in which a great mass of swarming drones overwhelms an opponent.[40] If China fields these swarms and resolves the command and
control problems that they present, US military officials will have to
develop successful anti-swarming countermeasures or fight a conflict
on China's terms: fast-moving, mobile, and lethal.

Blended Manned-Unmanned Missions

Until recently, a relatively sharp distinction has been drawn between
manned and unmanned aircraft, hinging around the presence or absence of a pilot. But this distinction is breaking down as the US military
has invested more in manned-unmanned teaming (MUMT) capability.
This capability, seen as critical for future Pentagon plans, would pair
complimentary systems and allow for more adaptive, flexible responses.
For example, the US Army has already been experimenting with having
pilots in Apache helicopters control Grey Eagle drones. This would
allow the Apache pilots, flying no more than 100 km from a potential target, to quickly destroy targets spotted by the Grey Eagle drones
without having to call in backup.[41] A similar effort by the US Air Force
would allow a pilot in a manned aircraft like an F-16 to control drones
as "loyal wingmen," deploying them as needed. Two models, the Mako

and Valkyrie, have been developed to this end.[42] In time, these expendable, low-cost UCAVs might fly ahead of manned aircraft to scout terrain, confuse or overwhelm radar defenses, or even swarm an adversary. The US Navy is also developing unmanned aerial refueling drones that would extend the range of manned aircraft like the F/A-18 Super Hornet and F-35 Joint Strike Fighter.[43] The United States is experimenting with a number of modes of remote command and control which would break the link between the drone and the ground control station thousands of miles away. For example, local ground troops might control sensors in the battlefield and seamlessly pass or distribute command to different actors, depending on the operational need, or even control or direct multiple unmanned and manned systems from a single device.[44] This would change the "remote" nature of drones and untether them from the ground control station that typically dictates their behavior.

Land and Sea Drones

Another evolution in drone technology that has the potential of furthering "drone thinking" and complicating decision-making is the expansion of drones to new domains, specifically land and sea. While drones have traditionally been designed for flight, the same principles lying beneath unmanned technology can be applied to ground forces and underwater vehicles. In its Unmanned Systems Integrated Roadmap, which projects developments until 2038, the Pentagon imagines a future with a proliferation of ground vehicles that are remotely controlled as today's drones are.[45] Some ground robotics models, such as the Packbot, are designed for tasks such as IED inspection and removal and have been used extensively in Afghanistan and Iraq for more than a decade.[46] But other models under development are also being considered for clearing dangerous vehicle routes, moving squads from place to place, and even combat.[47] In February 2018, the United States conducted its first set of tests with a ground robot shooting at targets on the battlefield.[48] The United States is not the only country experimenting with ground robotics. Russia has announced that it will be fielding tank-like ground robots in Syria and elsewhere for reconnaissance and close-fire support, although there are concerns about its reliability and effectiveness.[49]

The maritime domain is another area in which drones and associated robotic technology will play a growing role. The Pentagon has developed a number of drones, for surface and underwater use, that would be able to detect and disarm mines. The logic here is obvious: mines are extremely dangerous for both manned ships and submarines, and disabling them with drones drastically reduces the risks of losing sailors.[50] But other underwater drones will be used for ISR.[51] Especially if they have a wide range and can sustain a very long time in operations, underwater drones are ideal for picking up signals of submarines and other evidence of incoming naval vessels. Other Autonomous Underwater Vehicles (AUVs) are used for scouting the sea floor and detecting environmental hazards. Some unmanned maritime vehicles, such as the SeaFox, can also be used for port surveillance. Other countries are also experimenting with underwater drones, though they imagine much more aggressive uses than the United States has so far. For example, China is now using underwater drones to extend its claims in the South China Sea.[52] Russia has developed a nuclear armed underwater drone, termed the Status-6 or Kanyon, which has a range of 6,200 miles and can descend to 3,200 feet underwater.[53] At present, the United States is not developing a nuclear-armed drone, but it may eventually choose to do so if more of its rivals continue their development of the technology. An underwater arms race with drones is not out of the question if more countries follow suit.

Miniaturization

The final future trend through which drones may revolutionize political decision-making is miniaturization. To some extent, this is well-established trend, as drones have been getting smaller every year since the Predator made its debut. Such examples as the Raven, Black Hornet, and Switchblade already prove that soldiers on the battlefield can carry useful drones in their backpacks for rapid deployment. While the Raven and Black Hornet are largely reconnaissance drones, the Switchblade is now capable of carrying munitions and killing enemies by dive-bombing them up to 12 miles away from its launch. Other types of small quadcopter drones, already available on the commercial market, are being retrofitted by groups like the Islamic State to hold explosives and then be rammed into troops on the ground. We are already moving into a world where the drones doing the most damage, and attracting the most attention, are small drones rather than Reapers and Global Hawks.

This trend of miniaturization will continue with modern drone companies going even further to produce nano drones no bigger than birds or even insects. These are potential game-changers because they will be able to conduct surveillance without even being noticed. One model, the Nano Hummingbird developed by AeroVironment, weighs less than a pound and resembles a small bird. It has two flapping wings and could move seamlessly into populated areas without notice. Another model, the DragonflEye, is even smaller—it is a backpack equipped with energy, guidance, and navigation systems attached to living dragonflies, turning them into "cyborg drones."[54] This drone uses the natural appearance of the dragonfly as a camouflage and allows the insect to eat biomatter in its environment to sustain itself. In doing so, it cuts down on the energy that the drone element of the DragonflEye must sustain, while allowing its motions to be controlled remotely by its human pilot.[55]

A number of obstacles stand in the way of nano drones being put into use, not the least of which is their astronomical cost. But if the technical problems regarding their flight capacity, durability, and cost can be resolved, nano drones will usher in a revolution in what can be seen and heard. If a nano drone could fly into a building and perch itself in a corner, it could record video and audio of people and convey it back to intelligence officials without anyone noticing. This would be tremendously useful to spy agencies and soldiers in the field who are chasing terrorist leaders. For years, drones have been identifiable by their appearance and by the whirring noise they make when in the skies. The number of plausible uses for the technology will skyrocket if these factors are no longer in play. If a drone can resemble a bird or insect, it will automatically be in camouflage within the natural world and the risks of using it will plummet even further. If drones are as indistinguishable from their environment, as these models suggest is possible, they will be almost irresistible for future use in surveillance, reconnaissance, and spy craft, assuming their costs can be brought down to a manageable level.

Conclusion

All of these models of drones are not yet ready for widespread use, but may be in the next decade or so. The drone age is not one in which technology will sit still. With growing popular scientific and commercial interest in drones and a deep commercial base for their development, it is inevitable that drone technology will continue to develop by leaps and bounds, becoming cheaper and more capable with every passing year. If this trend continues, the technology itself will certainly outpace the contemporary legal and ethical frameworks associated with its use and throw up new dilemmas not yet imagined for political decision-making. Drones may allow us to see and know more about the world we live in, but it is far from obvious that they will make our choices clearer or our decisions easier.

One way to improve those decisions is to develop strong legal standards and norms for the use and sale of drones. To some extent, this is already underway with the humanitarian code of conduct under development and with some of the FAA's regulations for domestic use in the United States. But the political constraints around drone use are considerably weaker. Within the US government, although drone pilots are constrained by the rules of engagement, those authorizing targeted killings are much less constrained in making decisions. Under the Obama administration, the process for selecting targets for drone strikes was located entirely in the executive branch, with considerable latitude for discretion and little transparency, throughout President Obama's first term. Facing a possible defeat in the 2012 presidential elections, the Obama administration began to develop a secret drones "rulebook" to govern their use if Mitt Romney were to be elected president.[56] In 2013, President Obama laid out new standards for using drones for targeted killings and tightened the standards for permitting civilian casualties.[57] The Obama administration also developed tight export standards and launched a joint declaration with fifty other countries on the import and export of drones.[58] But President Donald Trump swept much of this away, loosening the standards for drone use and enabling exports to a wider variety of countries.[59] Today, the United States has a vast killing machine at its disposal, but little transparency or accountability for how it is used by the Trump administration.

All of this suggests that allowing countries to regulate themselves might be insufficient. Instead, we should look to the United Nations or other international organizations to regulate drones and keep their users honest. One option might be to back the development of an international regulatory mechanism, along the lines of the Convention on Certain Conventional Weapons (CCW), which could establish rules for how drones may be sold and used.[60] Alternatively, the United States could back the creation of a UN investigative body on drones which would help to collect information on how drones are used and shame those who use them carelessly or cruelly. A voluntary code of conduct for how drones might be sold and used, which the Obama administration explored, might be an idea worth reviving as drones spread around the world.

In the end, it is also important to realize that the drone age may be one in which everyone can take to the skies, but this accomplishment will not come without a cost. We will continue to risk turning our nightmares into reality until we come to understand how unmanned technology subtly shapes our behavior and the decisions that we make. Drone technology might open up new vistas for human accomplishment and allow many to follow their dreams, but it also alters the menu of choices that lie in front of us. As drone technology falls into the hands of more people, it enables shifts in risks and goals that are not always articulated or even understood by those using them. That drones are unmanned and thus less risky in terms of human life makes them seductive; it also makes things once considered too risky or too ambitious suddenly seem less so. To avoid the worst consequences that may flow from drones, we must learn to carefully measure how unmanned technology changes our strategic choices and to think through how a range of other actors—from repressive governments to terrorist organizations—will respond with their own use of the technology. Only by anticipating what drones do to ourselves and to others can we ensure that our embrace of unmanned technology does not come at the cost of our humanity.

NOTES

Chapter 1

1. The definitive account of the Awlaki case is Scott Shane, *Objective Troy: A Terrorist, A President and the Rise of the Drone* (New York: Tim Duggan Books, 2005), p. 164–172. Awlaki's family was warned that he might be the victim of a US drone strike if released. Jeremy Scahill, *Dirty Wars: The World is a Battlefield* (New York: Nation Books, 2013), p. 185–188.

2. The quote from unnamed US counterterrorism officials comes from Mark Mazetti, Charlie Savage, and Scott Shane, "How a U.S. Citizen Came to Be in America's Cross-Hairs," *New York Times*, 9 March 2013.

3. Patrick Symmes, "Anwar Awlaki: The Next Bin Laden," *GQ Magazine*, available at: http://www.gq.com/story/anwar-al-awlaki-profile. Accessed 19 December 2019.

4. Some have argued that the FBI even tried to turn Awlaki into an informant, although there is little hard evidence that this occurred and the FBI has never confirmed it. See Scahill, *Dirty Wars*, p. 31–47.

5. Jarret Brachman, "Anwar al Awlaki," in *The SAGE Encyclopedia of Terrorism*, ed. Gus Martin (London: Sage, 2011), p. 79–80.

6. Shane, *Objective Troy*, p. 250–251.

7. Executive Order 11905, full text available at: http://fas.org/irp/offdocs/eo11905.htm, accessed 15 August 2015.

8. This line was used by President Obama often, but see particularly "Remarks by the President on the Way Forward in Afghanistan," The White House, 22 June 2011, available at: https://obamawhitehouse.archives.gov/the-press-office/2011/06/22/remarks-president-way-forward-Afghanistan, accessed 1 August 2019.

9. The question is whether killing a US citizen in a foreign land would be a violation of the foreign murder statute under federal law, or whether this case would be different due to his involvement in armed attacks against the United States. See Shane, *Objective Troy*, p. 218–224.

10. See David Brooks, "Obama, Gospel and Verse," *New York Times*, 26 April 2007; John Blake, "How Obama's Favorite Theologian Shaped His First Year in Office," *CNN*, 5 February 2010

11. Mazetti, Savage, and Shane, "How a U.S. Citizen Came to Be in America's Cross-Hairs"; see also Shane, *Objective Troy*, p. 219–227.

12. The administration laid out its rationale in John O. Brennan, Assistant to the President for Homeland Security and Counterterrorism, "The Efficacy and Ethics of U.S. Counterterrorism Strategy," Woodrow Wilson Center, 30 April 2012; Harold Hongju Koh, "The Obama Administration and International Law," 25 March 2010, available at: http://www.state.gov/s/l/releases/remarks/139119.htm, accessed 28 September 2014; and Department of Justice White Paper, "Lawfulness of a Lethal Operation Against a U.S. Citizen Who Is a Senior Operational Leader of al Qai'da or an Associated Force," obtained by NBC News. For critiques of the legality of targeted killing, see UN Human Rights Council, "Report of the Special Rapporteur on Extrajudicial, Summary and Arbitrary Executions, Philip Alston, Addendum: Study on Targeted Killing," 28 May 2010, available at: http://www2.ohchr.org/english/bodies/hrcouncil/docs/14session/A.HRC.14.24.Add6.pdf, accessed 24 September 2012; Mary Ellen O'Connell, "Unlawful Killings in Combat Zones: The Case of Pakistan 2004–2009," Notre Dame Law School, Legal Studies Research Paper 09-43, July 2010, available at: http://papers.ssrn.com/sol3/papers.cfm?abstract_id=1501144, accessed 24 September 2012; and Michael J. Boyle, "The Legal and Ethical Implications of Drone Warfare," *International Journal of Human Rights* 19:2 (2015), p. 105–126.

13. Chris Woods, *Sudden Justice: America's Secret Drone Wars* (Oxford: Oxford University Press, 2015), p. 120. There is some debate as to where the drones were flown from. Woods reports a drone pilot saying that they were flown from an older US base in Djibouti, while Mazetti, Savage, and Shane note that a new, secret drone base in Saudi Arabia accelerated the hunt for Awlaki.

14. Jeremy Scahill, *Dirty Wars: The World Is a Battlefield* (New York: Nation Books, 2013), p. 454–455.

15. The account of this strike is drawn from Scahill, *Dirty Wars*, p. 455–456.

16. Shane, *Objective Troy*, p. 283.

17. Margaret Coker, Adam Entous, and Julian E. Barnes, "Drones Target Yemeni Cleric," *Wall Street Journal*, 7 May 2011.

18. Quoted in Scahhill, p. 456.

19. Scahill, *Dirty Wars*, p. 456, Shane, *Objective Troy*, p. 283.

20. Shane, *Objective Troy*, p. 284.

21. Scahill, *Dirty Wars*, p. 500.

22. Mazetti, Savage, and Shane, "How a U.S. Citizen Came to Be in America's Cross-Hairs."
23. Shane, *Objective Troy*, p. 291.
24. Tom Junod, "The Lethal Presidency of Barack Obama," *Esquire*, August 2012.
25. Scahill, *Dirty Wars*, p. 501.
26. Shane, *Objective Troy*, p. 291.
27. Mazetti, Savage, and Shane, "How a U.S. Citizen Came to Be in America's Cross-Hairs."
28. Scahill, *Dirty Wars*, p. 502.
29. Shane, *Objective Troy*, p. 291.
30. Shane, *Objective Troy*, p. 291; see also Jacquelyn Schneider and Julia Macdonald, "US Public Support for Drone Strikes," Center for New America Security (October 2016), available at: https://s3.amazonaws.com/files.cnas.org/documents/CNAS-Report-DronesandPublicSupport-Final2.pdf, accessed 31 October 2016.
31. Quoted in Jo Becker and Scott Shane, "Secret 'Kill List' Proves a Test of Obama's Principles and Will," *New York Times*, 29 May 2012.
32. Junod, "The Lethal Presidency of Barack Obama."
33. Nasser al-Awlaki, "The Drone That Killed My Grandson," *New York Times*, 17 July 2013.
34. The quote comes from General Michael Hayden, former CIA director under President George W. Bush. See Shane, *Objective Troy*, p. 75.
35. President Barack Obama, "Remarks by the President at the National Defense University," The White House, 23 May 2013, available at: https://obamawhitehouse.archives.gov/the-press-office/2013/05/23/remarks-president-national-defense-university, accessed 29 July 2019.
36. John Kaag and Sarah Kreps, *Drone Warfare* (Cambridge: Polity, 2014), p. 21.
37. Jay Stanley, "'Drones' vs. 'UAVs': What's in a Name?" ACLU Blog, 20 May 2013, available at: https://www.aclu.org/blog/drones-vs-uavs-whats-behind-name, accessed 27 July 2015.
38. Richard B. Gasparre, "The US and Unmanned Flight—Part I," *Air Force Technology*, 25 January 2008, available at: http://www.airforce-technology.com/features/feature1528/.
39. The US military retired Predator drones from service in mid-2018.
40. US Air Force, MQ-1B Predator, Fact Sheet, 20 July 2010, available at: http://www.af.mil/AboutUs/FactSheets/Display/tabid/224/Article/104469/mq-1b-predator.aspx, accessed 27 July 2015.
41. US Air Force, MQ-9 Reaper, Fact Sheet, 18 August 2010, available at: http://www.af.mil/AboutUs/FactSheets/Display/tabid/224/Article/104470/mq-9-reaper.aspx, accessed 27 July 2015.
42. DJI, "Phantom 3 Specs," http://www.dji.com/product/phantom-3/spec, accessed 27 July 2015.

43. This scale was replaced by another which focuses on UAS "groups" of systems in 2011. See Department of Defense, "Unmanned Aircraft Systems Airspace Integration Plan," March 2011, Appendix D.

44. The US Air Force designates this with the letter codes for their drones. Those labeled M, like the MQ-9 Reaper, are multipurpose, while those labeled R like the RQ-4 Global Hawk are for reconnaissance only.

45. Stephanie Carvin, "Getting Drones Wrong," *International Journal of Human Rights* 19:2 (2015), p. 127–141.

46. This characterization accepts the definition of David Hastings Dunn that a disruptive technology is "an innovative technology that triggers sudden and unexpected effects." See his "Drones: Disembodied Warfare and the Unarticulated Threat," *International Affairs* 89:5 (2013), p. 1238.

47. Andrew Cockburn, *Kill Chain: The Rise of the High Tech Assassins* (New York: Henry Holt, 2015), p. 184.

48. Kris Osborn, "Pentagon Plans to Cut Drone Budgets," *DOD Buzz*, 2 January 2014.

49. "Pentagon Eyes Sharp Increase in Drone Flights by 2019: Official," *Reuters*, 17 August 2015.

50. W.J. Hennigan, "Air Force Proposes $3 Billion Plan to Vastly Increase Its Drone Program," *Los Angeles Times*, 10 December 2015.

51. Nick Hopkins, "British Military has 500 Drones," *Guardian*, 6 May 2013.

52. Zachary Keck, "China Is Building 42,000 Military Drones: Should America Worry?" *National Interest Online*, 10 May 2015.

53. General Accounting Office, "Agencies Could Improve Sharing and End-Use Monitoring on Unmanned Aerial Vehicle Exports," GAO 12-536, July 2012, p. 9.

54. Elisa Catalano Ewers, Lauren Fish, Michael C. Horowitz, Alexandra Sander, and Paul Scharre, *Drone Proliferation: Policy Choices for the Trump Administration* (Washington, DC: Center for New American Security, 2017), p. 2.

55. An original estimate of twenty-three was produced from Lynn E. Davis, Michael J. McNerney, James Chow, Thomas Hamilton, Sarah Harting, and Daniel Byman, *Armed and Dangerous: UAVs and U.S. Security* (Washington, DC: RAND, 2014), p. 9. The report by Ewers et al. cited in note 54 suggests this increased to thirty.

56. Samuel Oakford, "Drones, Drones Everywhere: UN Ramping Up Peacekeeping Surveillance Flights," *al Jazeera America*, 27 August 2014, available at: http://america.aljazeera.com/articles/2014/8/27/united-nations-drones.html, accessed 20 August 2015.

57. Rachel Nuwer, "High Above, Drones Keep a Watchful Eye on Wildlife in Africa," *New York Times*, 13 March 2017.

58. Somini Segupta, "Rise of Drones in U.S. Drives Efforts to Limit Police Use," *New York Times*, 15 February 2013.

59. Jim Gold, "Poll: Americans OK with Some Domestic Drones—But Not to Catch Speeders," *NBC News*, 13 June 2012, available at: http://

usnews.nbcnews.com/_news/2012/06/13/12205763-poll-americans-ok-with-some-domestic-drones-but-not-to-catch-speeders?lite, accessed 12 October 2012).

60. "FAA Releases 2016–2036 Aerospace Forecast," Federal Aviation Administration, 24 March 2016, available at: https://www.faa.gov/news/updates/?newsId=85227, accessed 25 October 2016.

61. Cited in "Hostile Drones: The Hostile Use of Drones by Non-State Actors Against British Targets," *Remote Control*, January 2016, p. 19.

62. Cited in Segupta, "Rise of Drones in U.S. Drives Efforts to Limit Police Use."

63. "Jeff Bezos: Amazon Drones Will Be 'as Common as Seeing a Mail Truck,'" *CNN*, 16 August 2015.

64. Mike Murphy, "The First Successful Drone Delivery in the U.S. Has Taken Place," *Quartz*, 20 July 2015, available at: http://qz.com/458703/the-first-successful-drone-delivery-in-the-us-has-taken-place/, accessed 22 July 2015.

65. William Wan and Peter Finn, "Global Race to Match U.S. Drone Capabilities," *Washington Post*, 4 July 2011.

66. Cockburn, *Kill Chain*, p. 177–179.

67. Hendrick de Leeuw, *Conquest of the Air: The History and Future of Aviation* (New York: Vantage, 1960), p. 78.

68. De Leeuw, p. 84–88.

69. Carl Solburg, *Conquest of the Skies* (Boston: Little, Brown and Company, 1979), p. 13–29.

70. Solburg, p. 30–45.

71. Neil Postman, *Technopoly* (New York: Vintage, 1993), p. 9; see also Harold Innis, *The Bias of Communication* (Toronto: University of Toronto Press, 1999).

72. Ulrich Beck, *The Risk Society* (London: Sage, 1992); Christopher Coker, *War in an Age of Risk* (London: Polity, 2009).

73. Edward Luttwak, "Where Are the Great Powers? At Home with the Kids," *Foreign Affairs*, July/August 1994.

74. For a dissenting view, see Christopher Gelpi, Peter Feaver, and Jason Riefler, *Paying the Human Costs of War: American Public Opinion and Casualties in Military Conflicts* (Princeton, NJ: Princeton University Press, 1999).

75. Michael Ignatieff, *Virtual War: Kosovo and Beyond* (New York: Picador, 2001).

76. Rupert Smith, *The Utility of Force: The Art of War in the Modern World* (London: Penguin, 2006).

77. See the discussions in Christian Enemark, *Armed Drones and the Ethics of War: Military Virtue in a Post-Heroic Age* (London: Routledge, 2013); and Robert Sparrow, "War Without Virtue"; Bradley Jay Strawser, *Killing by Remote Control: The Ethics of an Unmanned Military* (Oxford: Oxford University Press, 2013), p. 84–105.

78. Yee-Kuang Heng, *War as Risk Management* (London: Routledge, 2006) and Coker, *War in an Age of Risk*, p. 8.

79. Douglas C. Lovelace Jr., quoted in James Igoe Walsh, "The Effectiveness of Drone Strikes in Counterinsurgency and Counterterrorism Campaigns," Strategic Studies Institute, September 2013, p. v. See also Michael J. Boyle, "The Costs and Consequences of Drone Warfare," *International Affairs*, 89:1 (2013), pp. 1–29 and Audrey Kurth Cronin, "Why Drones Fail," *Foreign Affairs*, July/August 2013.

80. See particularly "Text of John Brennan's Speech on Drone Strikes Today at the Wilson Center," Wilson Center, 30 April 2012, available at: https://www.lawfareblog.com/text-john-brennans-speech-drone-strikes-today-wilson-center, accessed 27 July 2019.

81. Michael Lewis, "Drones: Actually the Most Humane Form of Warfare Ever," *Atlantic*, 21 August 2013.

82. See Strawser and Matthew Fricker, Avery Plaw, and Brian Glyn Williams, "New Light on the Accuracy of the CIA's Predator Drone Campaign in Pakistan" *Terrorism Monitor* 8:41 (11 November 2010), p. 8–13. See also Scott Shane, "The Moral Case for Drones," *New York Times*, 14 July 2012.

83. James Dao, "Drone Pilots Are Found to Get Stress Disorders as Much as Those in Combat Do," *New York Times*, 22 February 2013; Sarah McCammon, "The Warfare May Be Remote but the Trauma Is Real," *NPR*, 24 April 2017.

84. Unnamed Drone Pilot at Nellis Air Force Base, Las Vegas, NV, "It Is War at a Very Intimate Level," in Peter Bergen and Daniel Rothenberg (eds.), *Drone Wars: Transforming Conflict, Law and Policy* (Cambridge: Cambridge University Press 2015), p. 116.

85. Description of symptoms and quotes from unnamed Air Force study, quoted in Eyal Press, "The Wounds of the Drone Warrior," *New York Times*, 13 June 2018, available at: https://www.nytimes.com/2018/06/13/magazine/veterans-ptsd-drone-warrior-wounds.html, accessed 31 May 2018.

86. Ruth Sherlock, "Islamic State Releases Drone Video of Kobane," *Telegraph*, 12 December 2014.

87. Michael R. Gordon, "U.S. Says It Shot Down Drone That Attacked Fighters in Syria," *New York Times*, 8 June 2017.

88. David Rohde, "The Drone Wars," *Reuters Magazine*, 26 January 2012, available at: https://www.reuters.com/article/us-david-rohde-drone-wars-idUSTRE80P11I20120126, accessed 3 August 2019.

89. Chavala Madlena, "We Dream About Drones, Said 13 Year Old Yemeni Before His Death in a CIA Drone Strike," *Guardian*, 10 February 2015. Mohammed Tuaiman was killed, just months after this interview, because of an alleged affiliation with AQAP.

90. Michael J. Boyle, "The Costs and Consequences of Drone Warfare," *International Affairs* 89:1 (2013), p. 1–29.

91. See particularly Kaag and Kreps, *Drone Warfare*, and Boyle, "The Costs and Consequences of Drone Warfare."

92. Boyle.

93. Micah Zenko, "U.S. Drones: The Counterinsurgency Air Force for Pakistan, Yemen and Somalia," Council on Foreign Relations, 27 November 2012, available at: https://www.cfr.org/blog/us-drones-counterinsurgency-air-force-pakistan-yemen-and-somalia, accessed 3 August 2019.

94. Michael C. Horowitz, Sarah Kreps, and Matthrew Fuhrmann, "Separating Fact from Fiction in the Debate over Drone Proliferation," *International Security* 41:2 (Fall 2016), p. 7–42.

95. Andrew Tilghman, "Army, SOCOM to Take On Daily Drone Missions," *Military Times*, 17 August 2015.

96. David Axe, "Air Force Drone Crews Got So Demoralized That They Booed Their Commander," *War Is Boring*, 29 September 2014, available at: https://medium.com/war-is-boring/air-force-drone-crews-got-so-demoralized-that-they-booed-their-commander-cfd455fca40f, accessed 4 August 2019.

97. Arthur Holland Michel, *Eyes in the Sky: The Secret Rise of Gorgon Stare and How It Will Watch Us All* (New York: Houghton Mifflin Harcourt, 2019).

98. See Bradley Jay Strawser, "Moral Predators: The Duty to Employ Uninhabited Aerial Vehicles," *Journal of Military Ethics* 9:4 (2010), p. 342–368. See also Strawser's debate with Asa Kasher in "Distinguishing Drones: An Exchange," in Bradley Jay Strawser, *Killing by Remote Control: The Ethics of an Unmanned Military* (Oxford: Oxford University Press, 2013), p. 47–65.

99. Jacques Ellul, *The Technological Society* (New York: Vintage, 1964), p. 97–100.

100. Postman, *Technopoly*, p. 8–9.

101. See Langdon Winner, "Do Artifacts Have Politics?" in Donald MacKenzie and Judie Wajcman, *The Social Shaping of Technology* (Berkshire: Open University Press, 1999). See also Majid Tehranian, *Technologies of Power: Information Machines and Democratic Prospects* (Westport: Praeger, 1990), which applied this insight to information technology.

102. Stuart E. Johnson and Martin C. Libicki, *Dominant Battlespace Knowledge: The Winning Edge* (Washington, DC: National Defense University, 1995).

103. See Grégoire Chamayou, *Manhunts: A Philosophical Inquiry* (Princeton, NJ: Princeton University Press, 2012); Grégoire Chamayou, *Drone Theory* (New York: Penguin, 2015); Bradley Peniston, "Army Warns Future War with Russia or China Would Be 'Extremely Lethal and Fast,'" *DefenseOne*, 4 October 2016, available at: http://www.defenseone.com/threats/2016/10/future-army/132105/, accessed 11 November 2016.

104. See William Arkin, *Unmanned: Drones, Data and the Illusion of Perfect Warfare* (New York: Little, Brown 2015).

105. Lorenzo Franceschi-Bicchierai, "Report: Russia Is Stockpiling Drones to Spy on Street Protests," https://www.cnn.com/2012/07/25/tech/innovation/russia-stockpiling-drones-wired/index.html, *CNN.com*, 25 July 2012.

106. Drew Hinshaw, "For African Generals, Drones Are the Latest Thing," *Wall Street Journal*, 27 September 2013.

107. Didi Kirsten Tatlow, "China Said to Deploy Drones After Unrest in Xinjiang," *New York Times*, 19 August 2014.

108. Kristine Bergtora Sandvik and Kjersti Lohne, "The Promise and Perils of 'Disaster Drones,'" *Humanitarian Practice Network*, Issue 58 (July 2013).

Chapter 2

1. Albert L. Weeks, "In Operation Aphrodite, Explosive-Laden Aircraft Were to Be Flown Against German Targets," *World War II* 15:1 (May 2000), p. 66–69.

2. Richard Hollingham, "V2: The Rocket That Launched the Space Age," BBC News, 8 September 2014, available at: http://www.bbc.com/future/story/20140905-the-nazis-space-age-rocket, accessed 6 September 2016.

3. Weeks, "In Operation Aphrodite." Estimates vary about the casualties of V-1 and V-2 rockets. The Imperial War Museum in the United Kingdom confirms the 30,000 casualties number. See: https://www.iwm.org.uk/history/the-terrifying-german-revenge-weapons-of-the-second-world-war, accessed 5 August 2019. But it is also important to acknowledge that there were as many 10,000 people laborers in concentration camps who died when Nazis forced them to work on the rockets. V-2 rockets were also directed at France, Belgium, and other targets.

4. The quote is from Waugh's novel *Unconditional Surrender*, cited in Dave Sloggett, *Drone Warfare: The Development of Unmanned Aerial Conflict* (New York: Skyhorse, 2014), p. 25.

5. Weeks, "In Operation Aphrodite."

6. Estimate of altitude in Ed Grabianowski, "The Secret Drone Mission That Killed Joseph Kennedy Jr." *Io9*, 21 February 2013.

7. Anthony Leviero, "Kennedy Jr. Died in Air Explosion," *New York Times*, 22 October 1945.

8. Quoted in Ann Rogers and John Hill, *Unmanned: Drone Warfare and Global Security* (London: Pluto, 2014), p. 18.

9. Richard Whittle, *Predator: The Secret Origins of the Drone Revolution* (New York: Henry Holt, 2014), p. 20.

10. Ed Grabianowski, "The Secret Drone Mission That Killed Joseph Kennedy Jr."

11. Weeks, "In Operation Aphrodite."

12. Alan Axelrod, *Lost Destiny: Joe Kennedy Jr. and the Doomed WWII Mission to Save London* (New York: Palgrave Macmillian, 2015), p. 183.

13. Weeks, "In Operation Aphrodite."
14. Whittle, *Predator*, p. 20.
15. Axelrod, *Lost Destiny*, p. 181–217. "Operation Aphrodite," Historic Wings, 12 August 2012, available at: http://fly.historicwings.com/2012/08/operation-aphrodite/, accessed 6 September 2016.
16. Axelrod, *Lost Destiny*, p. 181–217; "Operation Aphrodite."
17. Andrew Cockburn, *Kill Chain: The Rise of High-Tech Assassins* (New York: Henry Holt, 2015), p. 25.
18. Weeks, "In Operation Aphrodite." The navy had its own code name for the mission: Operation Anvil.
19. Axelrod, *Lost Destiny*, p. 243.
20. Quoted in Axelrod, p. 239.
21. Edward J. Renehan Jr., *The Kennedys at War* (New York: Doubleday, 2002), p. 302.
22. Axelrod, *Lost Destiny*, p. 248–249.
23. The quote is from the Navy Cross citation awarded to Kennedy after his death, cited at: https://valor.militarytimes.com/hero/21302, accessed 15 May 2019.
24. Renehan, *The Kennedys at War*, p. 304.
25. Weeks, "In Operation Aphrodite."
26. For example, according to author Jack Olson, the Kennedy family was not told many of the details of the attack including the facts that the PB4-Y plane had a faulty electrical system, that these faults were known to ground operators, and that the German rocket site was abandoned. See Jack Olson, *Aphrodite: Desperate Mission* (New York: G.P. Putnam, 1970).
27. For a critical take, see William Arkin, *Unmanned: Drones, Data and the Illusion of Perfect Warfare* (New York: Little & Brown, 2015), p. 51–53.
28. Lawrence R. "Nuke" Newcome, *Unmanned Aviation: A Brief History of Unmanned Aerial Vehicles* (Reston, VA: American Institute of Aeronautics and Astronautics, 2004), p. 11–14.
29. Newcome, p. 15–21.
30. Newcome, p. 20; Sloggett, *Drone Warfare*, p. 17.
31. Newcome, *Unmanned Aviation*, p. 20. There is some variance between sources about whether this distance was 1,000 yards or feet, but this chapter uses Newcome's estimate.
32. Rogers and Hill, *Unmanned: Drone Warfare and Global Security*, p. 14.
33. Major Bell, cited in David Hambling, *Swarm Troopers* (Archangel Publishing, 2015), p. 11.
34. Konstantin Kakaes, "From Orville Wright to September 11: What the History of Drone Technology Says About Its Future," in Peter L. Bergen and Daniel Rothenberg (eds.), *Drone Wars: Transforming Conflict, Law and Policy* (Cambridge: Cambridge University Press, 2015), p. 360.
35. Jimmy Stamp, "Unmanned Drones Have Been Around Since World War I," *Smithsonian Magazine*, 12 February 2013, available at: http://www.

smithsonianmag.com/arts-culture/unmanned-drones-have-been-around-since-world-war-i-16055939/?no-ist, accessed 9 September 2016.

36. Stamp.

37. Assertion that it was the first was from Thomas Mueller, quoted in Rogers and Hill, *Unmanned: Drone Warfare and Global Security*, p. 14. On the control, see John Edward Jackson, "A Robot's Family Tree: An Introduction and Brief History of Unmanned Systems," in Capt. John E. Jackson (ed), *One Nation Under Drones: Legality, Morality and Utility of Unmanned Combat Systems* (Annapolis, MD: Naval Institute Press, 2018), p. 2.

38. Production numbers from Kakaes, "From Orville Wright to September 11," p. 362, and estimate of total actual production from Stamp, "Unmanned Drones Have Been Around Since World War I."

39. John David Blom, *Unmanned Aerial Systems: A Historical Perspective*, Occasional Paper 37 (Combat Studies Institute Press, Fort Leavenworth, Kansas, 2010), p. 46.

40. "The Mother of All Drones," Vintage Wings of Canada, available at: http://www.vintagewings.ca/VintageNews/Stories/tabid/116/articleType/ArticleView/articleId/484/The-Mother-of-All-Drones.aspx, accessed 9 September 2016.

41. Rogers and Hill, *Unmanned: Drone Warfare and Global Security*.

42. Quote is from Rogers and Hill. Attention from the US Navy is reported in Abigail R. Hall and Christopher Coyne, "The Political Economy of Drones," *Defence and Peace Economics* (2013), p. 4.

43. Hall and Coyne, p. 4.

44. There is some debate over where the term "drone" comes from. Some argue that it is from the Queen Bee; others from the sounds of the aircraft itself. Still others from the fact that it acts like an insect does, mindlessly doing tasks. For a useful discussion of how the term conveys the ephemeral, disposable nature of the technology, see Chamayou, *Drone Theory*, p. 26–27. On the changing meaning of the word "drone," see Brian Benchoff, "A Brief History of 'Drone,'" Hackaday, 26 September 2016, available at: http://hackaday.com/2016/09/26/a-brief-history-of-drone/, accessed 28 September 2016.

45. Bill Yenne, *Birds of Prey: Predators, Reapers and America's Newest UAVs in Combat* (North Branch, MN: Specialty Press, 2010), p. 10.

46. Yenne, p. 10–11.

47. Peter W. Singer, *Wired at War: The Robotics Revolution and Conflict in the 21st Century* (New York: Penguin, 2009), p. 49.

48. Yenne, *Birds of Prey*, p. 11–12.

49. Blom estimates that the US Army and Navy put into service 50 OQ-1s, 600 OQ-2s, 5,822 OQ-3s, and 2,084 OQ-4s. See Blom, *Unmanned Aerial Systems*, p. 47.

50. Michael Beschloss, "Marilyn Monroe's World War II Drone Program," *New York Times*, 3 June 2014.

51. Singer, *Wired at War*, p. 50.
52. Newcome, *Unmanned Aviation*, p. 66.
53. Newcome, p. 66.
54. Newcome, p. 66.
55. Hall and Coyne, "The Political Economy of Drones," p. 5.
56. Blom, *Unmanned Aerial Systems*, p. 48.
57. Newcome, *Unmanned Aviation*, p. 66.
58. Rogers and Hill, *Unmanned: Drone Warfare and Global Security*, p. 16.
59. Newcome, *Unmanned Aviation*, p. 67–69.
60. Jackson, "A Robot's Family Tree," p. 3.
61. Hambling, *Swarm Troopers*, p. 10.
62. Hall and Coyne, "The Political Economy of Drones," p. 5.
63. Kakaes, "From Orville Wright to September 11," p. 365.
64. Kakaes, "From Orville Wright to September 11," p. 366.
65. The *New York Times* reported that they were only 8 miles from the explosion, but Kakaes ("From Orville Wright to September 11," p. 366) reports 25 miles.
66. Hanson W. Baldwin, "The 'Drone': Portent of a Push-Button War," *New York Times*, 25 August 1946.
67. Adam Rawnsley, "America Almost Had a Nuclear-Armed Drone Bomber," *War Is Boring*, 16 December 2014, available at: https://warisboring.com/america-almost-had-a-nuclear-armed-drone-bomber-e494e2e9a286#.r9bkxlkjc, accessed 23 September 2016.
68. Delmar J. Trester, Thermonuclear Weapon Delivery by Unmanned B-47: Project Brass Ring, p. 275.
69. Trester, p. 280.
70. See Trester and also Rawnsley, "America Almost Had a Nuclear-Armed Drone Bomber."
71. Blom, *Unmanned Aerial Systems*, p. 50.
72. Kakaes, "From Orville Wright to September 11," p. 367.
73. Blom, *Unmanned Aerial Systems*, p. 52.
74. Jackson, "A Robot's Family Tree," p. 5.
75. Sloggett, *Drone Warfare*, p. 64.
76. Sloggett, p. 65.
77. Sloggett, p. 65–71.
78. Newcome, *Unmanned Aviation*, p. 71.
79. Rogers and Hill, *Unmanned: Drone Warfare and Global Security*, p.20.
80. Thomas P. Ehrhard, *Air Force UAVs: The Secret History* (Arlington: Mitchell Institute, July 2010), p. 6.
81. Whittle, *Predator*, p. 21.
82. Ehrhard, *Air Force UAVs*, p. 56.
83. Ehrhard, p. 6
84. Whittle, *Predator*, p. 21.
85. Ehrhard, *Air Force UAVs*, p. 5.

86. Ehrhard, p. 56. See also Bill Grimes, *The History of Big Safari* (Bloomington, IN: Archway, 2014).
87. Rogers and Hill, *Unmanned: Drone Warfare and Global Security*, p. 21.
88. Ehrhard, *Air Force UAVs*, p. 8.
89. Whittle, *Predator*, p. 21. Rogers and Hill, *Unmanned: Drone Warfare and Global Security*, p. 21.
90. Blom, *Unmanned Aerial Systems*, p. 57.
91. Ehrhard, *Air Force UAVs*, p. 8.
92. See particularly William Wagner, *Lightning Bugs and Other Reconnaissance Drones* (Fallbrook, CA: Aero Publishing, 1982).
93. Ehrhard, *Air Force UAVs*, p. 8.
94. Rogers and Hill, *Unmanned: Drone Warfare and Global Security*, p. 21.
95. Rogers and Hill, p. 21.
96. Ehrhard, *Air Force UAVs*, p. 23.
97. Ehrhard, p. 9.
98. Ehhard, p. 9.
99. Blom, *Unmanned Aerial Systems*, p. 60.
100. These were known as the D-21 drone and Compass Arrow; neither came into full service due to technical problems and cost.
101. Rogers and Hill, *Unmanned: Drone Warfare and Global Security*, p. 22.
102. Ehrhard, *Air Force UAVs*, p. 23.
103. Ehrhard, p. 9.
104. Whittle, *Predator*, p. 22.
105. Ehrhard, *Air Force UAVs*, p. 24.
106. Blom, *Unmanned Aerial Systems*, p. 61.
107. Quote from Comptroller General of the United States in Katherine Hall Kindervater, "The Emergence of Lethal Surveillance: Watching and Killing in the History of Drone Technology" *Security Dialogue* 47:3 (2016), p. 228.
108. Ehrhard, *Air Force UAVs*, p. 58.
109. Cited in Ehrhard, p. 30. Ironically, Ryan would later become an advocate of drones and actually vastly increase the size of the fleet.
110. Ehrhard finds that the so-called "white scarf" syndrome was overstated and that most of the pilots and leaders from the US Air Force were supportive of unmanned aircraft, even if it reduced their cockpit numbers. See Ehrhard, p. 45.
111. R. Cargill Hall, "Reconnaissance Drones and Their First Use in the Cold War," *Air Power History* (Fall 2014), p. 24.
112. Ehrhard, *Air Force UAVs*, p. 25.
113. Blom, *Unmanned Aerial Systems*, p. 63.
114. Hall, "Reconnaissance Drones and Their First Use in the Cold War," p. 27.
115. Arkin, *Unmanned*, p. 55.
116. Ehrhard, *Air Force UAVs*, p. 29.

117. Kakaes, "From Orville Wright to September 11," p. 371.
118. Newcome, *Unmanned Aviation*, p. 105.
119. Jackson, "A Robot's Family Tree," p. 6.
120. Blom, *Unmanned Aerial Systems*, p. 66–71.
121. Kakaes, "From Orville Wright to September 11," p. 374.
122. Kakaes, p. 376.
123. Richard H. Van Atta and Michael J. Lippitz, with Jasper C. Lupo, Rob Mahoney, and Jack H. Nunn, "Transformation and Transition: DARPA's Role in Fostering an Emerging Revolution in Military Affairs," Institute for Defense Analysis (April 2003), p. 41.
124. Kakaes, "From Orville Wright to September 11," p. 371.
125. Kakaes, p. 372.
126. Blom, *Unmanned Aerial Systems*, p. 72.
127. Ehrhard, *Air Force UAVs*, p. 72.
128. Ralph Sanders, "UAVs: An Israeli Military Innovation," *Joint Forces Quarterly* (Winter 2002–2003), p. 115.
129. Sanders, p. 115.
130. Blom, *Unmanned Aerial Systems*, p. 72.
131. Newcome, *Unmanned Aviation*, p. 96.
132. Yenne, *Birds of Prey*, p. 26.
133. Yenne, p. 27.
134. Arkin, *Unmanned*, p. 46.
135. Cited in Arkin, p. 44.
136. Arkin, p. 43–47.
137. Whittle, *Predator*, p. 24.
138. Arkin, *Unmanned*, p. 47.
139. Whittle, *Predator*, p. 15.
140. Whittle, p. 14.
141. Van Atta et al., "Transformation and Transition," p. 43.
142. Bill Sweetman, "Drones: Invented and Forgotten," *Popular Science* 245:3 (1 September 1994).
143. Van Atta et al., "Transformation and Transition," p. 43.
144. Sweetman, "Drones: Invented and Forgotten."
145. Whittle, *Predator*, p. 58.
146. Whittle, p. 71, 82.
147. General Atomics had used the term "Predator" earlier to describe a less sophisticated model. There is still some debate as to whether it was named after an Arnold Schwarzenegger movie.
148. Arkin, *Unmanned*, p. 58.
149. Arkin, p. 58.
150. Arkin, p. 58.
151. Arkin, p. 58.
152. Frank Strickland, "The Early Evolution of the Predator Drone," *Studies in Intelligence* 57:1 (Extracts, March 2013), p. 6.

153. Arkin, *Unmanned*, p. 58.
154. Singer, *Wired at War*, p. 98–99.
155. Tom Simonite, "Moore's Lawls Dead. Now What?" *MIT Technology Review*, 13 May 2016, available at: https://www.technologyreview.com/s/601441/moores-law-is-dead-now-what/, accessed 5 October 2016.
156. Whittle, *Predator*, p. 142.
157. Whittle, p. 40.
158. Quoted in Whittle, p. 41.
159. Whittle, p. 102–105, 115.
160. Elizabeth Becker, "They're Unmanned, They Fly Low, They Get the Picture," *New York Times*, 3 June 1999.
161. Major Houston Cantwell, USAF, "Beyond Butterflies: Predator and the Evolution of Unmanned Aerial Vehicle in Air Force Culture," School of Advanced Air and Space Studies, Maxwell AFB, Alabama, June 2007, p. 24.
162. Newcome, *Unmanned Aviation*, p. 109.
163. Thomas P. Christie, "Operational Test and Evaluation Report of the Predator: Medium-Altitude Endurance Unmanned Aerial Vehicle," undated report. Quoted at length in James Dao, "A Nation Challenged: U.S. Is Using More Drones, Despite Concerns over Flaws," *New York Times*, 3 November 2001.
164. Whittle, *Predator*, p. 115.
165. The term is common, but for an illustrative use see Woods, *Sudden Justice*, p. 86.
166. Whittle, *Predator*, p. 169.
167. Whittle, p. 201.

Chapter 3

1. The description of the house as a fortress comes from Scott Gates and Kaushik Roy, *Unconventional Warfare in South Asia: Shadow Warriors and Counterinsurgency* (London: Ashgate, 2014), p. 125.
2. Mark Mazetti, *The Way of the Knife* (New York: Penguin, 2013), p. 104–105.
3. Mazetti, p. 104.
4. Brian Glyn Williams, *Predator: The CIA's Drone War on al Qaeda* (Washington, DC: Potomac Press, 2013), p. 46.
5. Mark Mazetti, "A Secret Deal on Drones, Sealed in Blood," *New York Times*, 6 April 2013.
6. Mazetti, *The Way of the Knife*, p. 104–105.
7. Chris Woods, *Sudden Justice: America's Secret Drone War* (Oxford: Oxford University Press, 2015), p. 100.
8. Mazetti, *The Way of the Knife*, p. 105 and Gates and Roy, *Unconventional Warfare in South Asia*, p. 124.
9. Gates and Roy, p. 124, and Mazetti, *The Way of the Knife*, p. 105.

10. Mazetti, p. 105.
11. Gates and Roy, *Unconventional Warfare in South Asia*, p. 125.
12. David Rohde and Mohammed Khan, "Militants' Defiance Puts Pakistan's Resolve in Doubt," *New York Times*, 10 June 2004.
13. Mazetti, *The Way of the Knife*, p. 106.
14. Woods, *Sudden Justice*, p. 101.
15. Mazetti, *The Way of the Knife*, p. 107.
16. Rohde and Khan, "Militants' Defiance Puts Pakistan's Resolve in Doubt."
17. Mazetti, *The Way of the Knife*, p. 107.
18. Woods, *Sudden Justice*, p. 101.
19. Steve Coll, "The Unblinking Stare: The Drone War in Pakistan," *New Yorker*, 24 November 2014.
20. Mazetti, *The Way of the Knife*, p. 108.
21. Cited in Mazetti, p. 109.
22. Woods, *Sudden Justice*, p. 101.
23. This description of the attack comes from eyewitnesses. See Alice K. Ross, "Drone Strikes in Pakistan: Ten Years On, Eyewitnesses Describe the Aftermath of the First Pakistan Drone Strike," Bureau of Investigative Journalism, 17 July 2014, available at: https://www.thebureauinvestigates. com/2014/06/17/ten-years-on-eyewitnesses-describe-the-aftermath-of-the-first-pakistan-drone-strike/, accessed 8 February 2016.
24. The casualty estimates are from Ross. The description of Nek Mohammed's injuries are from Mazetti, *The Way of the Knife*, p. 110.
25. Woods, *Sudden Justice*, p, 101.
26. See Ross, "Drone Strikes in Pakistan."
27. Mazetti, *The Way of the Knife*, p. 110.
28. "Drone Wars Pakistan: Analysis," New America Foundation, available at: http://securitydata.newamerica.net/drones/pakistan-analysis.html, accessed 8 February 2016. This excluded data for 2016.
29. According to the New America Foundation data, there were 122 drone strikes in Pakistan in 2010.
30. Quoted in Coll, "The Unblinking Stare."
31. Woods, *Sudden Justice*, p. 110–111.
32. Woods, p. 109–111.
33. Woods, p. 111.
34. Woods, p. 114.
35. An unnamed US official quoted in Glyn Williams, *Predator*, p. 83.
36. Quoted in Woods, *Sudden Justice*, p. 115. The original conversation is recorded in a Wikileaks cable dated 23 August 2008, entitled "Immunity for Musharraf Likely After Zadari's Election as President," available at: https://wikileaks.org/plusd/cables/08ISLAMABAD2802_a.html, accessed 9 February 2016.
37. Jane Mayer, "The Predator War," *New Yorker*, 26 October 2009.
38. Cited in Woods, *Sudden Justice*, p. 115.

39. Glyn Williams, *Predator*, p. 89–91.
40. Spencer Ackerman, "Victim of Obama's First Drone Strike: 'I Am a Living Example of What Drones Are.'" *Guardian*, 23 January 2016.
41. Daniel Klaidman, *Kill or Capture: The War on Terror and the Soul of the Obama Presidency* (New York: Houghton Mifflin Harcourt, 2012), p. 40.
42. Coll, "The Unblinking Stare."
43. Mazetti, *The Way of the Knife*, p. 226.
44. Jo Becker and Scott Shane, "Secret Kill List Proves a Test of Obama's Principles and Will," *New York Times*, 29 May 2012.
45. Described in Mark Bowden, "The Killing Machine: How to Think About Drones," *Atlantic*, September 2013.
46. Becker and Shane, "Secret Kill List Proves a Test of Obama's Principles and Will."
47. Becker and Shane.
48. Quoted in *CNN*, "US Airstrikes Called "Very Effective," 18 May 2009.
49. Peter Bergen, "DroneIs Obama's Weapon of Choice," *CNN*, 19 September 2012.
50. Peter Bergen and Jennifer Rowland, "Drone Wars," *Washington Quarterly* (Summer 2013), p. 10.
51. Quoted in Jane Perlez and Pir Zubair Shah, "Drones Batter Al Qaeda and Its Allies in Pakistan," *New York Times*, 4 April 2010.
52. Rosa Brooks, "Take Two Drones and Call Me in the Morning," *Foreign Policy*, 12 September 2012.
53. Woods, *Sudden Justice*, p. 160.
54. Report of the Special Rapporteur on Extrajudicial, Summary and Arbitrary Executions," UN General Assembly, New York, available at: https://www.justsecurity.org/wp-content/uploads/2013/10/UN-Special-Rapporteur-Extrajudicial-Christof-Heyns-Report-Drones.pdf, accessed 18 February 2016.
55. The quote is from Prime Minister Raza Gilani, "Drone Strikes Adding to Insurgency—PM," *The News* (Pakistan), 28 January 2012.
56. Pew Research Center, "Pakistani Public Opinion Ever More Critical of the U.S." 27 June 2012. Some scholars have argued that this opposition among the Pakistani public varies on education level and the degree of knowledge of the program. See C. Christine Fair, Karl Kalthenhaler, and William J. Miller, "Pakistani Opposition to American Drone Strikes," *Political Science Quarterly* 129:1 (2014), p. 1–33.
57. "Obama's Speech on Drones Policy," *New York Times*, 23 May 2013, available at: http://www.nytimes.com/2013/05/24/us/politics/transcript-of-obamas-speech-on-drone-policy.html, accessed 18 February 2016.
58. Spencer Ackerman, "Trump Ramped Up Drone Strikes in America's Shadow Wars," *Daily Beast*, 26 November 2018.
59. Charlie Savage and Eric Schmitt, "Trump Poised to Drop Some Limits on Drone Strikes and Commando Raids," *New York Times*, 21 September

2017, available at: https://www.nytimes.com/2017/09/21/us/politics/trump-drone-strikes-commando-raids-rules.html, accessed 1 June 2018.

60. This estimate is from B'Tselem, an Israeli human rights organization. Quoted in Daniel Byman, "Do Targeted Killings Work?" *Foreign Affairs* March–April 2006, p. 98.

61. For a discussion of this difference, see Steven R. David, "Israel's Policy of Targeted Killings," *Ethics and International Affairs* 17:1 (2003), p. 112.

62. There is a scholarly debate as to whether there is a meaningful distinction between assassinations of leaders in peacetime and the killing of combatants in warfare. For discussions, see David, "Israel's Policy of Targeted Killings" and Michael L. Gross, *Moral Dilemmas of Modern War: Torture, Assassination and Blackmail in an Age of Asymmetric Conflict* (Cambridge: Cambridge University Press, 2010), p. 100–121.

63. These arguments are well-summarized by Amos N. Guiora, who served as a legal advisor to the Israeli Defense Force. See Amos R. Guiora, *Legitimate Target: A Criteria-Based Approach to Targeted Killing* (Oxford: Oxford University Press, 2013).

64. Colonel Daniel Reisner of the international law office of the Military Advocate General's office for the Israeli Defense Force later acknowledged that Israel made a concerted effort to use precedent to shape international law in a way conducive to targeted killing. In a 2009 interview in *Haaretz*, he reflected that "what we are seeing now is a revision of international law . . . If you do something for long enough, the world will accept it. The whole of international law is now based on the notion that an act that is forbidden today becomes permissible if executed by enough countries . . . International law progresses through violations. We invented the targeted assassination thesis and we had to push it. At first there were protrusions that made it hard to insert easily into the legal moulds. Eight years later it is in the center of the bounds of legitimacy." Yotam Feldman and Uri Blau, "Consent and Advise," *Haaretz*, 29 January 2009, available at: http://www.haaretz.com/consent-and-advise-1.269127, accessed 8 March 2016.

65. Although Israel defended targeted killing publicly, it has not made all of its legal rationale for the policy clear. In 2006, the Israeli Supreme Court reviewed the policy and approved it, but put forward formal requirements about the involvement of the individual in hostilities and the impossibility of capture before strikes can take place. Yet Israel has made little of its legal reasoning behind its claim public and does not disclose how closely it follows these requirements. See Human Rights Council, "Report of the Special Rapporteur on Extrajudicial, Summary or Arbitrary Executions, Philip Alston, Addendum: Study on Targeted Killing," 28 May 2010, para 15–17.

66. Quoted in Christopher Kurtz, "How Norms Die: Torture and Assassination in American Security Policy," *Ethics and International Affairs* 28:4 (2014), p. 436–437.

67. Woods, *Sudden Justice*, p. 48.

68. The quote comes from the 9/11 Commission Report, quoted in Woods, p. 47. See also George Tenet (with Bill Hawlow), *At the Center of the Storm: My Years at the CIA* (New York: Harper Press, 2007), p. 160.

69. Richard Whittle, *Predator: The Secret Origins of the Drone Revolution* (New York: Henry Holt, 2014), p. 224.

70. Whittle, *Predator*, p. 223. Richard A. Clarke, *Against All Enemies: Inside America's War on Terror* (New York: Free Press, 2004).

71. Woods, *Sudden Justice*, p. 39. For an account of the pilot who saw bin Laden, see Scott Swanson, "War Is Not a Video Game—Not Even Remotely," *Breaking Defense*, 18 November 2014, available at: http://breakingdefense.com/2014/11/war-is-no-video-game-not-even-remotely/, accessed 6 April 2016.

72. Tenet, *At the Center of the Storm*, p. 160; Woods, *Sudden Justice*, p. 40.

73. Woods, *Sudden Justice*, p. 40.

74. Woods, p. 49.

75. Tenet, *At the Center of the Storm*, p. 160.

76. Whittle, *Predator*, p. 222–223.

77. Bob Woodward, "CIA Told to Do 'Whatever Necessary' to Kill Bin Laden," *Washington Post*, 21 October 2001.

78. Ward Thomas, "The New Age of Assassination," *SAIS Review of International Affairs* 25:1 (Winter–Spring 2005), p. 27–39; See also Woods, *Sudden Justice*, p. 54.

79. Quoted in Grégoire Chamayou, *Drone Theory* (trans: Janet Lloyd), (London: Penguin, 2015), p. 32.

80. Jennifer Daskal, "After the AUMF" *Harvard National Security Law Review* 5 (2014), p. 115–146.

81. Jeremy Scahill, *Dirty Wars: The World Is a Battlefield* (New York: Nation Books, 2013), p. 24_25.

82. Scahill, *Dirty Wars*, p. 25.

83. Woods, *Sudden Justice*, p. 50.

84. Michael J. Boyle, "The Legal and Ethical Implications of Drone Warfare," *International Journal of Human Rights* 19:2 (2015), p. 105–126.

85. Quoted in John Mearsheimer, "America Unhinged," *National Interest* (January/February 2014), p. 29.

86. Whittle, *Predator*, p. 242–254.

87. Woods, *Sudden Justice*, p. 56.

88. Gregory D. Johnsen, *The Last Refuge: Yemen, al Qaeda and America's War in Arabia* (New York: W.W. Norton, 2013), p. 119–123.

89. Woods, *Sudden Justice*, p. 60–61.

90. Johnsen, *The Last Refuge*, p. 123.

91. Woods, *Sudden Justice*, p. 57.

92. This exception is generally known as "hot pursuit" and would not apply to most targeted killings, which are conducted with planning and deliberation.

93. For a discussion of the difference between peacetime and wartime with respect to targeted killing, see Gabriella Blum and Philip B. Heymann, "Law of Policy of Targeted Killings," *Harvard National Security Journal* 145 (2010), p. 145–170.

94. Mary Ellen O'Connell, "Unlawful Killing with Combat Drones," Notre Dame Law School, Legal Studies Research Paper 09-43 (2010).

95. For extended critiques of their position, see Special Rapporteur on Extrajudicial, Summary and Arbitrary Execution Christof Heyns, "Note by the Secretary General: Extrajudicial, Summary or Arbitrary Executions," UN General Assembly 69th Session, 6 August 2014; and Philip Alston, "The CIA and Targeted Killings Beyond Border," *Harvard Law National Security Journal* (2012), p. 285–446.

96. John Kaag and Sarah Kreps, *Drone Warfare* (Cambridge: Polity, 2014), p. 26.

97. Andrew C. Cockburn, *Kill Chain: The Rise of High Tech Assassins* (New York: Henry Holt, 2015), p. 220. See also Greg Miller, "CIA Remains Behind Most Drone Strikes, Despite Effort to Shift Campaign to Defense," *Washington Post*, 25 November 2013.

98. Cockburn, *Kill Chain*, p. 220.

99. For a profile, see Greg Miller, "At CIA, A Convert to Islam Leads the Terrorism Hunt," *Washington Post*, 24 March 2012.

100. Mark Mazetti and Matt Apuzo, "Deep Support in Washington for CIA's Drone Missions" *New York Times*, 25 April 2015.

101. Greg Miller and Julian Tate, "CIA Shifts Focus to Killing Targets," *Washington Post*, 1 September 2011.

102. Miller and Tate.

103. Unnamed senior CIA official, quoted in Miller and Tate.

104. Woods, *Sudden Justice*, p. 17.

105. Woods, p. 17.

106. Naureen Shah, "A Move in the Shadows: Will JSOC's Control of Drones Improve Policy," in Peter L. Bergen and Daniel Rothenberg (eds.), *Drone Wars: Transforming Conflict, Law and Policy* (Cambridge: Cambridge University Press, 2015), p. 171–175.

107. Scahill, *Dirty Wars*, p. 150–154.

108. Shah, "A Move in the Shadows: Will JSOC's Control of Drones Improve Policy," p. 174.

109. Naureen Shah, "A Move in the Shadows: Will JSOC's Control of Drones Improve Policy," p. 163.

110. Scahill, *Dirty Wars*; see also Sean Naylor, *Relentless Strike: The Secret History of Joint Special Operations Command* (New York: St. Martin's, 2015).

111. Dana Priest and William M. Arkin, "Top Secret America: A Look at the Military's Joint Special Operations Command," *Washington Post*, 2 September 2011.

112. Shah, "A Move in the Shadows," p. 161.

113. Scahill, *Dirty Wars*, p. 251.

114. Quotes from Shah, "A Move in the Shadows," p. 161.

115. Miller, "CIA Remains Behind Most Drone Strikes."

116. Woods, *Sudden Justice*, p. 19–21.

117. Shah, "A Move in the Shadows," p. 164.

118. See particularly Scahill and Woods.

119. Shah, "A Move in the Shadows," p. 172.

120. Shah, p. 166.

121. Cited in Micah Zenko, "10 Things You Didn't Know About Drones," *Foreign Policy*, March/April 2012.

122. Ian Cobain, "Obama's Secret Kill List—The Disposition Matrix" *Guardian*, 14 July 2013.

123. The first speech by an administration official was given by Harold Hongju Koh to the American Society of International Law in 2010, but even he admitted this speech was not successful. See his "The Obama administration and International Law," 25 May 2010, available at: http://www.state.gov/s/l/releases/remarks/139119.htm, accessed 14 April 2016. For his later speech calling for transparency to "discipline drones" see Harold Hongju Koh, "How to End the Forever War," Oxford Union, 7 May 2013.

124. John Brennan, "The Ethics and Efficacy of US Counterterrorism Strategy," Address at the Woodrow Wilson Center, Washington, DC, 30 April 2012, available at: http://www.wilsoncenter.org/event/the-efficacy-and-ethics-us-counterterrorism-strategy, accessed 28 September 2014.

125. President Barack Obama, "Remarks by the President at the National Defense University," The White House, 23 May 2013, available at: http://www.whitehouse.gov/the-press-office/2013/05/23/remarks-president-national-defense-university, accessed 28 September 2014.

126. See the U.S. Department of Justice, "Lawfulness of Lethal Operations Against a U.S. Citizen Is a Senior Operational Leader of Al Qai'da or An Associated Force," and "Memorandum for the Attorney General Re: Application of Federal Criminal Laws and the Constitution to Contemplated Lethal Operations Against Shayk Anwar al-Aulaqi," 16 July 2010. For a discussion, see Scott Shane, *Objective Troy: A Terrorist, A President, and the Rise of the Drone* (New York: Tim Duggan Books, 2015), p. 218–227.

127. Presidential Policy Guidance, "Procedures for Approving Direct Action Against Terrorist Targets Located Outside the United States and Areas of Active Hostilities," 22 May 2013.

128. Presidential Policy Guidance, "Procedures for Approving Direct Action Against Terrorist Targets Located Outside the United States and Areas of Active Hostilities," p. 15.

129. Woods, *Sudden Justice*, p. 203.

130. Mark Mazetti, "Delays in Effort to Refocus CIA from Drone War," *New York Times*, 5 April 2014.

131. Karen De Young and Greg Miller, "White House Releases Its Count of Civilian Deaths in Counterterrorism Operations Under Obama," *Washington Post*, 1 July 2016, available at: https://www.washingtonpost. com/world/national-security/white-house-releases-its-count-of-civilian-deaths-in-counterterrorism-operations-under-obama/2016/ 07/01/3196aa1e-3fa2-11e6-80bc-d06711fd2125_story.html?utm_term=. a3c2349b4586, accessed 1 June 2018.

132. Tom LoBianco, "Donald Trump on Terrorists: Take Out Their Families," *CNN*, 3 December 2015, available at: https://www.cnn.com/2015/12/ 02/politics/donald-trump-terrorists-families/index.html, accessed 1 June 2019.

133. Savage and Schmitt, "Trump Poised to Drop Some Limits on Drone Strikes and Commando Raids."

134. Stephen Tankel, "Donald Trump's Shadow War," *Politico*, 9 May 2018.

135. Ackerman, "Trump Ramped Up Drone Strikes in America's Shadow Wars."

136. Dan De Luce and Sean D. Naylor, "The Drones Are Back," *Foreign Policy*, 26 March 2018, available at: http://foreignpolicy.com/2018/03/26/the-drones-are-back/, accessed 1 June 2018.

137. Insanullah Tipu Mehsud, "U.S. Drone Strike Kills Leader of Pakistani Taliban, Pakistan Says," *New York Times*, 15 June 2018.

138. Ackerman, "Trump Ramped Up Drone Strikes in America's Shadow Wars."

139. Ackerman.

140. Amanda Sperber, "Inside the Secretive US Air Campaign in Somalia," *The Nation*, 7 February 2019. Based on interviews with senior Somali government officials, Amnesty International reported that the United States was assuming all military-aged males in geographic proximity to al-Shabaab militants were combatants, though this was disputed by the US military. "The Hidden US War in Somalia: Civilian Casualties from Air Strikes in the Lower Shabelle," Amnesty International, 2019, p. 2.

141. Ackerman, "Trump Ramped Up Drone Strikes in America's Shadow Wars."

142. Charlie Savage, "U.S. Removes Libya from List of Zones with Looser Rules for Drone Strikes," *New York Times*, 20 January 2017.

143. Julian Borger, "U.S. Air Wars Under Trump: Increasingly Indiscriminate, Increasingly Opaque," *Guardian*, 23 January 2018, available at: https:// www.theguardian.com/us-news/2018/jan/23/us-air-wars-trump, accessed 1 June 2018.

144. Joe Penney, Eric Schmitt, Rukmini Callimachi, and Christoph Koetti, "CIA Drone Mission, Curtailed by Obama, Is Expanded in Africa Under Trump," *New York Times*, 9 September 2018; Kyle Rempfer, "New in

2019: Two New U.S. Air Bases in Africa Nearing Completion," *Air Force Times*, 3 January 2019.

145. Greg Jaffe, "White House Ignores Executive Order Requiring Count of Civilian Casualties in Counterterrorism Strikes," *Washington Post*, 1 May 2018.

146. Charlie Savage, "Trump Revokes Obama-Era Rule on Disclosing Civilian Casualties from Airstrikes Outside War Zones," *New York Times*, 6 March 2019.

147. Becker and Shane, "Secret Kill List Proves a Test of Obama's Principles and Will."

148. Cora Currier, "The Kill Chain," *The Intercept*, 15 October 2015, available at: https://theintercept.com/drone-papers/the-kill-chain/, accessed 24 March 2016.

149. Aki Peritz and Eric Rosenbach, *Find, Fix and Finish: Inside the Counterterrorism Campaigns That Killed Bin Laden and Devastated al Qaeda* (New York: Public Affairs, 2013).

150. Declan Walsh, "Mysterious 'Chip' Is CIA's Latest Weapon Against al Qaida Targets Hiding in Pakistan's Tribal Belt," *Guardian*, 31 May 2009.

151. Chamayou, *Drone Theory*, p. 52–59.

152. Currier, "The Kill Chain."

153. Tankel, "Donald Trump's Shadow War."

154. Tankel.

155. Woods, *Sudden Justice*, p. 12.

156. The term here comes from the historian Geoffrey Blainey, although he used it with a different meaning. See Geoffrey Blainey, *The Tyranny of Distance* (New York: Macmillan, 1975).

157. Declan Walsh, "Leading UN Official Criticizes CIA's Role in Drone Strikes," *Guardian*, 3 June 2010.

158. Ed Pilkington, "Life as a Drone Operator: 'Ever Step on Ants and Never Give It a Second Thought?'" *Guardian* (19 November 2015).

159. Chamayou, *Drone Theory*, p. 107.

160. Michael Haas, quoted in Pilkington, "Life as a Drone Operator."

161. Woods, *Sudden Justice*, p. 174–175.

162. See particularly John Kaag and Sarah Kreps, *Drone Warfare* (Cambridge: Polity, 2014), p. 107–113; Chamayou, *Drone Theory*, p. 187–189.

163. Michael Walzer, "Targeted Killings and Drone Warfare," *Dissent*, 11 January 2013, available at: https://www.dissentmagazine.org/online_articles/targeted-killing-and-drone-warfare, accessed 14 April 2016.

164. Michael J. Boyle, "The Costs and Consequences of Drone Warfare," *International Affairs* 89:1 (2013), p. 1–29.

165. See particularly Chamayou, *Drone Theory*.

166. T. Mark McCurley, "I Was a Drone Warrior for 11 Years. I Regret Nothing," *Politico*, 18 October 2015.

167. See Kenneth Anderson, "The Case for Drones," *Commentary* 135:6 (June 2013), p. 14–23.

168. Woods, *Sudden Justice*, p. 175.

169. Swanson, "WarIs Not a Video Game."

170. Quoted in "Drone Pilot, Nellis Air Force Base, Las Vegas, NV, 'It Is War at a Very Intimate Level,'" in Bergen and Rothenberg (eds), *Drone Wars*, p. 116.

171. Pilkington, "Life as a Drone Operator."

172. Matthew Power, "Confessions of a Drone Warrior," *GQ Magazine*, 22 October 2013.

173. Phil Stewart, "Overstretched Pilots Face Stress Risk," *Reuters*, 18 December 2011.

174. James Dao, "Drone Pilots Are Found to Get Stress Disorders as Much as Those in Combat Do," *New York Times*, 22 February 2013.

175. Chamayou, *Drone Theory*, p. 106–111.

176. The full letter is available at: https://www.documentcloud.org/documents/2515596-final-drone-letter.html, accessed 7 April 2016.

177. Jake Heller, "Former Drone Pilots Denounce 'Morally Outrageous' Program," *NBC News*, 7 December 2015, available at: http://www.nbcnews.com/news/us-news/former-drone-pilots-denounce-morally-outrageous-program-n472496, accessed 7 April 2016.

178. There is a vast literature on the effectiveness of targeted killings, but relatively few studies of the specific effectiveness of drones. See particularly Byman, "Do Targeted Killings Work?"; Aaron Mannes, "Testing the Snake Head Strategy: Does Killing or Capturing Its Leaders Reduce a Terrorist Group's Activity," *Journal of International Policy Solutions* 9 (Spring 2008), p. 40–49; Catherine Lotrionte, "When to Target Leaders," *Washington Quarterly* 26:3 (Summer 2003), p. 73–86; Mohammed M. Hafez and Joseph M. Hatfield, "Do Targeted Assassinations Work? A Multivariate Analysis of Israel's Controversial Tactic During Al-Aqsa Uprising," *Studies in Conflict and Terrorism* 29:4 (2006), p. 359–382; Jenna Jordan, "When Heads Roll: Assessing the Effectiveness of Leadership Decapitation," *Security Studies* 18 (2009), p. 719–755; Patrick B. Johnston, "Does Decapitation Work?" *International Security* 36:4 (Spring 2012), p. 47–79; and Stephanie Carvin, "The Trouble with Targeted Killing," *Security Studies* 21:3 (2012), p. 529–555. On the effectiveness of drones, see Patrick Johnston and Anoop K. Sarbahi, "The Impact of US Drone Strikes on Terrorism in Pakistan," *International Studies Quarterly* (January 2016), p. 203–219; Daniel Byman, "Why Drones Work," *Foreign Affairs*, July/August 2013, p. 32–43; Audrey Kurth Cronin, "Why Drones Fail," *Foreign Affairs*, July/August 2013, p. 44–54; and Boyle, "The Costs and Consequences of Drone Warfare." p. 1–29

179. Byman, "Why Drones Work."

180. Quoted in Miller and Tate, "CIA Shifts Focus to Killing Targets."

181. Perlez and Shah, "Drones Batter al Qaeda and Allies in Pakistan."
182. Greg Miller, "Al Qaeda Could Collapse, U.S. Officials Say," *Washington Post*, 26 July 2011.
183. Quoted in Jane Meyer, "The Predator War," *New Yorker*, 29 October 2009.
184. Jeh Johnson, Speech at the Oxford Union, 30 November 2012, available at: http://www.lawfareblog.com/2012/11/jeh-johnson-speech-at-the-oxford-union/#.UvAPyE2YaM8, accessed 19 December 2019.
185. Remarks by the President at the National Defense University, 23 May 2013, available at: http://www.whitehouse.gov/the-press-office/2013/05/23/remarks-president-national-defense-university, accessed 15 February 2014.
186. Michael V. Hayden, "To Keep America Safe, Embrace Drone Warfare," *New York Times*, 19 February 2016.
187. David Rohde, "My Guards Absolutely Feared Drones," in Bergen and Rothenberg (eds), *Drone Wars*, p. 9–12.
188. Quoted in Brian Glyn Williams, "The CIA's Covert Predator Drone War in Pakistan 2004–2010: The History of an Assassination Campaign," *Studies in Conflict and Terrorism* 33 (2010), p. 879.
189. Spencer Ackerman, Alan Yuhas, Jason Burke, and Jon Boone, "US Releases More Than 100 Documents Recovered from the Osama bin Laden Raid," *Guardian*, 20 May 2015.
190. Pam Benson, "Bin Laden Documents: Fear of Drones," *CNN.com*, 3 May 2012.
191. See also Daniel L. Byman, "Why Drones Work: The Case for Washington's Weapon of Choice," *Brookings*, 17 June 2013. http://www.brookings.edu/research/articles/2013/06/17-drones-obama-weapon-choice-us-counterterrorism-byman.
192. Jason Burke, "Bin Laden Letters Reveal al Qaida's Fears of Drone Strikes and Infiltration," *Guardian*, 1 March 2016.
193. See Rohde, "My Guards Absolutely Feared Drones," and Glynn Williams "The CIA's Covert Predator Drone War in Pakistan 2004-2010."
194. Johnston and Sarbahi, "The Impact of US Drone Strikes on Terrorism in Pakistan," p. 1–17.
195. Asfandyar Mir, "What Explains Counterterrorism Effectiveness? Evidence from the U.S. Drone War in Pakistan," *International Security* 43:2 (2018), p. 45–83.
196. Michael J. Boyle, "Is the US Drone War Effective?" *Current History* 113:762 (April 2014), p. 137–145.
197. Peter Bergen and Katherine Tiedemann, "Washington's Phantom War: The Effects of the U.S. Drone Program in Pakistan," *Foreign Affairs*, July/August 2011, p. 12–18.
198. Bergen and Tiedemann.
199. Former CIA Director Leon Panetta, quoted in "Obama's Middle East Strategy," *New Perspectives Quarterly* (Summer 2009), p. 33.

200. Tim Craig, "Karachi Residents Live in Fear as Pakistani Taliban Gains Strength," *Washington Post*, 3 February 2014.

201. *Living Under Drones: Death, Injury and Trauma from US Drone Practices in Pakistan*, Stanford Law School and NYU School of Law, September 2012, p. 133–134.

202. Hassan Abbas, "How Drones Create More Terrorists," *Atlantic*, 23 August 2013.

203. Aqil Shah, "Do Drone Strikes Cause Blowback? Evidence from Pakistan and Beyond," *International Security* 42:4 (Spring 2018), p. 47–84.

204. Ibrahim Mothana, "How Drones Help al Qaeda," *New York Times*, 12 June 2012.

205. Quote is from Anssaf Ali Mayo, head of a leading Islamic party, reported in Sudarasan Raghavan, "In Yemen, U.S. Airstrikes Breed Anger, and Sympathy for al Qaeda," *Washington Post*, 29 May 2012.

206. Raghavan.

207. Aside from a lack of good data on terrorist recruitment in these regions, we have no baseline to judge whether drones are making a difference in influencing an individual's decision to join a terrorist group. Some may be predisposed to do so and drones may be irrelevant; some may not be and drones may be dispositive in making them join up.

208. Quoted in Ed Pilkington, "Does Obama's 'Single Digit' Civilian Death Claim Stand Up to Scrutiny?" *Guardian*, 7 February 2013.

209. Quoted in David E. Sanger, *Confront and Conceal: Obama's Secret Wars and the Surprising Use of American Power* (New York: Broadway, 2013), p. 251.

210. Conor Friedersdorf, "Calling U.S. Drone Strikes 'Surgical' Is Orwellian Propaganda," *Atlantic*, 27 September 2012, available at: http://www.theatlantic.com/politics/archive/2012/09/calling-us-drone-strikes-surgical-is-orwellian-propaganda/262920/, accessed 19 December 2019

211. Remarks by the President at the National Defense University, 23 May 2013, available at: http://www.whitehouse.gov/the-press-office/2013/05/23/remarks-president-national-defense-university, accessed 15 February 2014.

212. Becker and Shane, "Secret 'Kill List' Proves a Test of Obama's Principles and Will."

213. Bergen and Tiedemann, "Washington's Phantom War."

214. Dennis C. Blair, "Drones Alone Are Not the Answer," *New York Times*, 14 August 2011.

215. Mark Mazetti and Scott Shane, "As New Drone Policy Is Weighed, Few Practical Effects Are Seen," *New York Times*, 21 March 2013.

216. Quoted in David Alexander, "Retired General Cautions Against Overuse of 'Hated' Drones," *Reuters*, 7 January 2013.

217. Mark Mazetti and Scott Shane, "Evidence Mounts for Taliban Role in Bomb Plot," *New York Times*, 5 May 2010.

218. Stanford Law School and NYU School of Law, *Living Under Drones: Death, Injury and Trauma from US Drone Practices in Pakistan*, September 2012.

219. Glyn Williams, "The CIA's Covert Predator Drone War in Pakistan 2004–2010," p. 879.

220. *Living Under Drones*, p. 81.

221. David Rohde, "The Obama Doctrine: How the President's Secret Wars Are Backfiring," *Foreign Policy* 192 (March/April 2012), p. 66.

222. Brian Glyn Williams argues that those rushing to the scene to do "recovery" after a drone strikes are often Taliban militants or civilians assisting the Taliban, and hence not civilians. See Brian Glyn Williams, "New Light on CIA 'Double Tap' Drone Strikes on Taliban 'First Responders' in Pakistan's Tribal Areas," *Perspectives on Terrorism* 7:3 (June 2013), p. 79–83.

223. *Living Under Drones*, p. 73–99.

224. Fair, Kalthenhalter, and Miller, "Pakistani Opposition to American Drone Strikes."

225. *Living Under Drones*, p. 99.

226. *Living Under Drones*, p. 100.

227. Coll, "The Unblinking Stare."

228. See Fair, Kalthenhalter, and Miller, "Pakistani Opposition to American Drone Strikes"; Christopher Swift, "The Drone Blowback Fallacy," *Foreign Affairs*, 1 July 2012. For an interesting rejection of these criticisms from Pakistan, see Pervez Hoodbhoy, "Drones: Theirs and Ours," *OpenDemocracy*, 3 November 2012.

229. For a full discussion of this dynamic, see Boyle, "The Costs and Consequences of Drone Warfare."

230. David Kilcullen and Andrew McDonald Exum, "Death from Above, Outrage Down Below," *New York Times*, 16 May 2009.

231. Quoted in "Watch: Pakistani FM Calls Drone Strikes 'Illegal,' Says 'This Has to Be Our War'," Asia Society, 27 September 2012, available at: http://asiasociety.org/blog/asia/watch-pakistan-fm-calls-drone-strikes-illegal-says-has-be-our-war, accessed 4 October 2012.

232. Boyle, "The Costs and Consequences of Drone Warfare."

233. Transcript of Remarks by John O. Brennan, Assistant to the President for Homeland Security and Counterterrorism, "The Efficacy and Ethics of U.S. Counterterrorism Strategy," Woodrow Wilson Center, 30 April 2012.

234. Jane Perlez, "Chinese Plan to Kill Drug Lord With Drone Highlights Military Advances," *New York Times*, 20 February 2013.

235. Ewan MacAskill, "Drone Killing of British Citizens in Syria Marks Major Departure for the UK," *Guardian*, 7 September 2015.

236. Michael J. Boyle, "Pakistani Drone Strikes Should Worry Obama," *The Conversation*, 30 September 2015.

237. Patrick Boehler and Gerry Doyle, "Use by Iraqi Military May Be a Boon for Chinese-Made Drones," *New York Times*, 17 December 2015.

238. W.J. Hennigan, "A Fast Growing Club: Countries That Use Drones for Killing by Remote Control," *Los Angeles Times*, 22 February 2016.

239. Rawan Sharif and Jack Watling, "How the UAE's Chinese Made Drone Is Changing the War in Yemen," *Foreign Policy*, 27 April 2018, available at: http://foreignpolicy.com/2018/04/27/drone-wars-how-the-uaes-chinese-made-drone-is-changing-the-war-in-yemen/, accessed 1 June 2018.

240. Chamayou, *Drone Theory*, p. 104.

Chapter 4

1. Barbara Tuchman, *The Guns of August* (New York: Ballatine Books, 1994), p. 119

2. Cited in Stephen Van Evera, "The Cult of the Offensive and the Origins of the First World War," *International Security* 9:1 (Summer 1984), p. 67.

3. Cited in Stephan Van Evera, *The Causes of War* (Ithaca: Cornell University Press, 1984), p. 206.

4. Tuchman, *The Guns of August*, p. 122.

5. Colonel Terrence J. Finnegan, *Shooting the Front: Allied Aerial Reconnaissance and Photographic Interpretation on the Western Front—World War I*, Joint Military Intelligence Center, undated, available at: http://ni-u.edu/ni_press/pdf/Shooting_the_Front.pdf, p. 10, accessed 19 December 2019.

6. Major Tyler Morton, "Manned Airborne Intelligence, Surveillance and Reconnaissance," *Air and Space Power Journal*, November–December 2012, p. 38.

7. Max Boot, *War Made New: Technology, Warfare and the Course of History 1500 to Today* (New York: Gotham Books, 2006), p. 208.

8. This quote is taken from John Warden, "Strategy and Air Power," Air and Space Power Journal, 25:1 (2011), p. 64. Full page reference, p. 64–77.

9. Editorial, "The First World War Aerial Photography: 1914," the *Photogrammic Record* 29:148 (December 2014), p. 381.

10. Finnegan, *Shooting the Front*, p. 11.

11. Unnamed observer, quoted in Finnegan, p. 13.

12. Walter J. Boyne, "The Influence of Airpower on the Marne," *Air Force Magazine*, July 2011, p. 69.

13. Boyne, "The Influence of Airpower on the Marne," p. 69.

14. Giulio Douhet, *Command of the Air*, trans. Dino Ferrari (Washington DC: Air Force Museum and History Program, 1998).

15. John H. Morrow Jr., "The Air War," in J. Horne (ed.), *A Companion to World War I* (Oxford: Wiley-Blackwell, 2010), p. 354.

16. Boyne, "The Influence of Airpower on the Marne," p. 70.

17. Morton, "Manned Airborne Intelligence, Surveillance and Reconnaissance," p. 37.

18. Finnegan, *Shooting the Front*, p. 21.
19. Finnegan, p. 21–22.
20. Boyne, "The Influence of Airpower on the Marne," p. 71; Finnegan, *Shooting the Front*, p. 23.
21. Finnegan, p. 27. The role of the taxis is exaggerated, as only a small number of troops traveled that way compared to the other modes of transport, but it has become part of the legend of this battle. See Holger W. Herwig, *The Marne, 1914: The Opening of World War I and the Battle That Changed the World* (New York: Random House, 2009), p. 262.
22. Figures from Annika Mombauer, "The Battle of the Marne: Myths and Reality of Germany's "Fateful Battle," *The Historian* 68:4 (2006), p. 755.
23. Herwig, *The Marne 1914*, p. xii.
24. Boyne, "The Influence of Airpower on the Marne," p. 71.
25. Mombauer, "The Battle of the Marne," p. 748–749.
26. Cited in Paul K. Saint-Amour, "Modernist Reconnaissance," *Modernism/Modernity* 10:2 (April 2003), p. 354.
27. "Orville Wright Says 10,000 Airplanes Would End the War Within Ten Weeks," *New York Times*, 1 July 1917.
28. Dan Gettinger, "The Ultimate Way of Seeing: Aerial Photography in World War I," Center for the Study of the Drone, Bard University, 28 January 2014, available at: http://dronecenter.bard.edu/wwi-photography/ , accessed 2 December 2016.
29. Paul Virilio, *War and Cinema: The Logistics of Perception* (London: Verso, 1989), p. 92.
30. The literature on the RMA is vast. See particularly Andrew F. Krepinevich, "Calvary to Computer: The Pattern of Military Revolutions," *National Interest* 30:13 (Fall 1994); Steven Metz and James Kievet, "Strategy and the Revolution in Military Affairs: From Theory to Policy," Carlisle, PA: Strategic Studies Institute, 27 June 1995. For a critique, see Stephen Biddle, *Military Power: Explaining Victory and Defeat in Modern Battle* (Princeton, NJ: Princeton University Press, 2004) and Colin S. Gray, *Another Bloody Century* (London: Phoenix, 2007).
31. Steven Metz and James Kievet, "Strategy and the Revolution in Military Affairs: From Theory to Policy," Carlisle, PA: Strategic Studies Institute, 27 June 1995.
32. Cited in Christopher Coker, *War in an Age of Risk* (Cambridge: Polity, 2009), p. 8.
33. Stuart E. Johnson and Martin E. Libicki, *Dominant Battlespace Awareness* (Washington, DC: National Defense University, 2000), Major Robert F. Piccerillo and David A. Brumbaugh, "Predictive Battlespace Awareness: Linking Intelligence, Surveillance and Reconnaissance to Effects-Based Operations," Headquarters Air Force, US Pentagon, 2004, available at: https://www.semanticscholar.org/paper/Predictive-Battlespace-Awareness%3A-Linking-and-to-Piccerillo-Brumbaugh/acb88

12fb69667695b3f29afa6be4f1c5e1b04fb, accessed 19 December 2019, and Coker, *War in an Age of Risk*, p. 8.

34. Joint Chiefs of Staff, "Joint Vision 2020: America's Military—Preparing for Tomorrow," 2000, available at: http://www.dtic.mil/dtic/tr/fulltext/u2/a526044.pdf, accessed 5 December 2016.

35. Arthur Cebrowski and John H. Garstka, "Network-Centric Warfare: Its Origin and Future," US Naval Institute, January 1998.

36. Peter Singer, *Wired for War: The Robotics Revolution and Conflict in the 21st Century* (New York: Penguin, 2009), p. 180.

37. Elizabeth Bone and Christopher Bolkcom, "Unmanned Aerial Vehicles: Background and Issues for Congress," Congressional Research Service, 25 April 2003, p. 2; Arkin, *Unmanned*, p. 45–47.

38. Bill Yenne, *Birds of Prey: Predators, Reapers and America's Newest UAVs in Combat* (North Branch, MN: Specialty Press, 2010), p. 33.

39. Bone and Bolkcom, "Unmanned Aerial Vehicles," p. 34.

40. Bone and Bolkcom, p. 33.

41. Arkin, p. 119.

42. Bone and Bolkcom, "Unmanned Aerial Vehicles," p. 37.

43. Arkin, p. 143.

44. Arkin, p. 123.

45. Interview with a Global Hawk pilot, 25 April 2016.

46. Interview with a USAF Reaper pilot, 25 April 2016.

47. Christopher Drew, "Military Is Awash in Data from Drones," *New York Times*, 10 January 2010

48. Arkin, p. 158.

49. Arkin, p. 174–176.

50. Sean D. Naylor, "The Army's Killer Drones: How aSecretive Special Ops Unit Decimated ISIS," Yahoo News, 7 March 2019.

51. Andrew Cockburn, *Kill Chain: The Rise of High-Tech Assassins* (New York: Henry Holt, 2015), p. 145.

52. Cockburn, p. 144.

53. Staff Sgt. Thomas J. Doscher, "Predator Passes 400,000 Hours" *Air Combat Command*, 29 August 2008, available at: http://www.acc.af.mil/News/ArticleDisplay/tabid/5725/Article/202000/predator-passes-400000-hours.aspx, accessed 13 December 2016.

54. Tech. Sgt. Daryl Knee, "MQ-1B, MQ-9 Flight Hours Fit 4 Million," *Air Combat Command Public Affairs*, 11 March 2019.

55. Doscher, "Predator Passes 400,000 Hours."

56. *Unmanned Aircraft Systems Flight Plan 2009–2047* (Washington, DC: United States Air Force), p. 28.

57. Hambling, *Swarm Troopers*, p. 59.

58. Office of the Secretary of Defense, *Unmanned Aircraft Systems Roadmap*, 2005, p. 27.

59. Hambling, *Swarm Troopers*, p. 69.

60. Charlie Osborne, "US Military Equipped with Tiny Spy Drones," ZDNet, 4 February 2019.

61. Osborne.

62. Joseph Trevithick, "The Pocket Sized Black Hornet Is About to Change Army Operations Forever," *The Drive*, 6 February 2019.

63. Trevithick.

64. Interview with former USAF Reaper pilot.

65. Cited in Jared Serbu, "Five Things You May Have Missed at the Air, Space and Cyber Conference," Federal News Radio, 26 September 2016, available at: http://federalnewsradio.com/air-force/2016/09/five-things-may-missed-air-space-cyber-conference/, accessed 9 February 2017.

66. Interview with senior US Air Force Official, 7 November 2016.

67. Interview with Lt. General David Deptula, Mitchell Institute for Aerospace Studies, 24 May 2016.

68. Interview with senior US Air Force Official.

69. Grégoire Chamayou, *Drone Theory* (trans: Janet Lloyd), (London: Penguin, 2015), p. 62

70. Anna Mulrine, "Pentagon Pushes for More Bandwidth, Citing 'National Security Needs,'" *Christian Science Monitor*, 20 February 2014.

71. Noah Shachtman, "Pentagon Paying China—Yes, China—to Carry Data," *Wired*, 29 April 2013.

72. Office of the Secretary of Defense, "Unmanned Aircraft Systems Roadmap," p. 50.

73. US Air Force, "Small Unmanned Aircraft Systems (SUAS) Flight Plan 2016–2036, p. p. 27.

74. Office of the Secretary of Defense, "Unmanned Aircraft Systems Roadmap," p. 50. Emphasis in the original.

75. Cited in Chamayou, *Drone Theory*, p. 40.

76. Dan Nostowitz, "Every Six Hours, the NSA Gathers as Much Data as Is Stored in the Entire Library of Congress," *Popular Science*, 10 May 2011.

77. David Axe, "U.S. Air Force Spy Planes Recorded 1,000 Hours of Video Every Day," War Is Boring, 2 January 2015, available at: https://medium.com/war-is-boring/u-s-air-force-spy-planes-recorded-1-000-hours-of-video-every-day-57a152506b14.

78. Woods, *Sudden Justice*, p. 268.

79. Axe, "U.S. Air Force Spy Planes Recorded 1,000 Hours of Video Every Day."

80. Quoted in Drew, "Military Is Awash in Data from Drones."

81. Drew.

82. Axe, "U.S. Air Force Spy Planes Recorded 1,000 Hours of Video Every Day."

83. Statistic cited in Pratap Chatterjee, "The Side of Drone Warfare No One Is Talking About," *The Nation*, 23 July 2015.

84. For examples, see Arkin *Unmanned*, Chamayou, *Drone Theory*, and Cockburn, *Kill Chain*.

85. Cited in Woods, *Sudden Justice*, p. 268.

86. Interview with Lt. General David Deptula.

87. Interview with a USAF drone pilot and trainer, 25 April 2016.

88. Craig Whitlock, "Gorgon Stare Gazes over War Zone," *Washington Post*, 29 April 2011.

89. David Cenciotti and David Axe, "This New Drone Sensor Can Scan a Whole City at Once," *War Is Boring*, available at: https://warisboring. com/the-new-sensor-on-this-drone-can-scan-a-whole-city-at-once-33c314d4c763#.6kozk8yuc, accessed 9 September 2014.

90. Loren Thompson, "Air Force's Secret 'Gorgon Stare' Program Leaves Terrorists Nowhere to Hide," *Forbes*, 10 April 2015.

91. Sharon Weinberger, "How ESPN Taught the Pentagon to Handle a Deluge of Drone Data," *Popular Mechanics*, 11 June 2012.

92. Daisuke Wakabayashi and Scott Shane, "Google Will Not Renew Pentagon Contract That Upset Employees," *New York Times*, 1 June 2018, available at: https://www.nytimes.com/2018/06/01/technology/google-pentagon-project-maven.html, accessed 2 June 2018.

93. Cockburn, *Kill Chain*, p. 36–37.

94. Richard P. Hallion, "Precision-Guided Munitions and the New Era of Warfare," Air Power Studies Center 53, Australia, 1995, available at: https://fas.org/man/dod-101/sys/smart/docs/paper53.htm, accessed 16 January 2017.

95. James E. Hickey, *Precision-Guided Munitions and Human Suffering in War* (New York: Routledge, 2016), p. 102.

96. See particularly Michael Ignatieff, *Virtual War: Kosovo and Beyond* (New York: Picador, 2001); Christopher Coker, *Humane Warfare* (New York: Routledge, 2001).

97. Neta Crawford, "Targeting Civilians and U.S. Strategic Bombing Norms," in Matthew Evangelista and Henry Shue (eds.), *The American Way of Bombing* (Ithaca: Cornell, 2014), p. 64–75.

98. Stephanie Carvin and Michael John Williams, *Law, Science, Liberalism and the American Way of Warfare: The Quest for Humanity in Conflict* (Cambridge: Cambridge University Press, 2015), p. 125–134.

99. Hallion, "Precision-Guided Munitions and the New Era of Warfare."

100. Hallion.

101. Carl Conetta, "The Wages of War: Iraqi Combatant and Non-Combatant Fatalities in the 2003 Conflict," Project for Defense Alternatives Research Mongraph #8, 20 October 2003, available at: http://www.comw.org/pda/0310rm8.html, accessed 10 February 2017.

102. Coker, *Humane Warfare*, p. 13

103. Brigadier General David A. Deptula, "Effects-Based Operations: Change in the Nature of Warfare," Aerospace Education Foundation, 2001; David A. Deptula, "Air Force Transformation: Past, Present and Future," *Aerospace Power Journal* 15:3 (Fall 2001), p. 85–91.

104. Deptula, "Effects-Based Operations," p. 17.
105. See particularly Stephen Wrage, *Immaculate Warfare: Participants Reflect on Air Campaigns over Kosovo, Afghanistan, and Iraq* (Westport, CT: Praeger, 2003).
106. Anonymous, "Precision Strike Weapons and Net-Centric Warfare," *Military Technology* 30:2 (2006), p. 47.
107. Max Boot, "The New American Way of War," *Foreign Affairs*, July 2003, p. 41–58. For a skeptical view, see Stephen Biddle, *Afghanistan and the Future of Warfare: Implications for Army and Defense Policy*, (Carlisle, PA: US Army War College, 2002).
108. Arkin, p. 142.
109. Alan J. Vick, Adam Grisson, William Rosenau, Beth Grill, and Karl P. Mueller, *Air Power in the New Counterinsurgency Era* (Santa Monica: RAND, 2006).
110. Charles J. Dunlap Jr., "Making Revolutionary Change: Airpower in COIN Today," *Parameters* (Summer 2008), p. 59.
111. Philip S. Meilinger, "A Matter of Precision," *Foreign Policy*, 18 November 2009.
112. Chamayou, *Drone Theory*, p. 33–35.
113. George A. Crawford, "Manhunting: Counter-Network Organization for Irregular Warfare," JSOU-Report 09-7, 2009, p. 1.
114. Mehrdad Vahabi, *The Political Economy of Predation* (Cambridge: Cambridge University Press, 2016), p. 97.
115. Interview with C, former USAF Predator Pilot, 26 April 2016.
116. Interview with senior USAF pilot, 7 November 2016.
117. Interview with a former Reaper pilot, 25 April 2016.
118. Interview with a former Reaper pilot.
119. Interview with Lt. General David Deptula.
120. Interview with senior US Air Force Official.
121. Cited in Micah Zenko and Amelia Mae Wolf, "Drones Kill More Civilians Than Pilots Do," *Foreign Policy*, 25 April 2016.
122. Cited in Zenko and Wolf.
123. "Obama's Speech on Drone Policy," *New York Times*, 23 May 2013.
124. Scott Shane, "The Moral Case for Drones," *New York Times*, 14 July 2012. For a critique, see Michael J. Boyle, "The Costs and Consequences of Drone Warfare," *International Affairs* 89:1 (2003), p. 1–29;
125. Micah Zenko, "Do Not Believe the U.S. Government's Official Numbers on Drone Strikes or Civilian Casualties," *Foreign Policy*, 5 July 2016.
126. Zenko and Wolf, "Drones Kill More Civilians Than Pilots Do."
127. Margaret Sullivan, "Middle East Civilian Deaths Have Soared Under Trump. And the Media Mostly Shrug," *Washington Post*, 18 March 2018, available at: https://www.washingtonpost.com/lifestyle/style/middle-east-civilian-deaths-have-soared-under-trump-and-the-media-mostly-shrug/2018/03/16/fc344968-2932-11e8-874b-d517e912f125_story.html?utm_term=.b1090f915350, accessed 6 June 2018.

128. Lolita C. Badour, "Pentagon to Increase Drone Flights by 50 Percent," *Associated Press*, 18 August 2015.

129. Daniel Goure, "Air Force Needs to Rethink Its Overall UAS Concept of Operations," *Real Clear Defense*, 29 April 2016.

130. Badour, "Pentagon to Increase Drone Flights by 50 Percent."

131. Dave Majumdar, "Exclusive: U.S. Drone Fleet at Breaking Point, Air Force Says," *Daily Beast*, January 4, 2015.

132. Center for the Study of the Drone, "Drone Geography: Mapping a System of Intelligence," 19 February 2015.

133. Hope Hedge Seck, "Pentagon Debuts 'R' Award Device for Drone Warfare to Mixed Reviews," Military.com, 6 January 2016, available at: http://www.military.com/daily-news/2016/01/09/pentagon-debuts-r-award-device-for-drone-warfare.html, accessed 9 February 2017.

134. Interview with a former drone pilot and instructor, 25 April 2016.

135. Interview with a former Predator pilot, 26 April 2016.

136. Interview with a former Predator pilot.

Chapter 5

1. This information is drawn from the final report on the Honshu earthquake prepared by the US National Geophysical Data Center at the National Oceanic and Atmospheric Administration, available at: https://www.ngdc.noaa.gov/hazard/data/publications/2011_0311.pdf accessed 28 July 2016.

2. The highest recorded wave in this tsunami was 38.9 meters.

3. Krista Mahr, "Fukushima Report: Japan Urged Calm While It Mulled Tokyo Evacuation," *Time Magazine*, 29 February 2012. The quote is from official testimony in the independent inquiry.

4. Mari Yamaguchi, "Japanese Police Say Man Arrested After He Admits Landing Droneat Prime Minister's Office," *US News and World Report*, 25 April 2015.

5. David Kravets, "Man Lands Drone Carrying Radioactive Sand on Japanese Prime Minister's Office," *Ars Technica*, 25 April 2015.

6. The prefecture is home to twelve of Japan's fifty reactors. See Yamaguchi, "Japanese Police Say Man Arrested After He Admits Landing Drone at Prime Minister's Office."

7. Sneha Shankar, "Japan Arrests Yasuo Yamamoto for Landing Radioactive Sand-Laced Drone on Shinzo Abe's Office," *International Business Times*, 25 April 2015.

8. "Tokyo Police Studying Technology to Detect, Capture Drones," *Japan Times*, 30 April 2015.

9. Robert Hutton, "Nuclear Drones from 'Dark Web' Cited by Obama in Terror Scenario," *Bloomberg Politics*, 1 April 2016.

10. Ben Riley-Smith, "ISIL Plotting to Use Drones for NuclearAttack on the West," *Telegraph*, 1 April 2016.

11. Sean Higgins, "FBI: Terrorists Expected to Use Drones 'Imminently'" *Washington Examiner*, 27 September 2017.

12. Nicholas Grossman, *Drones and Terrorism: Asymmetric Warfare and the Threat to Global Security* (London: I.B. Tauris, 2018), p. 95–98.

13. Kevin Poulsen, "Why the US Government Is Terrified of Hobbyist Drones," *Wired*, 5 February 2015.

14. Jack Nicas, "Criminals, Terrorists Find Uses for Drones, Raising Concerns," *Wall Street Journal*, 28 January 2015.

15. Jeff Stone, "Terrorist Drones Could Target Airports, Sensitive U.S. Sites, DHS Warns After ISIS Attack in Iraq," *International Business Times*, 4 August 2015.

16. Birmingham Police Commission, "The Security Impact of Drones: Challenges and Opportunities for the UK," Report, October 2014, p. 75.

17. Dan Goure, "Now That Terrorists Are Getting Air Forces, What Do We Do?" *National Interest*, 15 July 2016.

18. Larry Friese, with N.R. Jenzen-Jones, and Michael Smallwood, *Emerging Unmanned Threats: The Use of Commercially-Available UAVs by Armed Non-State Actors*, Armament Research Service, Special Report No. 2, February, 2016, p. 28.

19. The smaller drones—for example, the DJI Phantom models—cannot carry a payload of 5 kg. Other models, such as 3D Robotics Solo, X8, Aero-M, and the Skywalker Tech X-8 drones, can carry this payload. See Friese, Jenzen-Jones, and Smallwood, *Emerging Unmanned Threats*, p. 20, 28.

20. Brian Ballou, "Rezwan Ferdaus of Ashland Sentenced to 17 Years in Terror Plot; Plotted to Blow Up Pentagon, Capitol," *Boston Globe*, 1 November 2012.

21. Alex Brandon, "FBI: Man Plotted to Fly Drone-Like Toy Planes with Bombs into Schools," *CBS News*, 8 April 2014.

22. "Will ISIS Launch a Mass Drone Attack on a Stadium?" *Daily Beast*, 24 February 2016.

23. Chris Abbott, quoted in "Will ISIS Launch a Mass Drone Attack on a Stadium?"

24. "Committee on Counter-Unmanned Aircraft System (CUAS) Capability for Battalion and Below Operations," Board of Science and Technlogy, National Academy of Science, Washington, DC, 2018, p. 7.

25. David Wood, "Armed Drones Could Target President: Former U.S. Intelligence Chief," *Huffington Post*, 22 January 2013.

26. Michael Gips, "A Remote Threat," *Security Management Online*, October 2002.

27. Friese, Jenzen-Jones, and Smallwood, *Emerging Unmanned Threats*, p. 48–49.

28. Joseph Ax, "Apparent Attack in Venezuela Highlights Risk of Drone Strikes," *Reuters*, 5 August 2018.

29. "Saudi Security Shoots Down Toy Drone Near Royal Palace," *Reuters*, 22 April 2018.

30. Christoph Koettl and Barbara Marcolini, "A Closer Look at the Drone Attack on Maduro in Venezuela," *New York Times*, 10 August 2018.

31. Jeremy Kryt, "The Day of the Drone Assassin Has Arrived," *Daily Beast*, 8 August 2018.

32. Nick Waters, "Did Drones Attack Maduro in Caracas?" *Bellingcat*, 7 August 2018, available at: https://www.bellingcat.com/news/americas/2018/08/07/drones-attack-maduro-caracas/, accessed 3 July 2019.

33. Nicholas Casey, "Venezuela Seeks Arrest of Opposition Leader in Drone Episode," *New York Times*, 8 August 2018.

34. "Retired Colonel Accused in Maduro Drone 'Attack' Held in Venezuela: Wife," *Agence France Presse*, 30 January 2019.

35. Tyler Rogoway, "Was a New Type of Drone-Launched Weapon Used to Kill Al-Qaeda's #2 Man?" *The Drive*, 27 February 2017, available at: https://www.thedrive.com/the-war-zone/7914/was-a-new-type-of-drone-launched-weapon-used-to-kill-al-qaedas-2-man, accessed 3 July 2019.

36. It is worth noting that the US Army does not consider the Switchblade a drone but instead a loitering munition. While this illustrates the blurry lines between munitions, missiles, and drones, the Switchblade is called a drone here, in line with most conventional descriptions.

37. Michael Martinez, John Newsome, and Rene Marsh, "Handgun-Firing Drone Appears Legal, but FAA, Police, Probe Further," *CNN*, 21 July 2015, available at: http://www.cnn.com/2015/07/21/us/gun-drone-connecticut/, accessed 1 August 2016.

38. "Dad, Son Fight FAA over Gun-Firing, Flame-Throwing Drones," CBS News, 4 July 2016, available at: http://www.cbsnews.com/news/dad-son-fight-faa-gun-firing-flame-throwing-drones/, accessed 1 August 2016.

39. Friese, Jenzen-Jones, and Smallwood, *Emerging Unmanned Threats*, p. 28.

40. Robert J. Bunker, *Terrorist and Insurgent Unmanned Aerial Vehicles: Use, Potentials, and Military Implications*, US Army War College, August 2015, p. 8.

41. Dennis M. Gormley, "UAVs and Cruise Missiles as Possible Terrorist Weapons," Center for Nonproliferation Studies Occasional Paper #12, p. 3.

42. Jay Mandelbaum, James Ralston, Ivars Gutmanis, Andrew Hull, Christopher Martin, "Terrorist Use of Improvised or Commercially Available Precision-Guided UAVs at Stand-Off Ranges: An Approach for Formulating Mitigation Considerations, Institute for Defense Analysis D-3199 (October 2005).

43. Moazzam Begg allegedly confessed to a plot to use a drone to bomb the UK House of Commons with anthrax in 2003, but later claimed that this was done under duress and was a wild, fanciful plot made up to satisfy his interrogators. For the original report, see Severin Carell, "British Prisoner 'Confesses' Plot to Poison Bomb Parliament," *Independent*, 29 November 2003. Iraq also claimed to have disrupted a plot involving chemical weapons and toy planes, as well as plans to smuggle these materials to Europe. "Iraq Uncovers Al Qaeda 'Chemical Weapons Plot,'" BBC News, 1 June 2003.

44. Chris Smith, "Brussels Attackers Were Looking to Make Radioactive 'Dirty Bomb' BG, 28 May 2016, available at: http://bgr.com/2016/03/28/brussels-attack-dirty-bomb/, accessed 2 August 2016.

45. David Hambling, "Could ISIS Really Attack the U.S. With A Dirty Drone?" *Popular Mechanics*, 8 August 2016.

46. Ben Lerner, "Keep 'Dirty Drone' Threat in Perspective," *Homeland Security Today*, 8 May 2016.

47. Birmingham Policy Commission, "The Security Impact of Drones: Challenges and Opportunities for the UK," October 2014, p. 75.

48. Sarah Griffiths and Abigail Beall, "Could Amazon Drones Turn Hostile? Experts Warn UAVs May Be Hijacked by Terrorists and Hijackers," *Daily Mail*, 1 August 2016.

49. "Obama Says U.S. Has Asked Iran to Return Drone Aircraft," *CNN*, 13 December 2011.

50. Andy Greenberg, "Hacker Says He Can Hijack a $35K Police Drone a Mile Away," *Wired*, 2 March 2016.

51. Hannah Kuchler, "Cyber Experts Warn of the Hacking Capability of Drones," *Financial Times*, 31 July 2016.

52. Ewen MacAskill, "US Drones Hacked by Iraqi Insurgents," *Guardian*, 17 December 2009.

53. Grossman, *Drones and Terrorism*, p. 104–105.

54. David Axe, "How Islamic Jihad Hacked Israel's Drones," *Daily Beast*, 25 March 2016.

55. Doug Bolton, "Terrorists Could Use Drones to Attack Planes and Spread Propaganda, Government Security Advisor Warns," *Independent*, 6 December 2015.

56. Bunker, *Terrorist and Insurgent Unmanned Aerial Vehicles*, p. 8.

57. Dan Gettinger and Arthur Holland Michel, "Drone Sightings and Close Encounters: An Analysis," Center for the Study of the Drone, Bard College, 2015.

58. Bridget Brennan, "Drone Accidents Mostly Caused by Technical Problems, Not Operator Error, Research Shows," *ABC News (Australia)*, 23 August 2016, available at: http://www.abc.net.au/news/2016-08-24/drone-accidents-mostly-caused-by-faulty-equipment-research-shows/7780066, accessed 8 June 2017.

59. Mark Harris, "Near Misses Between Drones and Airplanes On the Rise in the US, FAA Says," *Guardian*, 25 March 2016.

60. Geoffrey Thomas, "Opinion: Tragic Drone Strike With Plane 'Inevitable,'" *CNN*, 18 April 2016, available at: http://www.cnn.com/2016/04/18/opinions/drones-planes-accidents/index.html, accessed 8 June 2017.

61. Samantha Masunaga, "Gatwick Airport Incident Shows the Threat of Rogue Drones in Commercial Airspace," *Los Angeles Times*, 20 December 2018.

62. General Accounting Office, "Small Unmanned Aircraft Systems: FAA Should Improve Its Management of Safety Risks," GAO-18-110, 24 May 2018.

63. Gettinger and Michel, "Drone Sightings and Close Encounters: An Analysis."

64. Patrick Benedict, "Commercial Drones aSerious Safety Concern?" *Scripps News Service*, 6 June 2013.

65. Jack Nicas, "FAA: U.S. Airliner Nearly Collided with a Drone in March," *Wall Street Journal*, 9 May 2014.

66. Thomas, "Opinion: Tragic Drone Strike with Plane 'Inevitable.'"

67. Air Line Pilots Association, International, "Unmanned Aircraft Systems: Challenges for Safely Operating in the National Aerospace System," White Paper, April 2011, p. 3, available at: http://www.alpa.org/portals/alpa/pressroom/inthecockpit/UASWhitePaper.pdf (accessed 18 October 2013).

68. Thomas, "Opinion: Tragic Drone Strike with Plane 'Inevitable.'"

69. "Virgin Atlantic in the 'Closest Ever' Near Miss with Drone on Approach to Heathrow," *Sky News*, 23 October 2018.

70. Exponent, UAS Safety Analysis, 16 December 2014, available at: http://www.uasamericafund.com/assets/micro-uav-safety-analysis.pdf, accessed 2 August 2016.

71. David Schneider, "What Might Happen if an Airliner Hit a Small Drone?" IEEE Spectrum, 11 March 2015, available at: http://spectrum.ieee.org/automaton/robotics/drones/what-might-happen-if-airliner-hit-small-drone, accessed 2 August 2016.

72. Birmingham Policy Commission, p. 75.

73. "Heathrow Plane Strike 'Not a Drone Incident,'" *Guardian*, 28 April 2016.

74. Michael Goldstein, "With the First Commercial Air Collision, Is a Drone Disaster Inevitable?" Forbes, 18 October 2017, available at: https://www.forbes.com/sites/michaelgoldstein/2017/10/18/with-first-commercial-air-collision-is-a-drone-disaster-inevitable/#61abb6b83855, accessed 5 March 2018.

75. Joan Lowy, "Drone Operator Faulted in NY Collision with Black Hawk," *Associated Press*, 14 December 2017.

76. Adam Lisberg, the corporate communications director at DJI Technology, quoted in Cody Derespina, "Drone Manufacturers Work to Combat Growing Terror Threat," *Fox News*, 20 April 2016.

77. Faine Greenwood, "DronesAre the New Flying Saucers," *Slate*, 16 January 2019.
78. Sam Blum, "We Still Know So Alarmingly Little About How a Drone Shut Down an Airport for Two Days," *Popular Mechanics*, 3 January 2019.
79. "Gatwick Drones Pair 'No Longer Suspects,'" BBC News, 23 December 2018.
80. "Police Think Gatwick Worker May Have Launched Drone Attack, Times Says," *Reuters*, 21 February 2019.
81. Richard Pérez-Peña, "Gatwick Airport Drone: Lots of Second-Guessing, but Not Many Answers," *New York Times*, 27 December 2018.
82. Gwyn Topham, "Gatwick Drone Disruption Cost Airport Just £1.4 million," *Guardian*, 18 June 2019; Hallie Detrick, "Gatwick's December Drone Closure Cost Airlines £64.5 million," *Fortune*, 22 January 2019;
83. Kanishka Singh, "Gatwick, Heathrow Airports Order Military-Grade Anti-Drone Equipment," *Reuters*, 3 January 2019.
84. Waseem Abassi, "Inmates Fly Mobile Phones, Drugs and Porn into Jail— Via Drone" *USA Today*, 15 June 2017, available at: https://www.usatoday.com/story/news/2017/06/15/inmates-increasingly-look-drones-smuggle-contraband-into-their-cells/102864854/, accessed 6 June 2018.
85. Stephan Dinan, "Drones Become Latest Tool Drug Cartels Use to Smuggle Drugs into U.S." *Washington Times*, 20 August 2017, available at: https://www.washingtontimes.com/news/2017/aug/20/mexican-drug-cartels-using-drones-to-smuggle-heroi/, accessed 3 July 2019.
86. "Great, Mexican Drug Cartels Now Have Weaponized Drones," *Vice*, 25 October 2017, available at: https://www.vice.com/en_us/article/j5jmb4/mexican-drug-cartels-have-weaponized-drones, accessed 3 July 2019.
87. Mandelbaum et al., "Terrorist Use of Improvised or Commercially Available Precision-Guided UAVs at Stand-Off Ranges," p. 2.
88. Grossman, *Drones and Terrorism*, p. 115.
89. Interview with a senior British intelligence official, London, UK, 13 July 2016.
90. Friese, Jenzen-Jones, and Smallwood, *Emerging Unmanned Threats*, p. 30–34.
91. Nicas, "Criminals, Terrorists Find New Uses for Drones, Raising Concerns."
92. "Hostile Drones: The Hostile Use of Drones by Non-State Actors Against British Targets," *Remote Control*, January 2016, p. 10–13; Friese, Jenzen-Jones, and Smallwood, *Emerging Unmanned Threats*, p. 36–50.
93. Michael D. Shear and Michael S. Schmidt, "White House Drone Crash Described as U.S. Worker's Drunken Lark," *New York Times*, 27 January 2015.
94. "Hostile Drones," p. 10–11.
95. "Hostile Drones," p. 10.

96. Matt O'Sullivan, "Drone Activity 'Raises Risk' for Pilots, Firefighters as Bush-Fire Season Nears," *Sydney Morning Herald*, 11 October 2015.
97. Friese, Jenzen-Jones, and Smallwood, p. 18.
98. Friese, Jenzen-Jones, and Smallwood, p, 18.
99. Friese, Jenzen-Jones, and Smallwood, p. 19.
100. Bunker, *Terrorist and Insurgent Unmanned Aerial Vehicles*, p. 8.
101. Bunker, p. 8. Friese, Jenzen-Jones, and Smallwood, *Emerging Unmanned Threats*, p. 47.
102. One estimate holds that the FARC had a membership of 18,000 in 1999. It began to decline during this period but was still powerful enough to kidnap Ingrid Betancourt in 2002. See the Mapping Militant Organizations project, http://web.stanford.edu/group/mappingmilitants/cgi-bin/groups/view/89, accessed 11 August 2016.
103. "Hostile Drones," p. 11.
104. Israeli Defense Forces, "Hamas Farj-5 Missiles and UAV Targets Severely Damaged," 17 November 2012, available at: https://www.idfblog.com/blog/2012/11/17/hamas-fajr-5-missiles-uav-targets-damaged/, accessed 16 August 2016.
105. Bunker, *Terrorist and Insurgent Unmanned Aerial Vehicles*, p. 11.
106. "Hostile Drones," p. 11.
107. "Who Killed Tunisian Drone Expert Mohammed al-Zawari," Al-Jazeera World, 12 December 2018, available at: https://www.aljazeera.com/programmes/aljazeeraworld/2018/12/killed-tunisian-drone-expert-mohammed-al-zawari-181204070318931.html, accessed 9 July 2018.
108. "Hostile Drones," p. 11.
109. Bunker, *Terrorist and Insurgent Unmanned Aerial Vehicles*, p. 12.
110. "Hostile Drones," p. 11.
111. Kyle Mizokami, "Terrorist Group Hezbollah Is Reportedly Using Drone Bombers," *Popular Mechanics*, 16 August 2016.
112. Bunker, *Terrorist and Insurgent Unmanned Aerial Vehicles*, p. 9.
113. David Axe, "Hezbollah Drone Is a Warning to the U.S.," *Daily Beast*, 17 August 2016.
114. Dan Gettinger and Arthur Holland Michel, "A Brief History of Hamas and Hezbollah's Drones," Center for the Study of the Drone, 14 July 2014, available at: http://dronecenter.bard.edu/hezbollah-hamas-drones/, accessed 16 August 2016.
115. Milton Hoenig, "Hezbollah and the Use of Drones as a Weapon of Terrorism," *Public Interest Report* 67:2 (Spring 2014).
116. Hoenig.
117. Quote is from Yochi Dreazen, "The Next Arab-Israeli War," *New Republic*, 26 March 2014. Quoted in Hoenig.
118. Alexander Smith, "Hamas Drone Program Will Not Worry Israel, Experts Say." NBC News, 15 July 2014.
119. Hoenig, "Hezbollah and the Use of Drones as a Weapon of Terrorism."

120. Avery Plaw, Elizabeth Santoro, "Hezbollah's Drone Program Sets Precedent for Non-State Actors," Jamestown Foundation, 10 November 2017, available at: https://jamestown.org/program/hezbollahs-drone-program-sets-precedents-non-state-actors/, accessed 9 July 2019.

121. Center for Arms Control, Energy and the Environment, Moscow Institute of Physics and Technology, "Terrorists Develop Unmanned Aerial Vehicles: On the Mirsad-1 Flight over Israel," 6 December 2004, available at: http://www.armscontrol.ru/UAV/mirsad1.htmm accessed 16 August 2016.

122. Hoenig, "Hezbollah and the Use of Drones as a Weapon of Terrorism."

123. Hoenig.

124. Hoenig.

125. "Hostile Drones," p. 11.

126. Anne Barnard, "Hezbollah Says It Flew Iranian-Designed Drone into Israel," New York Times, 11 October 2012.

127. Bunker, Terrorist and Insurgent Unmanned Aerial Vehicles, p. 11.

128. Versions of this quote are attributed to Vernon Law, a former Major League Baseball pitcher, and to the writer C.S. Lewis.

129. Bunker, Terrorist and Insurgent Unmanned Aerial Vehicles, p. 11.

130. Ian Austen, "Libyan Rebels Reportedly Used Tiny Canadian Surveillance Drone," New York Times, 24 August 2011.

131. Michelle Nichols, "U.N. Report Finds Likely Use of Armed Drone in Libya by Haftar or 'Third Party,'" Reuters, 8 May 2019.

132. "Ankara Warns Haftar over Seized Turks as LNA 'hits Turkish drone,'" Al Jazeera, 30 June 2019, available at: https://www.aljazeera.com/news/2019/06/ankara-warns-haftar-arrests-lna-hit-turkish-drone-190630192450014.html, accessed 9 July 2019.

133. A similar argument is made in Peter W. Singer, Wired at War: The Robotics Revolution and Conflict in the Twenty-First Century (New York: Penguin, 2009), p. 261–296.

134. Adam Rawnsley, "Ukraine Scrambles for UAVs, but Russia Owns the Skies," War Is Boring, 20 February 2015, available at: https://warisboring.com/ukraine-scrambles-for-uavs-but-russian-drones-own-the-skies-74f5007183a2#.txbgq11cf, accessed 17 August 2016; "Hostile Drones," p. 12.

135. Committee on Counter-Unmanned Aircraft System (CUAS) Capability for Battalion and Below Operations, Board of Science and Technlogy, National Academy of Science, Washington, DC, 2018, p. 5.

136. James Harvey, "Rebel Drones: UAV Overmatch in the Ukrainian Conflict," Foreign Military Studies Office, Fort Leavenworth, KS, January 2015.

137. General David Perkins, head of training and doctrine command for the US army, quoted in Sydney J. Freeburg Jr., "Russian Drone Threat: Army Seeks Ukraine Lessons," Breaking Defense, 14 October 2015, available

at: http://breakingdefense.com/2015/10/russian-drone-threat-army-seeks-ukraine-lessons/, accessed 17 August 2016.

138. Friese, Jenzen-Jones, and Smallwood, *Emerging Unmanned Threats*, p. 38.

139. Harvey, "Rebel Drones."

140. "In Ukraine, Tomorrow's Drone War Is Alive Today," *Defense One*, 9 March 2015, available at: http://www.defenseone.com/technology/2015/03/ukraine-tomorrows-drone-war-alive-today/107085/, accessed 17 August 2016.

141. Harvey, "Rebel Drones."

142. Friese, Jenzen-Jones, and Smallwood, *Emerging Unmanned Threats*, p. 38.

143. Friese, Jenzen-Jones, and Smallwood, *Emerging Unmanned Threats*, p. 38.

144. The People's Project, available at: http://www.peoplesproject.com/en/about/, accessed 17 August 2016.

145. Oriana Pawlyk, "U.S. Organization Sends Drones to Ukrainian Military," *Air Force Times*, 19 February 2015, available at: http://flightlines.airforcetimes.com/2015/02/19/u-s-organization-sends-drones-to-ukrainian-military/, accessed 17 August 2016.

146. Phil Stewart, "Exclusive: U.S. Supplied Drones Disappoint Ukraine at the Front Lines," *Reuters*, 21 December 2016.

147. Dan Lamothe, "Gotcha? Russian Military Base Reportedly Found in Ukraine, Detailed in Drone Video," *Washington Post*, June 2015.

148. Friese, Jenzen-Jones, and Smallwood, *Emerging Unmanned Threats*, p. 38.

149. Friese, Jenzen-Jones, and Smallwood, p. 46.

150. Friese, Jenzen-Jones, and Smallwood, p. 46.

151. Benedetta Argentieri, "Private Surveillance Drones Take Flight over Iraq," *War Is Boring*, 28 April 2015, available at: https://warisboring.com/private-surveillance-drones-take-flight-over-iraq-811d0f5f2a8f#.lot4cptj7m accessed 18 August 2016.

152. Friese, Jenzen-Jones, and Smallwood, *Emerging Unmanned Threats*, p. 46.

153. "Hostile Drones," p. 11.

154. Adam Saxton, "Drones: Russia's Hot New Propaganda Tool," *National Interest*, 29 April 2016, available at: http://nationalinterest.org/blog/the-buzz/drones-russias-hot-new-propaganda-tool-15987, accessed 9 March 2018.

155. Cody Poplin, "Look Who Else Has Drones: ISIS and al-Nusra," *Lawfare*, 24 October 2014, available at: https://www.lawfareblog.com/look-who-else-has-drones-isis-and-al-nusra, accessed 17 August 2016.

156. Friese, Jenzen-Jones, and Smallwood, *Emerging Unmanned Threats*, p. 45.

157. "Drone Films Fierce Battle for Syrian Village," Sky News, 8 May 2016, available at: http://news.sky.com/story/drone-films-fierce-battle-for-syrian-village-10271242, accessed 18 August 2016.

158. "ISIS Jihadists Put Out Hollywood-Style Propaganda Film," France 24 News, 14 June 2014.

159. Grossman, *Drones and Terrorism*, p. 118–119.

160. Friese, Jenzen-Jones, and Smallwood, *Emerging Unmanned Threats*, p. 40–43.

161. Friese, Jenzen-Jones, and Smallwood, p. 43; Bunker, *Terrorist and Insurgent Unmanned Aerial Vehicles*, p. 12.

162. "2 Held in Denmark Suspected of Buying Drones for IS," *Associated Press*, 26 September 2018.

163. Matthew L. Schehl, "ISIS Is Expanding the Reach and Sophistication of its Drone Fleet," *Marine Corps Times*, 17 April 2016.

164. Christopher Harmon, quoted in "Is ISIS Building a Drone Army?" *Daily Beast*, 18 March 2015.

165. David Hambling, "ISIS Is Reportedly Packing Drones with Explosives Now," *Popular Mechanics*, 16 December 2015.

166. Anthony Capacio, "Armed Drones Used by Islamic State Posing New Threat in Iraq," *Bloomberg*, 7 July 2016.

167. Joseph Trevithick, "New U.S. Army Manual Warns About Small Drones," *War Is Boring*, 18 August 2016.

168. "Islamic State Drone Kills Two Kurdish Fighter, Wounds Two French Soldiers," *Reuters*, 11 October 2016.

169. Eric Schmitt, "Papers Offer a Peek at ISIS' Drones, Lethal and Largely Off the Shelf," *New York Times*, 31 January 2017.

170. Thomas Gibbons-Neff, "Houthi Forces Appear to Be Using Iranian-Made Drones to Ram Saudi Air Defenses in Yemen, Report Says," *Washington Post*, 22 March 2017, available at: https://www.washingtonpost.com/news/checkpoint/wp/2017/03/22/houthi-forces-appear-to-be-using-iranian-made-drones-to-ram-saudi-air-defenses-in-yemen-report-says/?utm_term=.bbb2be328bbf, accessed 5 March 2018.

171. "Yemen's Houthis Say They Attacked Aramco Refinery in Riyadh with Drone," US News and World Report, *Reuters*, 18 July 2018; Jeremy Binnie, "Houthi UAVs Infiltrate Saudi Airspace," *Jane's*, 21 March 2019.

172. "Yemen's Rebels 'Attack' Abu Dhabi Airport Using a Drone," *Al Jazeera*, 27 July 2018.

173. Aaron Stein, "Low-Tech, High-Reward: The Houthi Drone Attack," Foreign Policy Research Institute, 11 January 2019.

174. Dan Reid, "A Swarm of Armed Drones Attacked a Russian Military Base in Syria," *CNBC*, 11 January 2018, available at: https://www.cnbc.com/2018/01/11/swarm-of-armed-diy-drones-attacks-russian-military-base-in-syria.html, accessed 9 March 2018.

175. Reid.

176. David Hambling, "Swarms of Cheap Drones Are Attacking Missile Defense Systems in Yemen," *New Scientist*, 8 March 2018.

177. Dionne Searcey, "Boko Haram Is Back. With Better Drones," *New York Times*, 13 September 2019.

178. Jen Judson, "Pentagon Asks for More Money to Counter ISIS Drones," *Defense News*, 9 July 2016.

179. Christopher Miller, "Commercial Drones Are Flying Off the Shelves and into the Hands of Armed Groups," *Mashable*, 18 February 2016.

180. David Morgan, "Exclusive: US Government, Police Working on Counter-Drone Systems," *Reuters*, 20 August 2015.

181. W.J. Hennigan and Brian Bennett, "They're 400,000 Strong and the Pentagon Sees Them as an Emerging Threat," *Los Angeles Times*, 1 April 2016.

182. Thomas Gibbons-Neff, "The U.S. Is Apparently Using Anti-Drone Rifles Against the Islamic State," *Washington Post*, 26 July 2016; Richard Sisk, "Battelle Fielding 100 Drone Zappers to Pentagon, Homeland Security," DefenseTech, 16 May 2016.

183. Richard Whittle, "Finmeccanica Unit Claims Counter-Drone Breakthrough," *Breaking Defense*, 19 January 2016.

184. Sydney J. Freeburg, "Army Vice Says Yes on Anti-Drone Tech, Maybe on Missiles, No on Iran Man," *Breaking Defense*, 21 June 2016.

185. Arthur Holland Michel, "Counter-Drone Systems," Center for the Study of the Drone, Bard College, February 2018, p. 1.

186. Poulsen, "Why the US Government Is Terrified of Hobbyist Drones."

187. Michel, "Counter-Drone Systems," p. 5–6.

188. David Shepardson, "U.S. Agency Seeks New Authority to Disable Threatening Drones," *Reuters*, 15 May 2018.

Chapter 6

1. This story was broken in Bloomsberg news and this account draws heavily from its reporting and subsequent updates. See Monte Reel, "Secret Cameras Record Baltimore's Every Move from Above," *Bloomsberg*, 23 August 2016, available at: https://www.bloomberg.com/features/2016-baltimore-secret-surveillance/, accessed 8 February 2018.

2. Reel. *A Review of the Baltimore Police Department's Use of Persistent Surveillance*, Police Foundation, 30 January 2017, p. 10.

3. Andrew deGrandpre and Andrew Tilghman, "Iran Linked to the Deaths of 500 U.S. Troops In Iraq, Afghanistan," *Military Times*, 14 July 2015, available at: https://www.militarytimes.com/news/pentagon-congress/2015/07/14/iran-linked-to-deaths-of-500-u-s-troops-in-iraq-afghanistan/, accessed 2 June 2018.

4. Gregg Zoroya, "How the IED Changed the US Military," *USA Today*, 18 December 2013, available at: https://www.usatoday.com/story/news/nation/2013/12/18/ied-10-years-blast-wounds-amputations/3803017/, accessed 8 February 2018.

5. Reel, "Secret Cameras Record Baltimore's Every Move from Above."

6. Joint Improvised Explosive Device Defeat Organization, Annual Report, 2010, p. 16, available at: http://www.dtic.mil/dtic/tr/fulltext/u2/a549409.pdf, accessed 8 February 2018.

7. There are critiques that the Joint Improvised Threat Defeat Organization (JIEDDO) was ineffective because it did not concentrate its efforts and

pursued projects on tangentially related to IEDs. See Kelsey D. Atherton, "When Big Data Went to War—and Lost," *Politico*, 12 October 2017, available at: https://www.politico.eu/article/iraq-war-when-big-data-went-to-war-and-lost/, accessed 13 April 2018.

8. See Reel, "Secret Cameras Record Baltimore's Every Move from Above."

9. *A Review of the Baltimore Police Department's Use of Persistent Surveillance*, p. 9.

10. See Reel, "Secret Cameras Record Baltimore's Every Move from Above."

11. Matthew Feeney, "Surveillance Takes Wing: Privacy in the Age of Police Drones," CATO Policy Analysis Number 807 (13 December 2016), p. 4.

12. Lily Hay Newman, "How Baltimore Became America's Laboratory for Spy Tech," *Wired*, 4 September 2016.

13. *A Review of the Baltimore Police Department's Use of Persistent Surveillance*.

14. Baltimore Community Support Program, Frequently Asked Questions, available at: https://www.baltimorecsp.com/faq, accessed 9 February 2014. It is also worth noting that police made the same claim about deterrence with a previous trial of cameras, though they acknowledged that surveillance may just make the criminals move elsewhere. See Nancy G. La Vigne, Samantha S. Lowry, Joshua A. Markman, and Alison M. Dwyer, "Evaluating the Use of Public Surveillance Cameras for Crime Control and Prevention—A Summary," Urban Institute, Justice Policy Center, September 2011, p. 2.

15. *A Review of the Baltimore Police Department's Use of Persistent Surveillance*.

16. Carl Messineo, "Why Baltimore's Covert Spy Plane Program Is a Major Battleground for Privacy and Free Speech," *National Memo*, 1 September 2016, http://www.nationalmemo.com/baltimores-covert-spy-plane-program-major-battleground-privacy-free-speech/, accessed 16 February 2018.

17. Matthew Feeney, "Baltimore Air Surveillance Should Cause Concerns," *The Hill*, 25 August 2016, available at: http://thehill.com/blogs/pundits-blog/civil-rights/293329-baltimore-police-drones-should-cause-concerns, accessed 16 February 2018.

18. Jay Stanley, "Baltimore Aerial Surveillance Program Retained Data Despite 45-Day Privacy Policy Limit," ACLU, 25 October 2016.

19. Baltimore Community Support Program, Frequently Asked Questions.

20. Kevin Rector and Luke Broadwater, "Report of Secret Aerial Surveillance by Baltimore Police Prompts Questions, Outrage," *Baltimore Sun*, 24 August 2016, available at: http://www.baltimoresun.com/news/maryland/baltimore-city/bs-md-ci-secret-surveillance-20160824-story.html, accessed 16 February 2018.

21. Juliet Linderman, "Baltimore Police Took 1 Million Surveillance Photos of the City," *Associated Press*, 7 October 2016.

22. Jay Stanley, "Baltimore Police Secretly Running Aerial Mass Surveillance Eye in the Sky," ACLU, 24 August 2016, available at: https://www.aclu.

org/blog/privacy-technology/surveillance-technologies/baltimore-police-secretly-running-aerial-mass, accessed 16 February 2018.

23. Kevin Rector, "Baltimore's Aerial Surveillance Program Goes Way Beyond Citiwatch, Experts Say," *Baltimore Sun*, 25 August 2016, available at: http://www.baltimoresun.com/news/maryland/crime/bs-md-ci-surveillance-differences-20160825-story.html, accessed 16 February 2018.

24. Rector. An important implication of this argument is that the Baltimore police department would be freed of the obligation to get a warrant for the collection of this video.

25. Kristin Bergtora Sandvik, "The Public Order Drone: Proliferation and Disorder in Civil Air Space," in Maria Gabrielsen Jumbert and Kristin Bergtora Sandvik (eds.), *The Good Drone* (London: Routledge, 2017), p. 115.

26. Ellen Rosen, "Skies Aren't Clogged with Drones Yet, But Don't Rule Them Out," *New York Times*, 19 March 2019.

27. David Koenig and Joseph Pisani, "Where Are the Drones? Amazon's Customers Are Still Waiting," *Associated Press*, 3 December 2018.

28. Alan Levin, "Groundbreaking U.S. Plan Would Permit Drone Flights over Crowds," *Bloomsberg*, 14 January 2019; Michael Laris, "Trump Administration Taps 10 Projects, Including Those in Virginia, Kansas and Nevada for Drone Pilot," *Washington Post*, 9 May 2019;

29. Austin Choi-Fitzpatrick et al., "Up in the Air: A Global Estimate of Non-Violent Drone Use 2009–2015," 14 April 2016, available at: http://digital.sandiego.edu/cgi/viewcontent.cgi?article=1000&context=gdl2016report, accessed 25 February 2018.

30. Dan Gettinger, "Drones at Home: Public Safety Drones," Center for the Study of the Drone, Bard College, New York, April 2017, p. 1.

31. Gettinger, p. 3.

32. Sandvik, "The Public Order Drone," p. 113.

33. Quoted in Sandvik, p. 113.

34. Sidney Fussell, "Kentucky Is Turning to Drones to Fix Its Unresolved-Murder Crisis," *Atlantic*, 6 November 2018; Cindy Chang, "LAPD Deploys Controversial Drone for the First Time," *Los Angeles Times*, 15 January 2019.

35. Jeff Brown, "Taking to the Air: Drones and Law Enforcement," *Dover Post*, Delaware, 14 December 2017, available at: http://www.govtech.com/em/disaster/Taking-to-the-Air-Drones-and-Law-Enforcement.html, accessed 25 February 2018.

36. John Villasenor, "The Drone Threat to Privacy," *Scientific American*, 14 November 2011.

37. William M. Welch, "At Nation's Doorstep, Police Drones Are Flying," *USA Today*, 28 January 2014.

38. Ron Nixon, "Drones, So Useful in War, May Be Too Costly for Border Duty," *New York Times* (2 November 2016).

39. "US Drones to Watch Entire Mexico Border from September 1," *Reuters*, 31 August 2010, available at: https://www.reuters.com/article/us-usa-immigration-security/u-s-drones-to-watch-entire-mexico-border-from-september-1-idUSTRE67T5DK20100830, accessed 24 March 2018.

40. Arthur Holland Michel, "Custom and Border Protection Drones," Center for the Study of the Drone, 7 January 2015, available at: http://dronecenter.bard.edu/customs-and-border-protection-drones/, accessed 24 March 2018.

41. Davie Bier and Matthew Feeney, "Drones on the Border: Efficacy and Privacy Implications," CATO Institute, 1 May 2018; Nixon, "Drones, So Useful in War, May Be Too Costly for Border Duty."

42. Office of the Inspector General, "U.S. Custom and Border Protection's Unmanned Aircraft Program Does Not Achieve Intended Results or Recognize All Costs of Operation," OIG-15-17, 24 December 2014, available at: https://www.oig.dhs.gov/assets/Mgmt/2015/OIG_15-17_Dec14.pdf, accessed 25 March 2018.

43. Office of the Inspector General, "CBP Has Not Ensured Safeguards for Data Collected Using Unmanned Aircraft Systems," OIG-18-79, 21 September 2018, available at: https://www.oig.dhs.gov/sites/default/files/assets/2018-09/OIG-18-79-Sep18.pdf, accessed 12 July 2019; See also Bier and Feeney, "Drones on the Border."

44. Brenda Fiegel, "Narco-Drones: A New Way to Transport Drugs," *Small Wars Journal*, available at: http://smallwarsjournal.com/jrnl/art/narco-drones-a-new-way-to-transport-drugs, accessed 25 March 2018.

45. David Axe, "Weaponized Drones," *Vice Magazine*, 25 October 2017, available at: https://motherboard.vice.com/en_us/article/j5jmb4/mexican-drug-cartels-have-weaponized-drones, accessed 25 March 2018.

46. Patrick Tucker, "DHS: Drug Traffickers Are Spoofing Border Drones," *DefenseOne*, 17 December 2015, available at: http://www.defenseone.com/technology/2015/12/DHS-Drug-Traffickers-Spoofing-Border-Drones/124613/, accessed 25 March 2018.

47. Feeney, "Surveillance Takes Wing," p. 3; details on the Leptron Avenger are available on its manufacturer's website: http://www.leptron.com/leptron_avenger_uas_helicopter.html, accessed 22 April 2018.

48. Electronic Frontier Foundation, "Miami-Dade PD Releases Information About Its Drone Program; Will the FAA Follow Suit?" 13 April 2012, available at: https://www.eff.org/deeplinks/2012/04/miami-dade-pd-releases-information-about-its-drone-program-will-faa-follow-suit, accessed 13 April 2018.

49. Feeney, "Surveillance.Takes Wing," p. 3.

50. Quote from Welch, "At Nation's Doorstep, Police Drones Are Flying."

51. American Civil Liberties Union, "Protecting Privacy from Aerial Surveillance: Recommendations for Government Use of Drone Aircraft," December 2011, p. 7.

52. Gettinger, "Drones at Home," p. 4.

53. "Poll: Catching Criminals Is Fine, But Don't Use Drones for Speeding Tickets, Americans Say," *CNN*, 13 June 2012.

54. Seth Jacobson, "Met to Become First UK Force to Deploy Drones to Monitor Road Users," *Guardian*, 8 July 2019, available at: https://www.theguardian.com/uk-news/2019/jul/08/met-become-first-uk-force-deploy-drone-monitor-road-users, accessed 12 July 2019.

55. Paul Lewis, "CCTV in the Sky: The Police Plan to Use Military-Style Spy Drones," *Guardian*, 23 January 2010.

56. Ciaran D'Arcy, "Drones to be Used in Crackdown on Illegal Dumping," *Irish Times*, 13 March 2017.

57. Malek Murison, "French Police Use Drones to Catch Dangerous Drivers," *Drone Life*, 14 November 2017, available at: https://dronelife.com/2017/11/14/french-police-drones/, accessed 16 April 2018.

58. Mark Scott and Natasha Singer, "How Europe Protects Your Online Data Differently Than the U.S.," *New York Times*, 31 January 2016; Sheera Frenkel, "Tech Giants Brace for Europe's New Data Privacy Rules," *New York Times*, 28 January 2018.

59. Villasenor, "The Drone Threat to Privacy."

60. Paul Lewis, "You're Being Watched: There's One CCTV Camera for Every 32 People in the UK," *Guardian*, 2 March 2011.

61. BBC, "The Most Spied Upon People in Europe," 28 February 2008.

62. Asaf Lubin, "A New Era of Mass Surveillance Is Emerging Across Europe," *Just Security*, 9 January 2017, available at: https://www.justsecurity.org/36098/era-mass-surveillance-emerging-europe/, accessed 16 April 2018.

63. Alissa M. Dolan and Richard M. Thompson II, "Integration of Drones into Domestic Airspace: Selected Legal Issues," Congressional Research Service, 4 April 2013, p. 13–14.

64. Barry Friedman, *Unwarranted: Policing Without Permission* (New York: Farrar, Straus and Giroux, 2017), p. 215–217.

65. Friedman, p. 217. In 1934, the Congress passed the Federal Communications Act, which placed limits on the ability of the US government to tap telephone lines.

66. Samuel D. Warren and Louis D. Brandeis, "The Right to Privacy," *Harvard Law Review* 4:5 (1890), p. 193–220.

67. Warren and Brandeis, p. 206.

68. Feeney, "Surveillance Takes Wing," p. 4–5.

69. Dolan and Thompson, "Integration of Drones into Domestic Airspace," p. 16.

70. Feeney, "Surveillance Takes Wing," p. 5.

71. Quoted in full in Feeney, p. 5.

72. Feeney, p. 5.

73. Gregory McNeal, *Drones and Aerial Surveillance: Considerations for Legislators*, Brookings Institute, November 2014, p. 8.

74. Cited in McNeal, p. 8.
75. Dolan and Thompson, "Integration of Drones into Domestic Airspace,"
    p. 7–8.
76. Dolan and Thompson, p. 9.
77. For a discussion see McNeal, *Drones and Aerial Surveillance*, p. 12–27.
78. National Conference of State Legislatures, "Current Unmanned Aircraft
    State Law Landscape," 10 September 2018, available at: http://www.
    ncsl.org/research/transportation/current-unmanned-aircraft-state-law-
    landscape.aspx, accessed 5 August 2019.
79. Arthur Holland Michel, "Local and State Drone Laws," Center for the
    Study of the Drone, Bard College, March 2017, p. 6–7.
80. Faine Greenwood, "Shooting Down Drones Isn't Funny or Brave,"
    Slate, 26 September 2016, available at: http://www.slate.com/articles/
    technology/future_tense/2016/09/shooting_down_drones_is_dangerous_
    and_stupid.html, accessed 8 May 2017.
81. Kelsey D. Atherton, "It Is a Federal Crime to Shoot Down a Drone,
    Says FAA," *Popular Science*, 15 April 2016, available at: https://www.
    popsci.com/it-is-federal-crime-to-shoot-down-drone-says-faa, accessed 8
    May 2018.
82. Faine Greenwood, "When Can the Government Shoot Down Civilian
    Drones?" *Slate*, 15 October 2018.
83. Avery E. Holton, Sean Lawson, and Cynthia Love, "Unmanned Aerial
    Vehicles: Opportunities, Barriers and the Future of "Drone Journalism"
    Journalism Practice 9:5 (2015), p. 634–650.
84. Jason Koebler, "Drone Journalist Faces 7 Years in Prison for Filming
    Dakota Pipeline Protests," *Vice*, 25 May 2017, available at: https://
    motherboard.vice.com/en_us/article/zmbdy5/drone-journalist-faces-
    7-years-in-prison-for-filming-north-dakota-access-pipeline-protests,
    accessed 14 May 2018.
85. Alleen Brown, "Ohio and Iowa Are the Latest to of Eight States to
    Consider Anti-Protest Bills Aimed at Pipeline Opponents," *The Intercept*,
    2 February 2018, available at: https://theintercept.com/2018/02/02/ohio-
    iowa-pipeline-protest-critical-infrastructure-bills/, accessed 14 May 2018.
86. Matt Novak, "Journalists Start Using Drones to View Immigrant
    Detention Camps After Government Blocks Entry," Gizmodo, 22
    June 2018.
87. Andrew Stobo Sniderman and Mark Hanis, "Drones for Human Rights,"
    New York Times, 30 January 2012.
88. Frederik Rosén, "Extremely Stealthy and Incredibly Close: Drones,
    Control and Legal Responsibility," DIIS Working Paper,
    Copenhagen, 2013.
89. Josh Lyons, "Drones in the Service of Human Rights," Human Rights
    Watch, 11 December 2017, available at https://www.hrw.org/news/2017/
    12/11/drones-service-human-rights, accessed 9 May 2018.

90. Witness, "Cameras Everywhere: Current Challenges and Opportunities at the Intersection of Human Rights, Video and Technology," Brooklyn, NY, 2011, p. 10.
91. Witness, p. 11.
92. Cyrus Farivar, "FBI and ATF Spent $2.1 Million on 23 Drones That Don't Work," *Ars Technica*, 36 March 2015.
93. Farivar, "FBI and ATF Spent $2.1 Million on 23 Drones That Don't Work."
94. Conor Friedersdorf, "The Rapid Rise of Federal Surveillance Drones Over America," *Atlantic*, 10 March 2016.
95. Tom Barry, "Drones over the Homeland: How Politics, Money and a Lack of Oversight Have Sparked Drone Proliferation and What We Can Do," International Policy Report, Center for International Police, April 2013, p. 7.
96. Gregg Zoroya, "Pentagon Report Justifies Deployment of Military Spy Drones over U.S.," *USA Today*, 9 March 2016, available at: https://www.usatoday.com/story/news/nation/2016/03/09/pentagon-admits-has-deployed-military-spy-drones-over-us/81474702/, accessed 14 May 2018.
97. Gregg Zoroya and Alan Gomez, "Pentagon Has 'Unique' Policy for Legal Use of Drones in U.S.," *USA Today*, 9 March 2016, available at: https://www.usatoday.com/story/news/nation/2016/03/09/pentagon-developed-unique-policy-ensure-drones-used-legally/81540412/, accessed 14 May 2018.
98. Kelsey D. Atherton, "The PentagonIs Flying More Drone Missions Along America's Border," *Defense News*, 10 February 2019.
99. Pentagon, "UAS Tracker," available at: https://www.defense.gov/Portals/1/Documents/Web%20site%20UAS%20Tracker.pd, accessed 14 May 2018.
100. Michael Peck, "The Pentagon's Plan: Use Drones to Feed Hurricane Victims," *National Interest*, 10 March 2019.
101. Zoroya, "Pentagon Report Justifies Deployment of Military Spy Drones over U.S."
102. Brian Bennett and Joel Rubin, "Drones Are Taking to the Skies in the U.S.," *Los Angeles Times*, 15 February 2013, available at: http://articles.latimes.com/2013/feb/15/nation/la-na-domestic-drones-20130216, accessed 14 May 2018.
103. Katie Mather, "LAPD Takes Another Step Toward Deploying Drones in Controversial Yearlong Test," *Los Angeles Times*, 9 January 2018, available at: http://www.latimes.com/local/lanow/la-me-ln-lapd-drone-money-20180109-story.html, accessed 14 May 2018,
104. Fran Spielman, "ACLU Sounds the Alarm About Bill Allowing Use of Drones to Monitor Protesters," *Chicago Sun Times*, 1 May 2018, available at: https://chicago.suntimes.com/politics/aclu-sounds-the-alarm-about-bill-allowing-use-of-drones-to-monitor-protesters/, accessed 14 May 2018.

105. Jeremy Bentham, *The Panopticon Writings*, ed. Milan Bozovic (London: Verso, 2011).

106. See particularly Michel Foucault, *Discipline and Punish: The Birth of the Prison* (New York: Vintage, 1995).

107. Charlie Savage, "Facial Scanning Is Making Gains in Surveillance," *New York Times*, 21 August 2013.

108. Timothy Williams, "Facial Recognition Software Moves from Overseas Wars to Local Police, " *New York Times*, 12 August 2015.

109. Center on Privacy and Technology, "The Perpetual Line-Up: Unregulated Police Face Recognition in American,"18 October 2016.

110. Drew Harwell, "FBI, ICE Find State Driver's License Photos Are a Gold-Mine for Facial Recognition Searches," *Washington Post*, 7 July 2019.

111. Lorenzo Franceschi-Bicchierai, "Report: Russia Is Stockpiling Drones to Spy on Street Protests," *CNN.com*, 25 June 2012, available at: http://www.cnn.com/2012/07/25/tech/innovation/russia-stockpiling-drones-wired/index.html, accessed 13 October 2012.

112. "Russian Guard Will Use Drones to Monitor Protests," *UA Wire*, 30 March 2017, http://uawire.org/news/russian-guard-will-use-drones-to-monitor-protesters, accessed 16 May 2018.

113. Pei Lei and Cate Cadell, "China Eyes 'Black Tech' to Boost Security as Parliament Meets," *Reuters*, 10 March 2018, available at: https://www.reuters.com/article/us-china-parliament-surveillance/china-eyes-black-tech-to-boost-security-as-parliament-meets-idUSKBN1GM06M, accessed 15 May 2018.

114. James A. Millward, "What It's Like to Live in a Surveillance State," *New York Times*, 3 February 2018.

115. Zhao Lei, "1,000 Drones Used by Police Across Country," *China Daily*, 19 June 2017, available at: http://www.chinadaily.com.cn/china/2017-06/19/content_29792454.htm, accessed 15 May 2018.

116. Yaniv Kubovich, "In First, Israeli Drones Drop Tear Gas to Disperse Palestinian Protesters," *Haaretz* (12 March 2018), available at: https://www.haaretz.com/israel-news/in-first-israeli-drones-drop-tear-gas-to-disperse-palestinian-protest-1.5892756, accessed 16 May 2018.

117. Declan Walsh, "Waves of Gazans vs. Israeli Tear Gas and Bullets: Deadliest Mayhem in Years," *New York Times*, 14 May 2018, available at: https://www.nytimes.com/2018/05/14/world/middleeast/gaza-israel-deadly-protest-scene.html, accessed 16 May 2018.

118. Tyler Rogoway, "Israel Uses Drone Racers to Down Incendiary Kites and Drones to Disperse Tear Gas over Gaza," *The Drive*, 14 May 2018.

119. Remote Control, Omega Research Foundation, and University of Bradford, "Tear-Gassing by Remote Control," December 2015, p. 22–28.

Chapter 7

1. National Society for Earthquake Technology—Nepal, "Gorkha Earthquake," available at: http://www.nset.org.np/eq2015/, accessed

30 June 2017. See also John P. Rafferty, "Nepal Earthquake of 2015," *Encyclopedia Brittanica*, 24 April 2017.

2. Gopal Sharma, "Armed with Drones, Aid Workers Seek Faster Response to Earthquakes, Floods," *Reuters*, 15 May 2016.

3. "Nepal Earthquake: Eight Million People Affected," BBC News, 28 April 2015, available at: http://www.bbc.com/news/world-asia-32492232.

4. Elijah Wolfson, "One Year After Devastating Earthquake, Nepal Is Still in Ruins," *Newsweek*, 21 April 2016, available at: http://www.newsweek. com/2016/04/29/nepal-earthquake-anniversary-2015-gorkha-kathmandu-450449.html, accessed 1 July 2017.

5. "Nepal Earthquake."

6. The estimate is from the Digital Archaeology Foundation, Nepal, available at: http://www.digitalarchaeologyfoundation.com/, accessed 1 July 2017.

7. Gardiner Harris, "Everest Climbers Are Killed as Nepal Quake Sets off Avalanche," *New York Times*, 25 April 2015.

8. David Hamilton quoted in Carole Cadwallr, "Nepal Quake: Everest Base Camp 'Looked Like It Had Been Flattened By a Bomb,'" *Guardian*, 26 April 2015, available at: https://www.theguardian.com/world/2015/apr/26/nepal-quake-everest-base-camp-looked-like-it-had-been-flattened-by-bomb, accessed 1 July 2017.

9. Thomas Fuller and Chris Buckley, "Earthquake Aftershocks Jolt Nepal as Death Toll Rises Above 3,400," *New York Times*, 26 April 2015.

10. "Nepal Earthquake."

11. Rafferty, "Nepal Earthquake of 2015."

12. Ellen Barry, "Weeks After Deadly Nepal Quake, Another Tremor Revives Fears," *New York Times*, 12 May 20165. Some later estimates put the death toll as 400. See Rafferty, "Nepal Earthquake of 2015."

13. Quoted in "Nepal Earthquake."

14. Faine Greenwood, "How Drones Can Help Nepal Recover from the Earthquake," *Slate*, 28 April 2015.

15. Nita Bhalla, "Relief Efforts After Nepal Quake Must be Scaled Up, $415 Million Needed—UN," *Reuters*, 29 April 2015.

16. "Nepal Earthquake."

17. The term "digital humanitarian" has been coined by Patrick Meier. See his *Digital Humanitarians: How Big DataIs Changing The Face of Humanitarian Response* (Boca Raton: CRC Press, 2015).

18. Sharma, "Armed with Drones, Aid Workers Seek Faster Response to Earthquakes, Floods."

19. Matt McFarland, "In Nepal, a Model for Using Drones for Humanitarianism Emerges," *Washington Post*, 7 October 2015.

20. Sharma, "Armed with Drones, Aid Workers Seek Faster Response to Earthquakes, Floods."

21. Alex Hern, "Nepal Moves to Limit Drone Flights Following Earthquake," *Guardian*, 6 May 2015.

22. Hannan Lewsley, "Eye in the Sky," *Nepali Times*, 4–10 December 2015, available at: http://nepalitimes.com/article/nation/nepal-government-crack-down-on-drones,2716, accessed 5 July 2017.

23. Sharma, "Armed with Drones, Aid Workers Seek Faster Response to Earthquakes, Floods."

24. Peter Rabley, property rights director for the Omidayar Network, quoted in Jennifer Hlad, "Drones: A Force for Good When Flying in the Face of Disaster," *Guardian*, 28 July 2015.

25. Dr. Robin Murphy, "Unmanned Systems and Hurricane Matthew: Lessons from 2010 Haiti Earthquake," CRASAR, Texas A&M University, 4 October 2016, available at: http://crasar.org/2016/10/04/unmanned-systems-and-hurricane-matthew-lessons-from-2010-haiti-earthquake/, accessed 16 July 2017. See also US Air Force, "Global Hawk Collects Reconnaissance Data During Haiti Relief Efforts," 15 January 2010, available at: http://www.af.mil/News/Article-Display/Article/118014/global-hawk-collects-reconnaissance-data-during-haiti-relief-efforts/, accessed 16 July 2017.

26. UN Institute for Training and Research (UNITAR), "UNOSAT Carries Out First UAV Mission for IOM in Haiti," 17 February 2012, available at: http://www.unitar.org/unosat-carries-out-first-uav-mission-iom-haiti, accessed 5 July 2017.

27. UAViators: Humanitarian UAV Network, Case Studies, p. 7–8, available at: http://uaviators.org/docs, accessed 13 July 2017.

28. Faine Greenwood and Konstantin Kakaes, *Drones and Aerial Observation: New Technologies for Property Rights, Human Rights, and Global Development: A Primer* (New York: New America Foundation, 2015), p. 16.

29. UN Institute for Training and Research (UNITAR), "UNOSAT Carries Out First UAV Mission for IOM in Haiti."

30. This is also known as Typhoon Yolanda in the Philippines.

31. Victoria Turk, "Drones Mapped the Philippines to Improve Typhoon Aid Efforts," *Vice*, 9 May 2014, available at: https://motherboard.vice.com/en_us/article/pgagwk/drones-mapped-the-philippines-to-improve-typhoon-aid-efforts, accessed 5 July 2017.

32. Andrew Schroeder and Patrick Meier, "Automation for the People: Opportunities and Challenges for Humanitarian Robotics," Humanitarian Practice Networks, April 2016, available at: http://odihpn.org/magazine/automation-for-the-people-opportunities-and-challenges-of-humanitarian-robotics/, accessed 12 July 2017.

33. New America Foundation, *Drones and Aerial Observation*, p. 16.

34. Meier, *Digital Humanitarians*, p. 87.

35. Meier, p. 90.

36. United Nations Office for the Coordinator for Humanitarian Affairs, "Unmanned Aerial Vehicles in Humanitarian Response," OCHA Policy and Studies Series (June 2014), p. 5.

37. Brian Clark Howard, "Vanuatu Puts Drones in the Sky to See Cyclone Damage," *National Geographic*, 8 April 2015, available at: http://news.nationalgeographic.com/2015/04/150406-vanuatu-cyclone-pam-relief-drones-uavs-crisis-mapping-patrick-meier/, accessed 7 July 2017. See also the case studies produced by the Swiss Foundation for Mine Action (FSD) for the Drones in Humanitarian and Environmental Applications project, available at: http://drones.fsd.ch/en/case-study-no-13-using-drones-to-inspect-post-earthquake-road-damage-in-ecuador/, accessed 12 July 2017.

38. UAViators: Humanitarian UAV Network, Case Studies, p. 7–8, available at: http://uaviators.org/docs, accessed 13 July 2017.

39. Schroeder and Meier, "Automation for the People: Opportunities and Challenges for Humanitarian Robotics." See also the full discussion of the project, available at: https://irevolutions.org/2015/08/19/world-bank-using-uavs/, accessed 12 July 2017.

40. See particularly the UAViators website, http://uaviators.org/, accessed 12 July 2017.

41. FSD, *Drones in Humanitarian Action: A Guide to the Use of Airborne Systems in Humanitarian Crises*, January 2017, p. 17, available at: http://drones.fsd.ch/en/homepage/, accessed 13 July 2017.

42. FSD, p. 17.

43. Steve Sweeney, "Drones Can Do a Lot of Good if Used Correctly and with Respect," *Drone 360*, 9 August 2016, available at: http://www.drone360mag.com/news-notes/2016/08/drones-can-do-a-lot-of-good-if-used-correctly-and-with-respect-humanitarians, accessed 7 July 2016. See also New America Foundation, *Drones and Aerial Observation*, p. 15.

44. Sweeney.

45. FSD, *Drones in Humanitarian Action: A Guide to the Use of Airborne Systems in Humanitarian Crises*, p. 28.

46. FSD, p. 29.

47. Faine Greenwood, "How to Make Maps with Drones," in New America Foundation, *Drones and Aerial Observation*, p. 35.

48. Greenwood, "How to Make Maps with Drones," p. 36.

49. FSD, *Drones in Humanitarian Action: A Guide to the Use of Airborne Systems in Humanitarian Crises*, p. 32.

50. Quoted in Sweeney, "Drones Can Do a Lot of Good if Used Correctly and with Respect.".

51. FSD, *Drones in Humanitarian Action: A Survey of Perceptions and Applications*, 2016, p. 5–6.

52. See the WeRobotics website, available at: http://werobotics.org/, accessed 14 July 2017.

53. Interview with WeRobotics cofounder Adam Klaptocz, "Drones for Good 2.0: How WeRobotics Is Redefining the Use of Unmanned Systems in Developing Countries," Waypoint, 15 February 2017, available at: http://waypoint.sensefly.com/drones-for-good-2-0-werobotics-developing-countries/, accessed 14 July 2017.

54. WeRobotics, "Building Cargo Drone Expertise in Papua New Guinea," iRevolutions, 25 February 2019, available at: https://blog.werobotics.org/2019/02/25/building-cargo-drone-expertise-in-papua-new-guinea/, accessed 16 July 2019.

55. FSD, *Drones in Humanitarian Action: A Guide to the Use of Airborne Systems in Humanitarian Crises*, p. 49.

56. Anthony Cuthbertson, "Firefighting, Criminal Chasing, and Rescue Drones Come to Europe," *Newsweek*, 8 April 2016, available at: http://www.newsweek.com/drones-rescue-dji-phantom-eena-eu-445467, accessed 17 July 2017.

57. FSD, *Drones in Humanitarian Action: A Guide to the Use of Airborne Systems in Humanitarian Crises*, p. 48.

58. "Migrant Crisis: Migration to Europe Explained in Seven Charts," BBC News, 4 March 2016, available at: http://www.bbc.com/news/world-europe-34131911, accessed 19 July 2017.

59. Cited in "Migrant Crisis: Migration to Europe Explained in Seven Charts."

60. Ben Quinn, "Migrant Death Toll Passes 5,000After Two Boats Capsize off Italy," Guardian, 23 December 2016.

61. Justin Stares, "EU Plans Drone Fleet to Track Migrants," *Politico*, 6 April 2016, available at: http://www.politico.eu/article/european-union-fleet-of-drones-to-track-migrants-refugees/, accessed 26 July 2017; Costas Paris and Robert Wall, "Europe Tries Out Sniffer Drones for Policing Ship Emissions," *Wall Street Journal*, 26 November 2015.

62. Stares, "EU Plans Drone Fleet to Track Migrants."

63. The Italian navy also used drones in its efforts to detect and rescue migrants in Operation Mare Nostrum in 2013–2014. See Maria Gabrielsen Jumbert, "Creating the EU Drone: Control, Sorting and Search and Rescue at Sea," in Kristen Bergtora Sandvik and Maria Gabrielsen Jumbert (eds.), *The Good Drone* (London: Routledge, 2016), p. 89.

64. European Maritime Safety Agency, *Maritime Surveillance in Practice: Using Integrated Maritime Services* (Lisbon: 2015).

65. European Commission, "European Commission Calls for Tough Standards to Regulate Civil Drones," 4 August 2014; Myrto Hatzigeorgeopoulos, "European Perspectives on Unmanned Aerial Vehicles," European Security Review (December 2012). See particularly the Riga Declaration on Remotely Piloted Aircraft, "Framing the Future of Aviation," 6 March 2015, and European Parliament, "Civil Drones in the European Union," Briefing, October 2015.

66. Interview with Leendert Bal, Director of "C" Operations, European Maritime Safety Agency, conducted by phone, 21 June 2016.

67. Interview with Leendert Bal.

68. Interview with Leendert Bal.

69. See the reports collected on Human Rights Watch, "Europe's Migration Crisis," https://www.hrw.org/tag/europes-migration-crisis, accessed 26 July 2017.

70. Jumbert, "Creating the EU Drone: Control, Sorting and Search and Rescue at Sea," p. 95.

71. Jumbert.

72. Mark Jacobsen, "The Promise of Drones," *Harvard International Review* 37:3 (Spring 2016); Dan Wang, "The Economics of Drone Delivery," Flexport, no date, available at: https://www.flexport.com/blog/drone-delivery-economics/, accessed 4 August 2017; the 0.88 cents estimate is from Tasha Keeney, "Amazon Drones Could Deliver a Package in Under Thirty Minutes for Under One Dollar," *Ark Invest*, 1 December 2015, available at: https://ark-invest.com/research/amazon-drone-delivery, accessed 4 August 2017.

73. Lora Kolodny, "Drone Delivery Start-Up Flirtey Raises $16 Million to Become Next-Gen UPS," *Techcrunch*, 18 January 2017, available at: https://techcrunch.com/2017/01/18/drone-delivery-startup-flirtey-raises-16-million-to-become-a-next-gen-ups/, accessed 28 July 2017.

74. FSD, *Drones in Humanitarian Action: A Guide to the Use of Airborne Systems in Humanitarian Crises*, p. 36.

75. FSD, p. 36.

76. FSD, p. 38.

77. United Nations Office for the Coordinator for Humanitarian Affairs, "Unmanned Aerial Vehicles in Humanitarian Response," p. 8.

78. Jacobsen, "The Promise of Drones"; United Nations Office for the Coordinator for Humanitarian Affairs, "Unmanned Aerial Vehicles in Humanitarian Response," p. 8.

79. FSD, *Drones in Humanitarian Action: A Guide to the Use of Airborne Systems in Humanitarian Crises*, p. 38.

80. Médecins Sans Frontières, "Papua New Guinea: Innovating to Reach Remote TB Patients and Improve Access to Treatment," 14 November 2014, available at: http://www.msf.org/en/article/papua-new-guinea-innovating-reach-remote-tb-patients-and-improve-access-treatment, accessed 28 July 2017.

81. Schroeder and Meier, "Automation for the People: Opportunities and Challenges for Humanitarian Robotics."

82. FSD, *Drones in Humanitarian Action: A Guide to the Use of Airborne Systems in Humanitarian Crises*, p. 37.

83. FSD, p. 37.

84. Interview with Gerald Poppinga and Mirjam Jansen Op De Haar, Drones for Development, Amsterdam, Netherlands, 23 May 2016, conducted by phone.

85. See the details on Zipline's website, http://flyzipline.com/product/index. html, accessed 7 August 2017.

86. See the details on Zipline's website: https://flyzipline.com/impact/, accessed 15 August 2019.

87. Interview with Will Hetzler, Cofounder, Zipline, 15 November 2016, conducted by phone.

88. Lora Kolodny, "Zipline Raises $25 Million to Deliver Medical Supplies by Drone," *Techcrunch*, 9 November 2016, available at: https://techcrunch. com/2016/11/09/zipline-raises-25-million-to-deliver-medical-supplies-by-drone/, accessed 7 August 2017; Riley de Leon, "Zipline Takes Flight in Ghana, Making It the World's Largest Drone Delivery Network," CNBC, 24 April 2019, available at: https://www.cnbc.com/2019/04/24/with-ghana-expansion-ziplines-medical-drones-now-reach-22m-people.html, accessed 13 August 2019.

89. John Markoff, "Drones Marshalled to Drop Lifesaving Supplies over Rwandan Terrain," *New York Times*, 4 April 2016.

90. Interview with Will Hetzler, Cofounder, Zipline.

91. Aryn Baker, "Zipline's Drones Are Saving Lives," *Time Magazine*, 31 May 2018.

92. Markoff, "Drones Marshalled to Drop Lifesaving Supplies over Rwandan Terrain.".

93. De Leon, "Zipline Takes Flight in Ghana, Making It the World's Largest Drone Delivery Network."

94. Interview with Will Hetzler, Cofounder, Zipline.

95. De Leon, "Zipline Takes Flight in Ghana, Making It the World's Largest Drone Delivery Network."

96. De Leon.

97. Austin Choi-Fitzpatrick, "Drones for Good: Technological Innovations, Social Movements, and the State," *Journal of International Affairs* 68:1 (Fall/Winter 2014), p. 19–36.

98. Gopal Sharma, "Nepal's Medicine Drones Bring Healthcare to the Himalayas," *Reuters*, 29 April 2018.

99. See the information on Redline drones available on the EPFL/Afrotech website, http://afrotech.epfl.ch/page-115280-en.html, accessed 7 August 2017.

100. Zoe Flood, "From Killing Machines to Agents of Hope: The Future of Drones in Africa," *Guardian* (27 July 2016).

101. Interview with Jonathan Ledgard, Redline/Afrotech/EFPL, conducted by Skype, 17 June 2016.

102. Interview with Jonathan Ledgard.

103. United Nations Office for the Coordinator for Humanitarian Affairs, "Unmanned Aerial Vehicles in Humanitarian Response," p. 8.

104. Jacobsen, "The Promise of Drones."

105. Jacobsen.

106. John Karlsrud and Frederik Rosén, "Lifting the Fog of War? Opportunities and Challenges of Drones in UN Peace Operations," in Kristen Bergtora Sandvik and Maria Gabrielsen Jumbert (eds), *The Good Drone*, p. 47.

107. Karlsrud and Rosén, p. 47.

108. Colum Lynch, "U.N. Wants to Use Dronesfor Peacekeeping Missions," *Washington Post*, 8 January 2013.

109. The UN was granted the right to do aerial surveillance in 2006, but did not use drones for this purpose until the Irish peacekeepers arrived in 2009. See John Karlsrud and Frederik Rosén, "In the Eye of the Beholder? The UN and the Use of Drones to Protect Civilians," *Stability: International Journal of Security and Development* 2:2 (2013), p. 2.

110. Karlsrud and Rosén, "Lifting the Fog of War? Opportunities and Challenges of Drones in UN Peace Operations," p. 47.

111. Karlsrud and Rosén, "In the Eye of the Beholder? The UN and the Use of Drones to Protect Civilians."

112. Fiona Blyth, "UN Peacekeeping Deploys Unarmed Drones to Eastern Congo," *Global Observatory*, 27 February 2013.

113. Joachim Koops, Norrie MacQueen, Thierry Tardy, and Paul D. Williams, *The Oxford Handbook of United Nations Peacekeeping Operations* (Oxford: Oxford University Press, 2015). p. 799.

114. Quoted in Lynch, "U.N. Wants to Use Drones for Peacekeeping Missions."

115. Kasaija Philip Apuuli, "The Use of Unmanned Aerial Vehicles (Drones) in United Nations Peacekeeping: The Case of the Democratic Republic of Congo," *American Society of International Law* 18:13 (13 June 2014), p. 325–342

116. Apuuli.

117. Blyth, "UN Peacekeeping Deploys Unarmed Drones to Eastern Congo."

118. Michelle Nichols, "U.N. Security Council Allows Drones for Eastern Congo," *Reuters*, 24 January 2013, available at: http://www.reuters.com/article/us-congo-democratic-un-idUSBRE90N0X720130124, accessed 16 August 2017.

119. Blyth, "UN Peacekeeping Deploys Unarmed Drones to Eastern Congo."

120. Apuuli, "The Use of Unmanned Aerial Vehicles (Drones) in United Nations Peacekeeping: The Case of the Democratic Republic of Congo."

121. Konstantin Kakaes, "The UN's Drones and Congo's War," in New America Foundation, *Drones and Aerial Observation*, p. 88; Better World Campaign, "The UN's Use of Unmanned Aerial Vehicles in the Democratic Republic of the Congo: U.S. Support and Potential Foreign Policy Advantages," p. 4.

122. Better World Campaign, "The UN's Use of Unmanned Aerial Vehicles in the Democratic Republic of the Congo: U.S. Support and Potential Foreign Policy Advantages," p. 4.

123. Better World Campaign, "The UN's Use of Unmanned Aerial Vehicles in the Democratic Republic of the Congo: U.S. Support and Potential Foreign Policy Advantages," p. 5.

124. "Are UN Drones the Future of Peacekeeping?" France 24 News, 9 April 2015.

125. Kakes, "The UN's Drones and Congo's War," p. 89.

126. Better World Campaign, "The UN's Use of Unmanned Aerial Vehicles in the Democratic Republic of the Congo: U.S. Support and Potential Foreign Policy Advantages," p. 6.

127. Better World Campaign, "The UN's Use of Unmanned Aerial Vehicles in the Democratic Republic of the Congo: U.S. Support and Potential Foreign Policy Advantages," p. 5.

128. "Are UN Drones the Future of Peacekeeping?"

129. Kakes, "The UN's Drones and Congo's War," in *New America Foundation, Drones and Aerial Observation*, p. 89.

130. Kakes, p. 91.

131. Nichols, "U.N. Security Council Allows Drones for Eastern Congo."

132. "Are UN Drones the Future of Peacekeeping?"

133. Kakes, "The UN's Drones and Congo's War," in *New America Foundation, Drones and Aerial Observation*, p. 92.

134. "Are UN Drones the Future of Peacekeeping?"

135. Interview with Dirk Druet, UN Department of Peacekeeping Operations (DPKO), by phone, 7 December 2016.

136. Interview with Dirk Druet.

137. There is an open question about whether the UN can effectively convey a deterrent threat, as this presumes the ability to enforce the threat against transgressions. Given limited resources and mandates for UN missions, especially observer missions, this is not always present.

138. United Nations, *Performance Peacekeeping: Final Report of the Expert Panel on Technology and Innovation in UN Peacekeeping* (New York: 22 December 2014).

139. Kakes, "The UN's Drones and Congo's War," in *New America Foundation, Drones and Aerial Observation*, p. 90.

140. Quoted In Siobhan O'Grady, "How a U.N. Drone Crashed in Congo and Was Promptly Forgotten," *Foreign Policy*, 10 September 2015.

141. Interview with Dirk Druet.

142. Samantha Power, *A Problem from Hell: America and the Age of Genocide* (New York: Basic Books, 2002).

143. Quoted in O'Grady, "How a U.N. Drone Crashed in Congo and Was Promptly Forgotten."

144. Karlsrud and Rosén, "In the Eye of the Beholder? The UN and the Use of Drones to Protect Civilians," p. 5.

145. Karlsrud and Rosén, p. 6.

146. Harvard Humanitarian Initiative, "The Signal Code: A Human Rights
     Approach to Information During Crisis," January 2017, available
     at: https://hhi.harvard.edu/publications/signal-code-human-rights-
     approach-information-during-crisis, accessed 3 September 2017.
147. Interview with Dirk Druet.
148. Karlsrud and Rosén, "In the Eye of the Beholder? The UN and the Use of
     Drones to Protect Civilians," p. 6.
149. Karlsrud and Rosén, "Lifting the Fog of War? Opportunities and
     Challenges of Drones in UN Peace Operations," p. 57.
150. "What Are Humanitarian Principles," OCHA, June 2012, available
     at: https://docs.unocha.org/sites/dms/Documents/OOM-
     humanitarianprinciples_eng_June12.pdf, accessed 15 August 2017.
151. "Unmanned Drones Used by UN Peacekeepers in the DRC," World
     Vision, 15 July 2014, available at: http://www.worldvision.org.uk/
     news-and-views/latest-news/2014/july/unmanned-drones-used-un-
     peacekeepers-drc/, accessed 3 September 2017.
152. See Kristin Bergtora Sandvik and Kjersti Lohne, "The Rise of the
     Humanitarian Drone: Giving Content to an Emerging Concept,"
     Millennium: Journal of International Studies 43:1 (2014), p. 145–164;
     Kristin Bergtora Sandvik and Kjersti Lohne, "The Promise and Perils of
     Disaster Drones," *Humanitarian Practice Network* Issue 58 (July 2013).
153. The phrase is taken from Karlsrud and Rosén, "Lifting the Fog of War?
     Opportunities and Challenges of Drones in UN Peace Operations."
154. See particularly Jack C. Chow, "Predators for Peace," *Foreign Policy*,
     27 April 2012; Jack C. Chow, "The Case for Humanitarian Drones,"
     OpenCanada, 12 December 2012.

Chapter 8

1. "ASDF Confirms Unidentified Drone Flying over East China Sea," *Asahi
   Shimbun*, 10 September 2013, available at: http://ajw.asahi.com/article/
   asia/china/AJ201309100069, accessed 4 April 2014.
2. Paul J. Smith, "The Senkaku/Diaoyu Island Controversy: A Crisis
   Postponed," *Naval War College Review* 66:2 (Spring 2013), p. 29.
3. Mira Rapp-Hooper, "The Battle for the Senkakus Moves to the Skies,"
   The Diplomat, 12 November 2013, http://thediplomat.com/2013/11/
   the-battle-for-the-senkakus-moves-to-the-skies/, accessed 3 April 2014.
   See also Smith, "The Senkaku/Diaoyu Island Controversy: A Crisis
   Postponed," p. 39.
4. Dan Gettinger, "An Act of War: Drones Are Testing China-Japan
   Relations," Center for the Study of the Drone, Bard College, 8 November
   2013, available at: http://dronecenter.bard.edu/act-war-drones-testing-
   china-japan-relations/, accessed 3 April 2014.
5. Quoted in Gettinger.

6. "Japan to Shoot Down Foreign Drones That Invade Its Airspace," *Japan Times*, 20 October 2013, http://www.japantimes.co.jp/news/2013/10/20/national/japan-to-shoot-down-foreign-drones-that-invade-its-airspace/#.Uz26mvldVLD, accessed 3 April 2014.

7. "China Warns Japan Against Shooting Down Drones over Islands," *Times of India*, 27 October 2013, http://timesofindia.indiatimes.com/world/china/China-warns-Japan-against-shooting-down-drones-over-islands/articleshow/24779422.cms?referral=PM, accessed 3 April 2014.

8. Naoya Yoshino, "Senkaku Showdown: US to Send in Drones," *Asian Review*, 26 November 2013, available at: http://asia.nikkei.com/Politics-Economy/International-Relations/Senkaku-showdown-US-to-send-in-drones, accessed 3 April 2014.

9. "China Bitterly Attacks Japanese Prime Minister over Air Zone Remarks," *Guardian*, 15 December 2013.

10. Smith, "The Senkaku/Diaoyu Island Controversy: A Crisis Postponed," p. 40.

11. "China to Set Up Drone Bases Along Coastal City," CRI English, October 23, 2012, available at: http://english.cri.cn/7146/2012/10/23/2702s728578.htm, accessed 14 March 2017.

12. Daniel A. Medina, "The Real Story Behind Japan's Drone Boom," *Quartz*, 21 July 2014, available at: https://qz.com/236440/the-real-story-behind-japans-drone-boom/, accessed 14 March 2017.

13. "Beijing Ponders Drones for Senkaku Surveillance Duty," *Japan Times*, 13 June 2015.

14. Medina, "The Real Story Behind Japan's Drone Boom."

15. "Japan Debates Shooting Down Chinese Drones over Contested Airspace," UAS Vision, 30 June 2015, available at: http://www.uasvision.com/2015/06/30/japan-debates-shooting-down-chinese-drones-over-contested-islands/, accessed 15 March 2017.

16. "Drone Joins Four Chinese Ships in Latest Senkaku Intrusion," *Japan Times*, 18 May 2017, available at: https://www.japantimes.co.jp/news/2017/05/18/national/drone-joins-four-chinese-ships-latest-senkaku-intrusion/#.XTHNSuhKiM8, accessed 19 July 2019.

17. The quote is from Acting Secretary of Defense Patrick Shanahan. See "US Sells 34 Drones to South China Allies," *Straits Times*, 5 June 2019.

18. Matthew Fuhrmann and Michael C. Horowitz, "Droning On: Explaining the Proliferation of Unmanned Aerial Vehicles," *International Organization* 71 (Spring 2017), p. 397–418.

19. This is probably an underestimate. See General Accounting Office, "Agencies Could Improve Sharing and End-Use Monitoring on Unmanned Aerial Vehicle Exports," GAO 12-536, July 2012, p. 9. See also Center for New American Security, "Proliferated Drones," available at: http://drones.cnas.org/, accessed 29 July 2019.

20. Ulrike Esther Franke, "The Global Diffusion of Unmanned Aerial Vehicles (UAVs) or Drones," in Mike Aaronson, Wali Aslam, Tom Dyson, and Regina Rauxloh (eds.), *Precision Strike Warfare and International Intervention: Strategic, Ethico-Legal and Decisional Implications* (London: Routledge, 2014).

21. See particularly Shashank Joshi and Aaron Stein, "Emerging Drone Nations," *Survival* 55:5 (2013), p. 53–78; Andrea Gilli and Mauro Gilli, "The Diffusion of Drone Warfare: Industrial, Organizational and Infrastructural Constraints," *Security Studies* 25 (2016), p. 50–84.

22. Joshi and Stein, "Emerging Drone Nations," p. 56–62.

23. Fuhrmann and Horowitz, "Droning On: Explaining the Proliferation of Unmanned Aerial Vehicles."

24. Teal Group, "Teal Group Predicts Worldwide UAV Market Will Total $89 Billion in Its 2013 UAV Market Profile and Forecast," 17 June 2013, available at: https://www.tealgroup.com/index.php/pages/press-releases/40-teal-group-predicts-worldwide-uav-market-will-total-89-billion-in-its-2013-uav-market-profile-and-forecast, accessed 19 December 2019.

25. Teal Group, "Teal Group Predicts Worldwide UAV Market Will Total $89 Billion in Its 2013 UAV Market Profile and Forecast," 17 June 2013, available at: https://www.tealgroup.com/index.php/pages/press-releases/40-teal-group-predicts-worldwide-uav-market-will-total-89-billion-in-its-2013-uav-market-profile-and-forecast, accessed 19 December 2019

26. Peter L. Bergen and Jennifer Rowland, "World of Drones: The Global Proliferation of Drone Technology," in Peter L. Bergen and Daniel Rothenberg (eds.), *Drone Wars: Transforming Conflict, Law and Policy* (Cambridge: Cambridge University Press, 2015), p. 300–301.

27. Wojciech Moskwa, "World Drone Market Seen Nearing $120 Billion in 2020, PwC Says," *Bloomberg Technology*, 9 May 2016, available at: https://www.bloomberg.com/news/articles/2016-05-09/world-drone-market-seen-nearing-127-billion-in-2020-pwc-says, accessed 26 March 2017.

28. "Drones: Reporting For Work," available at: http://www.goldmansachs.com/our-thinking/technology-driving-innovation/drones/, accessed 26 March 2017.

29. Aaron Mehta, "US Seeking Global Armed Drone Export Rules," *DefenseNews*, 25 August 2016.

30. Burak Ege Bekdil, "Export Potential: Turkey's Homemade Drones Could Boost Local Industry," *Defense News*, 22 April 2019, available at: https://www.defensenews.com/unmanned/2019/04/22/export-potential-turkeys-homemade-drones-could-boost-local-industry/, accessed 19 July 2019.

31. David Shepardson, "U.S. Commercial Drone Use to Expand Tenfold by 2021: Government Agency," *Reuters*, 22 March 2017, available at: http://www.reuters.com/article/us-usa-drones-idUSKBN16S2NM, accessed 26 March 2017.

32. Government Accountability Office, "Agencies Could Improve Sharing and End-Use Monitoring on Unmanned Aerial Vehicle Exports," GAO 12-536, July 2012, p. 11.

33. Dan Gettinger, "What You Need to Know About Drone Exports," Center for the Study of the Drone, Bard College, 6 March 2015.

34. Roderic Alley, "The Drone Debate: Sudden Bullet or Slow Boomerang," Centre for Strategic Studies, Victoria University of Wellington, No. 14 (2013), p. 32–33.

35. Government Accountability Office, "Agencies Could Improve Sharing and End-Use Monitoring on Unmanned Aerial Vehicle Exports," p. i.

36. Geoff Dyer, "US Drone Lobby Power Points to Revived Military-Industrial Complex," *Financial Times*, 9 October 2013.

37. W.J. Hennigan, "Drone Makers Urge the U.S. to Let Them Sell More Overseas," *Los Angeles Times*, 1 July 2012.

38. Jeremiah Getler, "U.S. Unmanned Aerial Systems," Congressional Research Service, 3 January 2012, p. 28.

39. "US Looks to Export Drone Technology to Allies," *Reuters*, 25 March 2010.

40. Doug Palmer and Jim Wolf, "Pentagon OK with Selling U.S. Drones to 66 Countries," *Reuters*, 6 September 2012.

41. Andrea Shalal and Emily Stephenson, "U.S. Establishes Policy for Exports of Armed Drones," *Reuters*, 18 February 2015.

42. "U.S. Sets Stringent Drone Sales Policy," *Arms Control Today*, April 2015, p. 31.

43. Sarah Knuckley, quoted in "U.S. Sets Stringent Drone Sales Policy," p. 31.

44. "U.S. Sets Stringent Drone Sales Policy," p. 31.

45. Richard Whittle, "Drone Dealer: The Case for Selling Armed UAVs," *National Interest*, 20 February 2015.

46. "A World of Drones," New America Foundation, available at: https://www.newamerica.org/in-depth/world-of-drones/1-introduction-how-we-became-world-drones/, accessed 12 April 2017.

47. Aaron Mehta, "White House Rolls Out Armed Drone Declaration," *Defense News*, 5 October 2016, http://www.defensenews.com/articles/white-house-rolls-out-armed-drone-declaration, accessed 12 April 2017.

48. Michael C. Horowitz and Joshua Schwartz, "A New U.S. Policy Makes It (Somewhat) Easier to Export Drones," *Washington Post*, 20 April 2018, available at: https://www.washingtonpost.com/news/monkey-cage/wp/2018/04/20/a-new-u-s-policy-makes-it-somewhat-easier-to-export-drones/?utm_term=.97c49cfc2b0cm, accessed 2 June 2018.

49. Lara Seligman, "Trump's Push to Boost Lethal Exports Reaps Few Rewards," *Foreign Policy*, 6 December 2018; see also George Nacouzi, J.D. Williams, Brian Dolan, Anne Stickells, David Luckey, Colin Ludwig, Jia Xu, Yuliya Shokh, Daniel M. Gerstein and Michael H. Decker, "Assessment of the Proliferation of Certain Remotely Piloted Aircraft

Systems," Santa Monica: RAND, 2019, available at: https://www.rand.org/pubs/research_reports/RR2369.html .

50. Seligman, "Trump's Push to Boost Lethal Exports Reaps Few Rewards.";
    Natasha Turak, "Pentagon Is Scrambling as China 'Sells the Hell Out of'
    Armed Drones to US Allies," *CNBC*, 21 February 2019.

51. "Israel and the Drone Wars: Examining Israel's Production, Use and
    Proliferation of UAVs," *Drone Wars UK*, January 2014, p. 5.

52. "Israel and the Drone Wars: Examining Israel's Production, Use and
    Proliferation of UAVs."

53. J.R. Wilson, *UAV Round Up*, (Bay City: Aerospace America, July/August
    2013), p. 32.

54. Jefferson Morley, "Israel's Drone Dominance," *Salon*, 15 May 2012.

55. "Israel and the Drone Wars: Examining Israel's Production, Use and
    Proliferation of UAVs," p. 18.

56. Uri Sadot, "A Perspective on Israel," Center for New American Security,
    Proliferated Drones Series, 2015. Original estimate from Frost and
    Sullivan.

57. Sadot "A Perspective on Israel."

58. Josh Meyer, "Why Israel Dominates Global Drone Exports," *Quartz*,
    10 July 2013, available at: https://qz.com/102200/why-israel-dominates-
    global-drone-exports/, accessed 15 April 2017.

59. Frost and Sullivan, "Israel is Top Global Exporter of Unmanned
    Aerial Systems with a Continued Positive Outlook Ahead," 20 May
    2013, available at: https://www.frost.com/prod/servlet/press-release.
    pag?docid=278664709, accessed 15 April 2017.

60. "Israel and the Drone Wars: Examining Israel's Production, Use and
    Proliferation of UAVs," p. 23.

61. Ora Coren, "Israel Aerospace Industries Moving Production to
    Mississippi," *Haaretz*, 17 November 2009.

62. "Israel and the Drone Wars: Examining Israel's Production, Use and
    Proliferation of UAVs," p. 21.

63. Edward Wong, "Hacking U.S. Secrets, China Pushes for Drones,"
    New York Times, 20 September 2013.

64. "Report: Israel Frozen Out of F-35 Development," *Defense Industry Daily*,
    19 April 2005.

65. William Wan and Peter Finn, "Global Race to Match U.S. Drone
    Capabilities," *Washington Post*, 4 July 2011.

66. "Israel and the Drone Wars: Examining Israel's Production, Use and
    Proliferation of UAVs," p. 23.

67. Yaakov Katz and Amir Bohbot, "How Israel Sold Russia Drones to
    Stop Missiles from Reaching Iran," *Jerusalem Post*, 3 February 2017; Ben
    Sullivan, "Israel Gave Russia a Farming Drone and Now Everyone's
    Pissed," *Vice*, 17 November 2016.

68. Kavitha Surana, "Israel Freezes Export of Suicide Drone to Azerbaijan After Allegation of Abuse," *Foreign Policy*, 30 August 2017, available at: https://foreignpolicy.com/2017/08/30/israel-freezes-export-of-suicide-drone-to-azerbaijan-after-allegation-of-abuse-azerbaijan-armenia/, accessed 19 July 2019.

69. Kimberly Hsu, "China's Military Unmanned Aerial Vehicle Industry," U.S.-China Economic and Security Review Commission Staff Research Backgrounder, 13 July 2013, p. 3.

70. Hsu, p. 3.

71. *The Role of Autonomy in DoD Systems* (Washington, DC: Defense Science Board, U.S. Department of Defense, 2012), p. 71.

72. Hsu, "China's Military Unmanned Aerial Vehicle Industry," p. 8.

73. Zhao Lei, "China Gaining Market Share in Military Drones," *China Daily*, 26 September 2013.

74. Wong, "Hacking U.S. Secrets, China Pushes for Drones."

75. Wong.

76. Wan and Finn, "Global Race to Match U.S. Drone Capabilities."

77. Cited in Sarah Kreps and Micah Zenko, "The Next Drone Wars: Preparing for Proliferation," *Foreign Affairs*, March/April 2015, p. 68–79.

78. The quote is from Wong, "Hacking U.S. Secrets, China Pushes for Drones."

79. Ian M. Easton and L.C. Russell Hsiao, "The Chinese People's Liberation Army's Unmanned Aerial Vehicle Project: Organizational Capacities and Operational Capacities," Project 2049, 13 March 2013, p. 2.

80. Easton and Hsiao, p. 5.

81. Hsu, "China's Military Unmanned Aerial Vehicle Industry," p. 14.

82. John Reed, "Photo of the Week: China's Got a Stealth Drone," *Foreign Policy*, 14 May 2013.

83. Jeffrey Lin and P.W. Singer, "Meet China's Sharp Sword, a Stealth Drone That Can Likely Carry Two Tons of Bombs," *Popular Science*, 18 January 2017, available at: http://www.popsci.com/china-sharp-sword-lijian-stealth-drone, accessed 15 May 2017.

84. Patrick Tucker, "China's Beating US to Market on Combat Drones, by Copying US Technology," *Defense One*, 6 November 2018.

85. Jeffrey Lin and P.W. Singer, "China Flies Its Largest Ever Drone: The Divine Eagle," *Popular Science*, 6 February 2015, available at: http://www.popsci.com/china-flies-its-largest-ever-drone-divine-eagle, accessed 15 May 2017.

86. Quoted in Wan and Finn, "Global Race to Match U.S. Drone Capabilities."

87. Wong, "Hacking U.S. Secrets, China Pushes for Drones."

88. "China Makes Gains in Military-Drone Market, China Daily Says," *Bloomberg News*, 26 September 2013.

89. Ian Armstrong, "What's Behind China's Big New Drone Deal," *The Diplomat*, 20 April 2017.

90. Peter Apps, "Chinese, Local Drones Reflect Changing Middle East," *Reuters*, 7 March 2019.

91. Adam Rawnsley, "Meet China's Killer Drones," *Foreign Policy*, 14 January 2016.

92. Michael C. Horowitz, Sarah E. Kreps, and Matthew Fuhrmann, "Separating Fact from Fiction in the Debate over Drone Proliferation," *International Security* 41:2 (Fall 2016), p. 12.

93. Rawnsley, "Meet China's Killer Drones."

94. The estimate range of the Caihong-5 is 6,500 km. See Saibal Dasgupta, "China Set to Export Advance Combat Drones to Pakistan, Others," *Times of India*, 1 November 2016.

95. Clay Dillow, "China: A Rising Drone Weapons Dealer to the World," CNBC, 5 March 2016, available at: http://www.cnbc.com/2016/03/03/china-a-rising-drone-weapons-dealer-to-the-world.html, accessed 15 May 2017.

96. Rawnsley, "Meet China's Killer Drones."

97. Wendell Minnick, "Report: China's UAVs Could Challenge Western Dominance," *Defense News*, 25 June 2013.

98. UAV Round Up, Aerospace America, p. 35.

99. "Russia Planning to Buy Aerial Drones from the UAE," Pakistan Defense, 18 July 2013 available at: http://www.defence.pk/forums/arab-defence/265292-russia-planning-buy-aerial-drones-uae.html, and "Russia to Fund Drone Development Through 2025," *Voice of Russia*, 2 August 2013, available at: http://voiceofrussia.com/news/2013_08_02/Russia-to-fund-drone-development-program-through-2025-0127/

100. Isabelle Facon, "Proliferated Drones: A Perspective on Russia," Center for New American Security .

101. The 800 estimate comes from Michael Pearson, "Russia's Resurgent Drone Program," CNN (16 October 2015), and see also Isabelle Facon, "Proliferated Drones: A Perspective on Russia," Center for New American Security (CNAS).

102. Patrick Tucker, "US Army Racing to Catch Up to Russia on Battle Drones," Defense One, 28 September 2016, available at: http://www.defenseone.com/technology/2016/09/us-army-racing-catch-russia-battle-drones/131936/, accessed 16 May 2017.

103. Bill Gertz, "Russia to Deploy Long-Range Attack Drone by 2016," *Washington Free Beacon*, 13 November 2013.

104. Facon, "Proliferated Drones: A Perspective on Russia."

105. Karl Soper, "New UAVs Completing Russian State Testing," *Jane's 360*, 14 May 2018; Vladimir Karnozov, "Russia Prepares to Flight-Test the Sukhoi-70 UCAV," AIN Online, 25 February 2019, available at: https://

www.ainonline.com/aviation-news/defense/2019-01-25/russia-prepares-flight-test-sukhoi-s-70-ucav, accessed 19 July 2019.

106. Jason Le Miere, "Russia Developing New 'Swarm of Drones' in New Arms Race with U.S., China," *Newsweek*, 15 May 2017.

107. Liz Sly, "The Kalashnikov Assault Rifle Changed the World. Now There's a Kalashnikov Kamikaze Drone," *Washington Post*, 23 February 2019.

108. "Iran, Russia Sharing Drone Technology," *Tehran Times*, 23 July 2017.

109. Daniel A. Medina, "How Japan Fell in Love with America's Drones," *Defense One*, 21 July 2014.

110. Lieutenant General Masayuki Hironaka, "A Perspective on Japan," Center for New American Security.

111. Captain Sukjoon Yoon, "A Perspective on South Korea," Center for New American Security.

112. "Turkey," European Forum on Armed Drones, no date, available at: https://www.efadrones.org/countries/turkey/, accessed 19 July 2019.

113. Johanna Polle, "MALE-Drone Proliferation in Europe: Assessing the Status Quo Regarding Acquisition, Research and Development, and Employment," IFSH/IFAR Working Paper #21, November 2018.

114. Quoted in Glennon J. Harrison, "Unmanned Aircraft Systems (UAS): Manufacturing Trends," Congressional Research Service, 30 January 2013.

115. George Arnett, "The Numbers Behind the Worldwide Trade in Drones," *Guardian*, 16 March 2015.

116. Ulrike Esther Franke, "A Perspective on Germany," Center for New American Security.

117. Interview with Ludwig DeCamps, NATO Headquarters, Belgium, 5 July 2016; see also Michael C. Sirak, "NATO's New Eyes in the Sky," *Air Force Magazine*, September 2013, p. 50–54.

118. "Future Combat Air System: UK and France Move Ahead," AirForce Technology, 17 May 2016, available at: http://www.airforce-technology.com/features/featurefuture-combat-air-system-uk-and-france-move-ahead-4893221/.

119. Pierre Tran, "Eurodrone to Launch This Summer," *Defense News*, 15 March 2015, available at: http://www.defensenews.com/story/defense/air-space/isr/2015/03/15/euro-drone-studies-to-be-launched-this-summer/70276960/, accessed 16 May 2017.

120. Thomas C. Schelling, *Arms and Influence* (New Haven: Yale University Press, 1966), p. 7–9.

121. Michael C. Horowitz, Sarah E. Kreps, and Matthew Fuhrmann, "Separating Fact from Fiction in the Debate over Drone Proliferation," *International Security* 41:2 (Fall 2016), p. 7–42.

122. Horowitz, Kreps, and Fuhrmann, p. 9.

123. John Reed, "Predator Drones 'Useless' in Most Wars, Top Air Force General Says," *Foreign Policy*, 19 September 2013.

124. Horowitz, Kreps, and Fuhrmann, "Separating Fact from Fiction in the Debate over Drone Proliferation."

125. Pradip R. Sagar, "49 Drones to Keep an Eye on China and Pakistan Borders," *DNA*, 30 September 2013, available at: http://www.dnaindia.com/india/1895756/report-dna-exclusive-49-drones-to-keep-an-eye-on-china-and-pakistan-borders, accessed 13 October 2013.

126. Craig Whitlock, "U.S. Military Drone Surveillance Expanding to Hot Spots Beyond Declared Combat Zones," *Washington Post*, 20 July 2013.

127. See Horowitz, Kreps, and Fuhrmann, "Separating Fact from Fiction in the Debate over Drone Proliferation."

128. Quoted in Christopher Drew, "Military Is Awash in Data from Drones," *New York Times*, 10 January 2010.

129. Michel Tuan Pham, Tom Meyvis, and Rongrong Zhou, "Beyond the Obvious: Chronic Vividness in Imagery and the Use of Information in Decision-Making," *Organizational Behavior and Human Decision Processes* 84:2 (2001), p. 226–253; Keren Yarhi-Milo, *Knowing the Adversary: Leaders, Intelligence and the Assessment of Intentions in International Politics* (Princeton: Princeton University Press, 2014).

130. Bill Roggio, "Indian Commandos Strike Terrorists inside Pakistan-Occupied Kashmir," *Long War Journal*, 29 September 2016.

131. Quoted in M. Ilyas Khan, "India's 'Surgical Strikes' in Kashmir: Truth or Illusion?" BBC News, 23 October 2016, available at: https://www.bbc.com/news/world-asia-india-37702790, accessed 25 July 2019.

132. Irfan Ghauri, "Indian Drone Shot Down over Pakistan's Airspace," *Express Tribune*, 20 November 2016, available at: http://tribune.com.pk/story/1237093/indian-drone-shot-pakistans-airspace/, accessed 19 December 2019.

133. "Russia Developing Anti-Drone Defence," Voice of Russia, 17 November 2013.

134. John Horgan, "Unmanned Flight," *National Geographic*, March 2013, available at: http://ngm.nationalgeographic.com/2013/03/unmanned-flight/horgan-text?rptregcta=reg_free_np&rptregcampaign=20131016_rw_membership_n1p_us_se_w#.

135. "Advanced System to Guard Russia from Hi-Tech Surveillance, Drone Attacks," RT.News, 18 November 2013.

136. Ishaan Tharoor, "China Unveils a New Anti-Drone Laser, but It's the Growing Chinese Drone Fleet That Matters," *Washington Post*, 6 November 2014.

137. See Immanuel Kant, *Perpetual Peace* (Indianapolis: Hackett, 1983); Grégoire Chamayou, *Drone Theory* (London: Penguin, 2015), p. 182–187.

138. Cited in Hugh Gusterson, *Drone: Remote Control Warfare* (Cambridge, MA: MIT Press, 2016), p. 22.

139. David Hastings Dunn, "Drones: Disembodied Aerial Warfare and the Unarticulated Threat," *International Affairs* 89:5 (2013), p. 1238.

140. Jason Koebler, "Obama: Administration Saw Drone Strikes as a 'Cure All' for Terrorism," *US News*, 23 May 2013, available at: https://www.usnews.com/news/articles/2013/05/23/obama-administration-saw-drone-strikes-as-cure-all-for-terrorism, accessed June 1, 2017. See also Horowitz, Kreps, and Fuhrmann,"Separating Fact from Fiction in the Debate over Drone Proliferation," p. 14.

141. See the discussion in James Igoe Walsh and Marcus Schulzke, "The Ethics of Drone Strikes: Does Reducing the Cost of Conflict Encourage War," Strategic Studies Institute, U.S. Army War College, September 2015; Jacquelyn Schneider and Julia MacDonald, "Presidential Risk Orientation and Force Employment Decisions: The Case of Unmanned Weaponry," *Journal of Conflict Resolution* 61:3 (2017), p. 511–536; Jacquelyn Schneider and Julia MacDonald, "U.S. Public Support for Drone Strikes," Center for New American Security, 20 September 2016, available at: https://www.cnas.org/publications/reports/u-s-public-support-for-drone-strikes, accessed 6 June 2017.

142. Aniseh Bassiri Tabrizi and Justin Bronk, "Armed Drones in the Middle East," Royal United Services Institute, Occasional Paper, December 2018.

143. On these strategies, see Alexander L. George and William E. Simons, *The Limits of Coercive Diplomacy*, 2nd ed., (Boulder: Westview Press, 1994). See also Alexander L. George, *Forceful Persuasion: Coercive Diplomacy as an Alternative to War* (Washington, DC: US Institute of Peace, 2009).

144. Zafar Malik, "India Drone Violates Pak Airspace," *The Nation (Pakistan)*, 4 August 2013.

145. "India Deploys U.S.-Made Surveillance Drones Near LoC," *Express Tribune*, 8 March 2017.

146. Salman Siddiq, "Pakistani Company Produces Country's Second Drone Technology," *Express Tribune*, 22 November 2016.

147. Jitendra Bahadur Singh, "Pakistan Deploys Weaponized Drones Across LoC Border," *India Today*, 21 March 2019, available at: https://www.indiatoday.in/mail-today/story/pakistan-deploys-weaponised-drones-across-loc-border-1483153-2019-03-21, accessed 23 July 2019.

148. Stella Kim, Paula Hancocks, and Madison Park, "South Korea Investigates Two North Korean Drones," *CNN*, 2 April 2014, available at: http://www.cnn.com/2014/04/02/world/asia/korea-drone-tensions/, accessed 6 June 2017.

149. Kyle Mizohami, "What's in North Korea's Drone Arsenal," *Popular Mechanics*, 22 January 2016, available at: http://www.popularmechanics.com/military/weapons/a19090/inside-north-koreas-unmanned-aerial-vehicle-arsenal/, accessed 6 June 2017.

150. Van Johnson, "Kim Jong Un's Tin Can Air Force," *Foreign Policy*, 12 November 2014, available at: http://foreignpolicy.com/2014/11/12/kim-jong-uns-tin-can-air-force/, accessed 2 June 2017.

151. "Seoul Fires Warning Shots at 'North Korea Drone'" *Sky News*, 13 January 2016, available at: http://news.sky.com/story/seoul-fires-warning-shots-at-north-korea-drone-.10128538, accessed 19 December 2019.

152. Ju-Min Park and Christine Kim, "South Korea Fires at Suspected Drone amid Missile Crisis," *Reuters*, 23 May 2017.

153. Safya Khan-Ruf, "North Korea 'Developing More Sophisticated Aerial Drones' Warns South as Tensions Simmer," *Independent (UK)*, 20 December 2016. http://www.independent.co.uk/news/world/asia/north-korea-developing-advanced-aerial-drones-says-south-a7486526.html

154. "N.K. Estimated to Have 1,000 drones: Report," Yonhap News Agency, 29 March 2017, available at: http://english.yonhapnews.co.kr/news/2017/03/29/0200000000AEN20170329002100315.html, accessed 6 June 2017.

155. Siobhan Gorman, Yochi J. Dreazen, and August Cole, "Insurgents Hack U.S. Drones," *Wall Street Journal*, 17 December 2009; Scott Petersen, "Exclusive: Iran Hijacked US Drone, Says Iranian Engineer," *Christian Science Monitor*, 15 December 2011.

156. Michael D. Shear, Helene Cooper, and Eric Schmitt, "Trump Says He Was 'Cocked and Loaded' to Strike Iran, but Pulled Back," *New York Times*, 21 June 2019.

157. "U.S. May Have Downed More Than One Iranian Drone, Centcom Commander Gen. Kenneth McKenzie Says," CBS News, 23 July 2019, available at: https://www.cbsnews.com/news/centcom-commander-kenneth-mckenzie-suggests-us-brought-down-more-than-one-iranian-drone/, accessed 23 July 2019.

158. Ben Hubbard, Palko Karasz, and Stanley Reed, "Two Major Oil Installations Hit by Drone Strike, and U.S. Blames Iran," *New York Times*, 14 September 2019.

159. Geoff Brumfiel, "What We Know About the Attack on Saudi Oil Facilities," NPR, 19 September 2019.

160. Amy Zegart, "Deterrence in the Drone Age," Working Group on Foreign Policy and Grand Strategy, Hoover Institute, Stanford University, November 20, 2014, available at: http://www.hoover.org/sites/default/files/fw_hoover_foreign_policy_working_group_unconventional_threat_essay_series/201411%20-%20Zegart.pdf, accessed 6 June 2017.

161. Micah Zenko and Sarah Kreps, *Limiting Armed Drone Proliferation* (New York: Council on Foreign Relations Press, 2014).

162. See particularly Schelling, *The Strategy of Conflict*.

163. Horowitz, Kreps, and Fuhrmann, "Separating Fact from Fiction in the Debate over Drone Proliferation," p. 29.

164. Horowitz, Kreps, and Fuhrman, p. 31.

165. Amy Zegart, "Cheap Flights, Credible Threats: The Future of Armed Drones and Coercion," *Journal of Strategic Studies* (2018), p. 17–18.
166. Zegart, p. 18–22.
167. Amy Zegart, "The Coming Revolution of Drone Warfare," *Wall Street Journal*, 18 March 2015.
168. Zegart, "Cheap Flights, Credible Threats."
169. Michael Mayer, "The New Killer Drones: Understanding the Strategic Implications of Next-Generation Unmanned Combat Aerial Vehicles," *International Affairs* 91:4 (2015), p. 765–780.
170. Melissa Salyk-Virk, "Drone Warfare Could Enter the Kashmir Conflict," New America, 4 December 2018, available at: https://www.newamerica. org/international-security/blog/drone-warfare-could-enter-kashmir-conflict/, accessed 23 July 2019.
171. On the need for urgency, see George, *Forceful Persuasion.*
172. David Zucchino, "War Zone Drone Crashes Add Up," *Los Angeles Times*, 6 July 2010.
173. Getler, "U.S. Unmanned Aerial Systems," p. 18.
174. Craig Whitlock, "When Drones Fall from the Sky," *Washington Post*, 20 June 2014.
175. Kelsey D. Atherton, "What Causes So Many Drones Crashes?" *Popular Science*, 4 March 2013.
176. Whitlock, "When Drones Fall from the Sky."
177. Geoffrey Ingersoll, "Frightening Drone Footage Shows Near Miss with Airbus A300 over Kabul," *Business Insider*, 3 June 2013. Footage revealing this incident was only recently declassified.
178. Craig Whitlock, "Drone Crashes Mount at Civilian Airports," *Washington Post*, 30 November 2012.
179. Craig Whitlock and Greg Miller, "U.S. Moves Drone Fleet from Camp Lemonnier to Ease Djibouti's Safety Concerns," *Washington Post*, 24 September 2013.

Chapter 9

1. Roger Berkowitz, "Drones and the Question of the 'Human,'" *Ethics and International Affairs* 28:2 (2014), p. 159–169.
2. "An Open Letter to the United Nations Convention on Certain Conventional Weapons," accessed at: https://www.dropbox.com/s/g4ijcaqq6ivq19d/2017%20Open%20Letter%20to%20the%20United%20Nations%20Convention%20on%20Certain%20Conventional%20Weapons.pdf?dl=0, accessed 18 May 2018.
3. Quoted in Catherine Clifford, "Hundreds of A.I. Experts Echo Elon Musk, Stephen Hawking in Calling for a Ban on Killer Robots," *CNBC*, 8 November 2017, available at: https://www.cnbc.com/2017/11/08/

ai-experts-join-elon-musk-stephen-hawking-call-for-killer-robot-ban.
html, accessed 18 May 2018.

4. The literature on nuclear weapons and crisis behavior is vast. See
    particularly Thomas Schelling, *Arms and Influence*, rev. ed. (New
    Haven: Yale University Press, 2008); Richard K. Betts, *Nuclear
    Blackmail and Nuclear Balance* (Washington, DC: Brookings, 1987);
    Robert Jervis, *The Meaning of the Nuclear Revolution: Statecraft and the
    Promise of Armageddon* (Ithaca, NY: Cornell University Press, 1989);
    Robert Powell, *Nuclear Deterrence Theory: The Search for Credibility*
    (Cambridge: Cambridge University Press, 1990); Todd Sechser
    and Matthew Fuhrmann, *Nuclear Weapons and Coercive Diplomacy*
    (Cambridge: Cambridge University Press, 2017).

5. See Glenn Snyder, "The Balance of Power and the Balance of Terror,"
    in P. Seabury (ed), *The Balance of Power* (San Francisco: Chandler,
    1965); Robert Jervis, *The Illogic of America's Nuclear Strategy* (Ithaca,
    NY: Cornell University Press, 1984); and Jervis, *The Meaning of the
    Nuclear Revolution*.

6. This is the problem of moral hazard, discussed particularly in John Kaag
    and Sarah Kreps, *Drone Warfare* (Cambridge: Polity, 2014).

7. Neil Postman, *Technopoly* (New York: Vintage, 1993), p. 61.

8. See Grégoire Chamayou, *Drone Theory* (trans: Janet Lloyd),
    (London: Penguin, 2015) and Hugh Gusterson, *Drone: Remote Control
    Warfare* (Cambridge, MA: MIT Press, 2016), p. 22 for articulation of this
    argument regarding the use of drones.

9. Stephanie Carvin, "Getting Drones Wrong," *International Journal of
    Human Rights* 19:2 (2015), p. 127–141.

10. Jacques Ellul, *The Technological Society* (New York: Vintage, 1964).

11. Michael C. Horowitz, "Artificial Intelligence, International
    Competition and the Balance of Power," *Texas National Security Review*
    (May 2018), available at: https://tnsr.org/2018/05/artificial-intelligence-
    international-competition-and-the-balance-of-power/#_ftn14, accessed
    23 May 2018.

12. Darrel M. West, "Will Robots and AI Take Your Job? The Economic
    and Political Consequences of Automation," Brookings Institute, 18
    April 2018, available at: https://www.brookings.edu/blog/techtank/2018/
    04/18/will-robots-and-ai-take-your-job-the-economic-and-political-
    consequences-of-automation/, accessed 23 May 2018.

13. Henry Kissinger, "How the Enlightenment Ends," *Atlantic*, June 2018,
    available at: https://www.theatlantic.com/magazine/archive/2018/06/
    henry-kissinger-ai-could-mean-the-end-of-human-history/559124/,
    accessed 23 May 2018.

14. James Vincent, "Putin Says the Nation That Leads in AI 'will be the
    ruler of the world.'" *Verge*, 4 September 2017, available at: https://www.

theverge.com/2017/9/4/16251226/russia-ai-putin-rule-the-world, accessed 24 May 2018.

15. Seth Figerman, "Elon Musk Predicts World War III," *CNN*, 5 September 2017, available at: http://money.cnn.com/2017/09/04/technology/culture/elon-musk-ai-world-war/index.html, accessed 30 May 2018.

16. Jake Swearingen, "A.I. Is Flying Drones (Very, Very Slowly)," *New York Times*, 26 March 2019.

17. Scott Shane, Cade Metz, and Daisuke Wakabayashi, "How a Pentagon Contract Became an Identity Crisis for Google," *New York Times*, 30 May 2018.

18. "Artificial Intelligence at Google: Our Principles," available at: https://ai.google/principles/, accessed 30 July 2019; Kelsey D. Atherton, "Targeting the Future of the DoD's Controversial Project Maven Initiative," C4ISRNET, 27 July 2018.

19. Horowitz, "Artificial Intelligence, International Competition and the Balance of Power."

20. Paul Scharre, *Army of None: Autonomous Weapons and the Future of War* (New York: W.W. Norton, 2018), p. 43–48.

21. Paul Scharre, "Autonomous Weapons and Operational Risk," Center for New American Security (February 2016), https://www.files.ethz.ch/isn/196288/CNAS_Autonomous-weapons-operational-risk.pdf, accessed 24 May 2018.

22. Paul Scharre, "Robotics on the Battlefield, Part II," Center for New American Security (October 2014), p. 46.

23. U.S. Department of Defense, Directive 3000.09, 21 November 2012.

24. Darrell M. West, "Brookings Survey Finds Divided Views on Artificial Intelligence for Warfare, but Support Rises if Adversaries Are Developing It," Brookings Institute, 29 August 2018.

25. Billy Perrigo, "A Global Arms Race for Killer Robots Is Transforming the Battlefield," *Time*, 9 April 2018.

26. Greg Williams, "Why China Will Win the Global Race for Complete AI Dominance," *Wired*, 16 April 2018, available at: http://www.wired.co.uk/article/why-china-will-win-the-global-battle-for-ai-dominance, accessed 24 May 2018. See also Jonathan Ray, Katie Atha, Edward Francis, Caleb Dependahl, Dr. James Mulvenon, Daniel Alderman, and Leigh Ann Ragland-Luce, "China's Industrial and Military Robotics Development," Center for Intelligence Analysis and Research, October 2016, available at: https://www.uscc.gov/sites/default/files/Research/DGI_China's%20Industrial%20and%20Military%20Robotics%20Development.pdf, accessed 24 May 2018.

27. Bradley Peniston, "Army Warns That Future War with Russia or China Would Be 'Extremely Lethal and Fast,'" *DefenseOne*, 4 October 2016,

available at: https://www.defenseone.com/threats/2016/10/future-army/
132105/, accessed 30 July 2019.

28. Scharre, "Robotics on the Battlefield, Part II," p. 10.

29. Evan Ackerman, "DARPA's Semi-Disposable Gremlin Drones Will Fly
in 2019," IEEE Spectrum, 9 May 2018, available at: https://spectrum.ieee.
org/automaton/robotics/military-robots/darpas-semidisposable-gremlin-
drones-will-fly-by-2019, accessed 26 May 2018.

30. David Hambling, "Change in the Air: Disruptive Developments in
Armed UAV Technology," UNIDIR, 6 June 2017, p. 3.

31. Kelsey D. Atherton, "LOCUST Launcher Fires a Swarm of Navy
Drones" Popular Science, 24 May 2016.

32. Alexis C. Madrigal, "Drone Swarms Are Going to Be Terrifying and Hard
to Stop," Atlantic, 7 March 2018.

33. Strategic Capabilities Office, U.S. Department of Defense, "Perdix Fact
Sheet," 2017, available at: https://www.defense.gov/Portals/1/Documents/
pubs/Perdix%20Fact%20Sheet.pdf, accessed 26 May 2018.

34. Paul Scharre, Robotics on the Battlefield, Part II: The Coming Swarm,
(Washington, DC: Center for New American Security, October
2014), p. 14.

35. Jason Kehe, "Drone Swarms As You Know Them Are Just an Illusion—
for Now," Wired, 14 August 2018.

36. Daniel Cebul, "Raytheon, DARPA Developing Technology to Control
Drone Swarms," Defense News, 26 March 2018, available at: https://www.
defensenews.com/unmanned/2018/03/26/raytheon-darpa-developing-
technology-to-control-drone-swarms/, accessed 26 May 2018.

37. Zachary Kallenborn, "The Era of the Drone Swarm Is Coming, and We
Need to Be Ready for It," Modern War Institute, 25 October 2018.

38. Patrick Tucker, "The US Military's Drone Swarm Strategy Just Passed a
Key Test," DefenseOne, 21 November 2018.

39. Cebul, "Raytheon, DARPA Developing Technology to Control Drone
Swarms."

40. Hambling, "Change in the Air: Disruptive Developments in Armed UAV
Technology," p. 4–6.

41. Lt. Col. Stephan G. Van Riper, "Apache Manned-Unmanned Teaming
Capability," Association of the United States Army, September 2014,
available at: https://www.ausa.org/sites/default/files/apache-manned-
unmanned-teaming.pdf, accessed 26 May 2014.

42. Valerie Insinna, "Unmanned: Check Out This New Video with the Air
Force's Coolest Future Tech," Defense News, 26 March 2018, available
at: https://www.defensenews.com/unmanned/2018/03/26/check-out-this-
new-video-with-the-air-forces-coolest-future-tech/, accessed 26 May 2018.

43. Aaron Gregg, "Boeing Wins $805 Million Contract to Build Navy's MQ-
25 Stingray Drone," Washington Post, 30 August 2018.

44. Mark Pomerleau, "Army's Multidomain Battle Brings Manned-Unmanned Teaming to the Fore," C4ISRNET, 31 August 2017, available at: https://www.c4isrnet.com/unmanned/uas/2017/08/31/armys-multidomain-battle-brings-manned-unmanned-teaming-to-the-fore/, accessed 26 May 2018.

45. US Department of Defense, "Unmanned Systems Integrated Roadmap, 2013–2038," Publication Number 14-S-0553, p. 7.

46. See particularly Singer, *Wired for War*, p. 19–29.

47. US Department of Defense, "Unmanned Systems Integrated Roadmap, 2013–2038," p. 7.

48. Patrick Tucker, "U.S. Army Now Holding Drills with Ground Robots That Shoot," *Defense One*, 8 February 2018.

49. Samuel Bendett, "Is Russia Building an Army of Robots?" *National Interest* (19 March 2018), available at: http://nationalinterest.org/blog/the-buzz/russia-building-army-robots-24969, accessed 26 May 2018.

50. US Department of Defense, Unmanned Systems Integrated Roadmap, 2013–2038, p. 8.

51. For a full list of maritime drones, see http://auvac.org/, accessed 26 May 2018.

52. Nyshka Chandran, "Beijing Is Using Underwater Drones in South China Sea to Show Off Its Might," *CNBC*, 12 August 2017, available at: https://www.cnbc.com/2017/08/12/china-uses-underwater-drones-in-south-china-sea.html, accessed 29 May 2018.

53. Valerie Insinna, "Russia's Nuclear Underwater Drone Is Real and Is in the Nuclear Posture Review," *Defense News*, 12 January 2018, available at: https://www.defensenews.com/space/2018/01/12/russias-nuclear-underwater-drone-is-real-and-in-the-nuclear-posture-review/, accessed 26 May 2018.

54. Rory Jackson, "Small Is Beautiful: Nano Drone Tech Is Advancing," *Defense IQ*, 20 July 2017, available at: https://www.defenceiq.com/defence-technology/articles/nano-drone-tech-is-advancing, accessed 26 May 2018.

55. Sophie Weiner, "This Genetically-Modified Cyborg Dragonfly Is the Tiniest Drone," *Popular Mechanics*, 1 June 2017, available at: https://www.popularmechanics.com/flight/drones/a26729/genetically-modified-cyborg-dragonfly/, accessed 26 May 2018.

56. Scott Shane, "Election Spurred a Move to Codify U.S. Drone Policy," *New York Times*, 24 November 2012.

57. "Obama's Speech on Drone Policy," *New York Times*, 23 May 2013.

58. Dennis M. Gormley, "Norms Needed to Regulate the Use of Armed Drones," Middle East Institute, 2 August 2017.

59. "The President Is Making It Easier to Order Lethal Drone Strikes," *The Economist*, 18 May 2017; Mike Stone and Matt Spetalnick, "Exclusive: Trump to Boost Exports of Lethal Drones to More

U.S. Allies—Sources," *Reuters*, 20 March 2018, available at: https://
www.reuters.com/article/us-usa-arms-drones-exclusive/exclusive-
trump-to-boost-exports-of-lethal-drones-to-more-u-s-allies-sources-
idUSKBN1GW12D, accessed 30 May 2018.

60. Michael J. Boyle, "The Costs and Consequences of Drone Warfare,"
     International Affairs 89:1 (2013), p. 28–29.

INDEX

———⟫◆⟪———